D1569794

MACRO-ENGINEERING

Water Science and Technology Library

VOLUME 54

The titles published in this series are listed at the end of this volume.

MACRO-ENGINEERING

A Challenge for the Future

Edited by

Viorel Badescu

Polytechnic University of Bucharest, Romania

Richard B. Cathcart

Geographos, Burbank, California, U.S.A.

and

Roelof D. Schuiling

Utrecht University, The Netherlands

A C.I.P. Catalogue record for this book is available from the Library of Congress.

ISBN-10 1-4020-3739-2 (HB)
ISBN-13 978-1-4020-3739-9 (HB)
ISBN-10 1-4020-4604-9 (e-book)
ISBN-13 978-1-4020-4604-9 (e-book)

Published by Springer,
P.O. Box 17, 3300 AA Dordrecht, The Netherlands.

Cover Image: Solar reflectors in the Earth orbit.
By the courtesy of Aerospace Systems Ltd. See also Fig. 12, page 241.

Printed on acid-free paper

This book is dedicated to all the brave individuals who dream of a better world, but in order to realize their ideas have to fight bureaucracies and the stubborn prejudices of the masses.

TABLE OF CONTENTS

CONTRIBUTING AUTHORS

Viorel Badescu

Candida Oancea Institute, Polytechnic University of Bucharest, Spl. Independentei 313, 79590, Bucharest, Romania. E-mail: badescu@theta.termo.pub.ro

Govindasamy Bala

L-103 Energy and Environment Directorate, Lawrence Livermore National Laboratory, 7000 East Avenue, CA 94550, Livermore, USA. E-mail: bala1@llnl.gov

Alexander A. Bolonkin

C & R Co., 1310 Avenue R, #6-F, NY 11229, Brooklyn, USA. E-mail: abolonkin@juno.com

Ken Caldeira

Carnegie Institution, Stanford, CA, USA.

Richard Brook Cathcart

Geographos, 1300 West Olive Avenue, CA 91506, Burbank, USA. E-mail: rbcathcart@charter.net

Milan M. Ćirković

Astronomical Observatory of Belgrade, Volgina 7, 11160, Belgrade, Serbia. E-mail: arioch@eunet.yu

Christopher Grassotti

Atmospheric and Environmental Research, Inc., 131 Hartwell Avenue, MA 02421-3126, Lexington, USA.

John M. Henderson

Atmospheric and Environmental Research, Inc., 131 Hartwell Avenue, MA 02421-3126, Lexington, USA.

Ross N. Hoffman

Atmospheric and Environmental Research, Inc., 131 Hartwell Avenue, MA 02421-3126, Lexington, USA. E-mail: rhoffman@aer.com

Richard LaRosa

Sealevelcontrol.com, 317 Oak Street, NY 11550-7713, South Hempstead, USA. E-mail: RLaRosa331@aol.com

S. Mark Leidner

Atmospheric and Environmental Research, Inc., 131 Hartwell Avenue, MA 02421-3126, Lexington, USA.

Evert H. du Marchie van Voorthuysen

GEZEN Foundation, Nieuwe Kerkhof 30a, 9712 PW, Groningen, The Netherlands. E-mail: voorthuysen@home.nl

Colin R. McInnes

Department of Mechanical Engineering, University of Strathclyde, G1 1XJ, Glasgow, UK. E-mail: colin.mcinnes@strath.ac.uk

George D. Modica

Atmospheric and Environmental Research, Inc., 131 Hartwell Avenue, MA 02421-3126, Lexington, USA.

Thomas Nehrkorn

Atmospheric and Environmental Research, Inc., 131 Hartwell Avenue, MA 02421-3126, Lexington, USA.

Gerrit Oudakker

Tetradon bv, Taling 5, 1721 DC, Broek op Langedijk, The Netherlands.

Roelof Dirk Schuiling

Institute of Geosciences, Utrecht University, PO Box 80021, 3508 TA, Utrecht, The Netherlands. E-mail: schuiling@geo.uu.nl

PREFACE

Macro-engineering involves the large-scale modification and manipulation of natural systems for the benefit of mankind. Some of the major applications of macro-engineering are power production, land reclamation, food production, climate change, environment, water, transport, coastal protection and the protection against (future) disasters. If we go back into the history of civilization, one might consider the evolution of agriculture to be the biggest macro-engineering change of the face of the earth, an evolution that was drawn out over several millennia.

Macro-engineering is not only concerned with earth-bound projects, however, but also ventures into extraterrestrial problems.

This book takes account of the fact that Macro-engineering is certain to become a more widespread profession. Its practitioners will remake, unmake and extend our infrastructure located – mainly – in planet Earth. Macro-engineers do not claim that everything is possible; but we do claim that Macro-engineering can design biosphere benignity to be constructed with minimal energy and raw materials while little (if any) harmful by-products and waste are produced that require costly disposal.

Galileo Galilei (1564–1642) participated already in a wide variety of common Earthly macro-projects (drainage, flood-control in Italy) and introduced economics, physics and mathematics into machine making. His 'Two New Sciences' can be considered the first textbook in the modern discipline of Macro-engineering since it blended all the sciences of anthropogenic artifacts.

Most macro-projects originate in the minds of a few inspired individuals. From 1944 until 1950, Geoffrey Dobbs in the UK published a series of articles that were compiled in 1951 into a book, 'On planning the Earth' (K.R.P. Publications Ltd., Belfast, Northern Ireland). Dobbs dramatically illuminated the biggest drawback for Macro-engineering:

"Planning implies control.... As history has repeatedly shown, the ultimate evil of centralized planning is that it is the stealing of choices."

Will Ley (1906–69), a well-remembered popularizer of space travel, exotic zoology and Macro-engineering penned "Engineers' Dreams: Great Projects that could come true" (The Viking Press: New York, April 1954). Not long after Ley's 1954 book was first printed – it went through multiple unrivaled editions – a small cadre of trained and otherwise qualified personnel gathered around so that the modern-day field could organize more effectively.

The word 'Macro-engineering' evidently had its first print appearance in the UK weekly science round-up magazine, New Scientist, on 12 March 1964 at page 685. There, columnist "Geminus", in "It seems to me", elegantly alleged: "The real cause of our attachment to macro-engineering is at once more subtle and more profound". Frank P. Davidson, in "Macro-engineering: A Capability in Search of a Methodology", Futures, December 1968, at p. 153 probably used 'macro-engineering' for the first time in a USA publication. The New York Times, often cited as one of America's daily newspapers of record, may have first employed 'Macro-engineering' in print in its 19 February 1978 issue.

Between 1978 and 1998, eight books appeared with 'Macro-engineering' in their titles:

- Frank P. Davidson, L.J. Giacoletto and Robert Salkeld (Eds.) "Macro-engineering and the Infrastructure of Tomorrow" (Westview Press, Boulder CO, 1978);
- Frank P. Davidson, C. Lawrence Meador and Robert Salkeld (Eds.), "How Big and Still Beautiful? Macro-engineering Revisited" (Westview Press, Boulder CO, 1980);
- Robert Salkeld, Frank P. Davidson and C. Lawrence Meador (Eds.), "Macro-Engineering: The Rich Potential" (American Institute of Aeronautics and Astronautics, NY, 1981);
- Frank P. Davidson and C. Lawrence Meador (Eds.), "Macro-engineering and the Future: A Management Perspective" (Westview Press, Boulder CO, 1982);
- Frank P. Davidson and John Stuart Cox, "MACRO: A Clear Vision of How Science and Technology Will Shape Our Future" (William Morrrow and Company, NY, 1983);
- Frank P. Davidson and C. Lawrence Meador (Eds.), "Macro-engineering: Global Infrastructure Solutions" (Ellis Horwood, NY, 1992);
- Frank P. Davidson, Ernst G. Frankel and C. Lawrence Meador (Eds.), "Macro-engineering: MIT Brunel Lectures on Global Infrastructure" (Ellis Horwood, Chichester UK, 1997) and lastly,
- Uwe Kitzinger and Ernst G. Frankel (Eds.), "Macro-engineering and the Earth: World Projects for the Year 2000 and Beyond" (Ellis Horwood, Chichester UK, 1998).

The lead editor of most of the printed book titles mentioned above, Frank P. Davidson (born 1918), is a distinguished expert in International Law and most contributors were international lawyers and financiers.

Between 2003 and 2004 three book-length treatments of the *economics* of macro-projects were published:

- Bent Flyvbjerg, Nils Bruzelius and Werner Rothengatter, "Megaprojects and Risk: An Anatomy of Ambition" (Cambridge University Press, Cambridge UK, 2003),
- Alan Altshuler and David Luberoff "Mega-Projects: The Changing Politics of Urban Public Investment" (The Brookings Institution, Washington DC, 2003), and

- Henry Petroski, "Pushing the Limits: New Adventures in Engineering" (Alfred A. Knopf, NY, 2004), a book that in its final chapter reviews Willy Ley's 1954 book.

What makes our book different from those previously published is its scope (chiefly physics and engineering), its multiple contributors (each enjoying worldwide fame as macro-engineers already) and its up-to-date data and futuristic information. Our book is also intended to foster a needed professional spirit, because Macro-engineering's creative and scientific activity transforms global Nature [Mars, for instance] to serve the needs and wants of large numbers of persons – we create what did not exist. Therefore, the principal audience target consists of students and researchers involved in large-scale projects. The text might be used for professional reference and some chapters can be used as teaching material in faculties of natural sciences and engineering.

In this book a number of macro-engineering projects or project proposals will be presented, which cover the range between daring proposals that can be executed with existing technologies to space projects that still exist only in the imagination. Macro-engineering planners design "macro-projects" that may cause a fundamental change in humanity's planetary outlook. Some macro-projects are beneficial or essential infrastructure. Most earthbound macro-engineering projects require massive funding, significant manpower, large-scale equipment and millions of tons of material. Macro-engineering projects extend the state-of-the-art of technology and may take place in difficult and sometimes hostile environments.

Worldwide macro-engineering changes, like projects aiming at changing the climate have the negative aspect that they imply control over almost everything and everyone. Not all macro-engineering projects have global dimensions, though, but some may affect only specific regions. Such projects mainly aim to improve the economic status and the living conditions of particular groups of people in a region.

To conclude, macro-projects have defining characteristics:

a) complex and trying engineering and management problems must be solved even before the main project is undertaken;
b) significant public and private commitments are necessary, both in terms of money and societal dedication;
c) sometimes scientific and purely technical problems of inordinate complexity arise that must be deftly macro-managed with favorable final outcomes; and
d) macro-projects usually impact the planetary biosphere or large regions of the Earth's human-inhabited volume.

Our book attempts to bring honest appraisals of some macro-project plans that can make Earth a better place in which to live. It also touches on technologies that will permit humanity to move from Earth into Solar System and its surrounding Universe, the ultimate "Environment".

The book is structured along the following lines. After an introductory chapter by R.B. Cathcart and V. Badescu, first a number of earth-bound, somewhat unusual projects, but still within the sphere of present-day technology are presented. Sequestration of CO_2 in mineral form to avoid excessive greenhouse effects and recovery at the same time of large amounts of the energy of carbonation and hydration (R.D. Schuiling, chapter 2) forms a transition to particular macro-engineering projects oriented to large scale energy production and to control and modify the climate. The first two chapters in this series are related to the large-scale use of solar energy in deserts (E. du Marchie van Voorthuysen, chapter 3) and to the large-scale energy production in wind farms at sea (R.D. Schuiling, chapter 4). The series continues with chapter 6 by G. Bala and K. Caldeira on climate control, flanked by two chapters on hurricane control (chapter 5 by R. LaRosa and chapter 7 by R. Hoffmann and co-workers).

From this point on the chapters become more space-oriented. However, the transition between Earth-based macro-projects to space-based application is smooth. The editors have allowed the authors a certain freedom of speculation, but in the resulting mix of science and fiction we have tried to stick to scientific standards, while allowing bold proposals. Chapter 8, by A.A. Bolonkin, proposes two new macro-engineering projects: the inflatable pneumatic high altitude towers (height up to 100 km) and the kinetic cable space towers (height up 160,000 km). A second contribution by A.A. Bolonkin (chapter 10) refers to the simple cable anti-gravitator and to electrostatic levitation and artificial gravity. Some implications of the radical macro-engineering efforts in medium-to-long-term future of humanity are discussed in chapter 9, by M. Ćirković and R.B. Cathcart. Chapter 11, by C. McInnes, focuses on two particular cases where orbiting solar reflectors may be used: the active cooling of the climate to mitigate against anthropogenic climate change due to the increase of the carbon dioxide concentration in the Earth's atmosphere and the active heating of the climate to mitigate against an advance of the polar ice sheets of a magnitude comparable to that induced by the Milankovitch cycles. In chapter 12, by V. Badescu and R.B. Cathcart, the concept of stellar engine is defined and one theorizes on various ways of controlling the Sun movement in the Galaxy. Finally, in chapter 13, by M. Ćirković, the problem of the detectability of macro-engineering projects over interstellar distances is discussed, in the context of Search for ExtraTerrestrial Intelligence (SETI).

The Editors

FOREWORD

Make no small plans for they have not the power to stir men's blood.

Niccolo Machiavelli

Man is the engineering animal. Since prehistoric times, we have used our skills at engineering to adapt our environment to suit ourselves. We have leveled mountains, engineered man-made lakes and drained swamps, reversed the course of rivers, and made deserts bloom. The stone monoliths of Stonehenge were an engineering feat unrivaled for its time. Later, the great pyramids boasted in stone to the gods of Egypt that we humans, too, could make mountains. In more modern times, Gustav Eiffel's tower – at the time the tallest building in the world - was the marvel of the age, a poem in steel.

For better or for worse, as our technology becomes more powerful, our engineering feats are moving to planetary scales. The United States interstate highway system, a network of concrete cumulatively almost a hundred thousand kilometers long, is an engineering structure of continent-spanning scale. By burning fossil fuels and increasing the carbon dioxide content of the atmosphere – now creating global effects on the climate of Earth – we are even now altering the climate by our own efforts and returning the planet to the warmer environment that had prevailed in the age of the dinosaurs. This human-induced global climate change is, so far, an unplanned act of planetary engineering, but actions of similar consequence could be accomplished deliberately, if we want to do so. We can engineer the environments of entire worlds, or at least we will soon be able to do so, if we so choose. In the near term, we will want to engineer the climate of our planet deliberately, and not accidentally. If the warming due to the greenhouse effect is not to our liking, we can and will learn how to change it.

And this process will, in the long term, be not only a choice, but also a necessity. Over the time span of a hundred thousand years, the cycle of ice ages bringing kilometer-thick sheets of ice onto the continents will be something we will want to engineer away. On a time scale of a billion years, the sun itself is gradually increasing in luminosity. If life itself is to survive, we will have to learn to effect planetary-scale climate engineering, either by regulating the temperature of the

V. Badescu et al. (eds.), Macro-Engineering: A Challenge for the Future, 1–2.

Earth with albedo changes or solar shield, or by moving the planet itself to a more suitable location further from the sun, as well as terraforming the other planets and moons of our solar system into habitable places for life. And on the time scale of five billion years, the sun will exhaust its nuclear fuel and swell into a red giant. This will be the ultimate challenge to macro-engineering. By this time, we will learn to tame stars. In the long view, space flight and planetary engineering are not luxuries, but a necessity. Humanity may well have spread life into the galaxy, and be looking at the great gulfs between galaxies.

We are engineering our planet now, getting ready for the greening of the solar system.

Engineering is at the very heart of what makes us human. Other animals adapt to live in the world; we adapt the world to ourselves, building our hopes and our dreams in concrete and steel. Daniel Burnham, the great builder of parks and buildings, expanded upon the advice of Machiavelli: “Make big plans! ” said Burnham. “Aim high in hope and work, remembering that a noble, logical diagram once recorded will not die, but long after we are gone be a living thing, asserting itself with ever-growing insistence”.

I will end with a final quote, this time from Michelangelo:

“The greater danger for most of us is not that our aim is too high and we miss it, but that it is too low and we reach it. ”

Let us resolve to make no small plans.

Geoffrey A. Landis

ACKNOWLEDGEMENTS

In preparing this volume the editors received help and guidance from Drs. Petra van Steenbergen, Maria Jonckheere and Ria Balk (Springer NL) and Prof. Stanislaw Sieniutycz (Warsaw University of Technology).

A critical part of writing any book is the review process, and the authors and editors are very much obliged to the researchers who patiently helped them read through subsequent chapters and who made valuable suggestions: Dr. Rob N.J. Comans (Energy Research Centre of the Netherlands), Prof. Steven J. Dick (NASA History Division), Dr. James Edward Oberg (Houston), Dr. Titus Filipas (University of Craiova), Dr. Jim Hansen (Massachussets Institute of Technology), Dr. Mark Hempsell (Aerospace Engineering, Bristol), Dr. P. Hoyng (Foundation of Space Research, Netherlands), Dr. Wouter J.J. Huijgen (Energy Centre of the Netherlands), Dr. Harry D. Kambezidis (National Observatory of Athens), Dr. Geoffrey Landis (NASA John Glenn Research Center), Dr. Neville Marzwell (NASA Jet Propulsion Laboratory), Dr. Chandrakant B. Panchal (Argonne National Laboratory), Dr. Jerome Pearson (Star Technology and Research, Inc. Mount Pleasant), Prof. Robert Pitz-Paal (Deutsches Zentrum fur Luft-und Raumfahrt), Dr. Peter Siegmund (Royal Netherlands Meteorological Institute), Prof. M. Sigmond (University of Toronto), Dr. Richard L.S. Taylor (Probability Research Group, London), Dr. James R. Underwood, jr.(Austin, Texas), Dr. Luis A. Vega (Pacific International Center for High Technology Research, Honolulu), Dr. Max Whisson (Perth, Western Australia).

One of the editors (VB) is particularly indebted to Dr. Ross Hoffman and Prof. Colin McInnes for further help with the reviewing process. The editors thank JSC Aerospace systems, Moscow, Russia (www.spacer.ru) for permission to use the figure on the front cover.

The editors, furthermore, owe a debt of gratitude to all authors. Collaborating with these stimulating colleagues has been a privilege and a very satisfying experience.

V. Badescu et al. (eds.), Macro-Engineering: A Challenge for the Future, 3.

ACKNOWLEDGEMENTS

CHAPTER 1

GEO-ENGINEERING AND ENERGY PRODUCTION IN THE 21ST CENTURY

RICHARD BROOK CATHCART[1] AND VIOREL BADESCU[2]

[1] *Geographos, 1300 West Olive Avenue, Burbank, CA 91506, USA*

[2] *Candida Oancea Institute, Polytechnic University of Bucharest, Spl. Independentei 313, Bucharest, 79590, Romania*

Abstract: This is a survey of some thought-provoking macro-engineering project ideas. They involve geographically large-scale changes to the Earth's surface or energy balance. A subjective assessment of the theoretical progress accomplished in Macro-engineering is given and suggestions are offered as to where the most fruitful avenues of modern macro-project investigation may lie during the early 21st Century for this class of projects

Keywords: macro-engineering, geo-engineering Earth-surface, future technology

1. INTRODUCTION

If one examines the literature of Environmental Macro-engineering (EM) published during the last century (Fogg, 1995), or the testimony on EM before review conferences of the early 21st Century, one must conclude that most of the statements made during the last half of the 20th Century can be repeated today. Hoffert (2002) presented a synopsis of what the planet's surface would become if certain patented and public domain inventions could be applied to the Earth-surface. This summary was inspired, in part, by the sequence of "new technologies" assessments described and illustrated in *Time* (19 September 1960 through 8 September 2003) as well as by the "Key to the Global Thermostat: A Symposium on the Architect's Role in Global Warming", a 16 October 2003 conference (Hawthorne, 2003).

EM combines Geo-engineering, including some less-than-global macro-projects, with Terraforming in an over-arching new professional viewpoint on the Earth-surface. EM, therefore, is a visionary system of scientific and engineering ideas, principles, laws, and hypotheses. The adventurous spirit of "What if?" pervades

V. Badescu et al. (eds.), Macro-Engineering: A Challenge for the Future, 5–20.

EM (Defries, 2002), and brings together the understanding of Macro-engineering's current and potential arsenal. Future technologies will redefine the meaning of the Earth-surface and may progress from Geo-engineering to Terraforming of Mars.

This chapter focuses on two main ideas. First, that an appropriate usage of the regional geographic factors can enable large scale power production. Second, that various beneficial effects (like mitigation of green house gas climate warming) may be obtained by managing the earth's energy balance. A number of macro-engineering projects are presented to support these ideas.

2. POWER PRODUCTION

2.1 Heliohydroelectric Power Generation

The concept of heliohydroelectric (HHE) power generation utilizes the conversion of solar energy of evaporation to hydraulic energy, and then to electricity. When topographical and hydrological conditions are favorable, a hydraulic head can be established and maintained by the evaporation of water. A system of two reservoirs is required to establish the hydraulic head: one reservoir, an "infinite" constant level source of water, the other, a "sink" for evaporation. There are two ways in which such a system can be developed and exploited. Natural depressions well below sea level exist in many parts of the world. Those in the proximity of open seas and situated in regions of abundant sunshine, offer excellent prospect for an HHE system. Seas that are almost enclosed bodies of water offer a means of "artificially creating a depression". If the Red Sea is chosen to be the site of such a system, construction of a dam across the Strait of Bab-Al-Mandab between Arabia and Africa, would transform the entire Red Sea into a "sink" reservoir. The system (Red Sea–Indian Ocean) thus formed, after the effect of evaporation has lowered the level of the Red Sea, may be used for HHE energy conversion, the Indian Ocean being the "infinite" reservoir.

The Red Sea occupies a fault separating Africa from Arabia. It is completely closed, except for a narrow strait, formed in geologically recent times, linking it to the Indian Ocean and, since 1869 to the Mediterranean Sea via the Suez Canal. The sea extends from 14°30′N to 26°N latitude over a 1900 km distance. The average width of the sea is 230 km, the widest distance (about 350 km) being in the south between Jizan (Saudi Arabia) and Massawa (Ethiopia). This distance narrows down toward the Bab-Al-Mandab in the South and the Gulfs of Suez and Aqaba in the North. Bab-Al-Mandab is divided into two channels separated by Perim Island. The small channel between Perim and the Arabian coast has a 2.7 km width; the large channel between Perim and Africa is 16.7 km wide. It appears quite feasible to transform the entire Red Sea into an evaporation reservoir by building a dam across Bab-Al-Mandab. The total area enclosed would be about 438000 km^2. One of the critical elements in the Red Sea HHE Project is the dam. It is larger and more ambitious than any dam which has been constructed anywhere in the world. It will have a piled rubble/sealed core width, and at 1 km wide, about 100 km long and more than 150 m high will have a maximum emplaced structural volume of $\sim 15 \cdot 10^9$ m^3. [Peixoto and Kettani (1973) describe a similar but much smaller project for Bahrain.]

The optimization of the power produced by an HHE system of this type has been discussed before (Kettani and Scott, 1974). If a physical head H_e is established between the Red Sea and the Indian Ocean, the potential energy formed by the discharge Φ_s falling through H_e, can be transformed into electrical power P_e. For a steady state operation, and neglecting precipitation, Φ_s equals the amount of water evaporated by the sun at every moment. If the rate of evaporation is λ_s (m/s), and S is the area of water presented to the sun for evaporation, the power output is:

$$P_e = (g\eta)(\rho_l \lambda_s)(SH_e), \tag{1}$$

where g is gravity acceleration, ρ_l is the density of the sea water, and η is the efficiency of the hydroelectric plant. The first quantity in the left hand side of Eq. (1) is practically constant since g is given and η is close to unity (say 0.9). The second quantity depends basically on evaporation. Note that ρ_l is about $1.03\,\text{kg/m}^3$ for sea water at Bab-Al-Mandab and evaporation from the Red Sea was estimated to about 3.5 m/year by Kettani and Scott (1974). However, a more recent estimate of the evaporation excess is 2.1 m/year. The third quantity should be a maximum. As $H_e = 0$, S is maximum. When H_e is maximum, there will be no water left and $S = 0$. The surface area S decreases as the head H_e increases. Therefore, there will be a value of H_e for which the product (SH_e) is a maximum. This optimum efficiency of the power plant is achieved when equilibrium prevails between the amount of water flowing in and the amount evaporating by the sun's energy into the atmosphere. Early calculations show this happens for $H_{em} = 500$ m (Kettani and Scott, 1974). Thus, the potential capacity of the Red Sea HHE system using Eq. (1) is $P_{em} = 67000$ MW and the average energy that could be produced yearly is $W_e = 57 \cdot 10^{10}$ kWh/y. For a 500 meter head the remaining area left for evaporation would be about $133000\,\text{km}^2$.

The seawater evaporation rate depends of course on the incident solar irradiance, which in turn is a function of latitude. Therefore, the level of solar irradiance is not the same on the whole Red Sea surface. However, to give a rough idea, one may notice that at Aswan (situated not far away from Red Sea shore, at 23.9°N latitude and 32.8°E longitude) the yearly average solar global irradiation is 23.97 $\text{MJm}^{-2}\text{day}^{-1}$. The time necessary to reach a 500 meter head at a yearly evaporation rate of 2.1 meters is 238 years. It is not reasonable to wait this long. It is possible to start operations as soon as the dam across Bab-Al-Mandab has been built, but at a rate below the maximum possible.

The economics of the project depends on the way in which the evaporation is managed. For the first ten years after the dam is constructed it is expected that no power is generated and that the full evaporation of the Red Sea is used to lower the water level and to produce a hydraulic head. At the end of 10 years it then becomes possible to generate 5,000 MW of power. From that time one part of the evaporation may be used to generate power and part of it to lower the water level still further. After 100 years the water level might be kept constant and the power generation increased to 40,000 Megawatts. Alternatively, evaporation might be allowed to lower the water level of the 500 meter optimum head. This would occur after 240 years and if all the evaporation is then used to generate power about 67,000 MW could be produced.

In practice, the energy produced by the Red Sea HHE plant will be dictated by both the demand and not solely by the energy available. Any excess energy can be used for seawater desalination, and one should bring energy intensive industries to Bab-Al-Mandab, like aluminium or magnesium production, or even hydrogen production. The overall average over several years of consumption should equal the average energy capacity of the plant. More specifically, consider production at the optimum value of 570 billion kWh a year. Energy could be produced at say ten times this nominal rate, reaching 5,700 billion kWh/year; the only effect would be a rise of the sea level within the reservoir by about 35 meters corresponding to an insignificant loss in efficiency. HHE power generation affords therefore a unique way of storing solar energy and leveling out its diurnal and seasonal variations. The energy produced could be made even larger for any number of years. Indeed, the hydraulic head could vary between 320 meters and 800 meters with a loss of efficiency lower than 10 percent.

The HHE Red Sea power plant has a finite life-span just like any other project. After years of evaporation, salt deposition would eventually fill out the depression completely until no surface is left for evaporation. This "silting" of the evaporation basin is in no way different from the silting of water reservoirs behind dams in regular hydroelectric plants. The number of years necessary for the basin to be filled with salt would be equal to the ratio of the volume remaining after the optimum head has been reached over the volume of salt deposited yearly. A small calculation shows that the Red Sea would be completely filled to practically ocean level after 59,000 years of operation (Schuiling et al., 2006). [But well before that time the Red Sea will become a huge solar lake that could provide additional energy generation analogous to the Dead Sea project discussed below.]

Because the amount of water in the atmosphere cannot be increased, the effect of creating such an artificial depression would be to rise the level of all the oceans of the world by a height equal to the volume of the depression (here 149,800 km^3) divided by the area of the oceans. This rise would then be 30 cm (Schuiling et al., 2006). The amount of water involved annually in the water cycle is estimated to be 423,000 km^3. The rise of the ocean by 30 cm would correspond to tapping the hydraulic energy of all the rain falling on Earth across a 30 cm hydraulic head. This amounts to about 52,000 MW or about the capacity of the Red Sea HHE Plant. In other words, HHE power generation taps hydraulic energy which is otherwise impossible to tap.

2.2 Qattara Depression Power Plant

In the Libyan Desert only 80 km away from the Mediterranean, a vast depression 300 km long and 150 km wide can be found, the floor of which at its lowest point lies 135 m below sea level. It is called Qattara depression. If seawater is conducted to the northern edge of this depression by an open channel or an underground supply line, the difference of elevation between the Mediterranean and the bottom of the depression can be used for generating power in water turbines (Ball, 1933). Although the depression has no outlet, the utilized water is subjected to very high evaporation from this area (about 1800 mm per annum). In the Qattara depression optimal efficiency will be achieved as soon as the water level has reached −60 m, corresponding to a surface area of 12000 km^2 and to an annual evaporation volume of more than 20000 million m^3. Thus a seawater discharge of at least 650 m^3/s can be continuously fed into the turbines of the power plant without the level in the new Qattara basin rising any further. The capacity attainable with this hydro-solar scheme combined with pumped storage reaches 4000 MW peak load energy (Bassler, 1972).

[Peixoto and Kettani (1973) argue that part of the Nile should be diverted to make the Qattara depression an agricultural area.]

2.3 Dead Sea Solar Lake

A scheme aimed to modify the Dead Sea into a solar lake is outlined in Assaf (1976). The size of the sea and the possible application to other salt lakes and semi-closed basins makes the potential of the scheme for power production and desalinization of sea water very attractive. The Dead Sea is divided by the Lisan Strait into two basins, the northern basin, which is up to 330 m deep and has an area of 720 km^2, and a southern shallow basin (a few meters depth) with an area of 230 km^2. The Dead Sea level is between 392 and 399 m below sea level. The scheme assumes that the solar lake will occupy the northern basin. The seawater near the bottom will become hotter as a result of solar radiation absorption and will serve as a heat reservoir for a thermal power plant. A salt concentration gradient will prevent to some extent mixing of the bottom hot water with the surface coldest fluid. The basic idea is to reduce even more mixing over the solar lake, by floating wind breaks which might consist of a partly submerged net of a few cm height and a whole size less than 1 m. Computations show the optimum hot bottom water temperature is about 160 °C but realistic values should fall between 80° and 100 °C. The overall power generation efficiency is about 8% and a mean power of 4000 to 7000 MW is estimated.

2.4 Ocean Thermal Energy Conversion

Ocean thermal energy conversion (OTEC) systems utilize the temperature difference between water masses in the ocean to generate power. OTEC is independent of the diurnal variation of insolation and this is an important advantage among solar energy conversion alternatives. The temperature difference between the surface and a depth of 1000 m in the equatorial and lower temperate zone latitudes ranges from 25 to 18 °C. The thermal capacity of the warm upper layers of the sea in suitable regions (like Gulf of Mexico, Gulf Stream, Hawaii) is vast and OTEC systems can contribute significantly to world energy needs.

Preliminary estimates of the available ocean thermal energy of the Gulf Stream off the southern coast of the United States were reported by Mangarella and Heronemus (1979). The extractable power P of a closed OTEC thermal cycle is given by

$$P = \eta c_p \dot{m} \Delta T_h \left(1 - \frac{T_c}{T_h}\right), \qquad (2)$$

where η is cycle efficiency, c_p is sea water specific heat, $\dot{m}$ is the mass flow rate of water through the power plant evaporators, ΔT_h is the temperature drop of ambient hot water through evaporators and T_h and T_c are the temperatures of the working fluid on hot-side and cold-side of the heat engine, respectively. Equation (2) was applied to the Florida current whose importance as an OTEC power source lies in the fact that it is part of Gulf Stream and therefore derives much of its surface thermal energy from the solar energy exchange processes taking place over the broad expanse of the sub-tropical North Atlantic. Consequently, the energy carried by the Florida current is several times larger than the solar energy absorbed by the sea on the relatively small surface area of the Straits of Florida. The north-south volume transport within the top 60 m is about $6 \cdot 10^6$ m^3/s and the current average speed is 1.8 m/s. The mean summer surface currents range from 1.0 to 1.5 m/s while in winter the surface currents are 80–90% of the summer values. For an OTEC plant about 50 km east of Virginia Key with a 700 m intake depth a power capacity of 150 MW could be attained if the evaporators process the flow within an area 10 m wide by 60 m deep. For the total ocean thermal energy in the Florida current the average extractable energy is about $3.3 - 3.5 \cdot 10^{12}$ kWh/y depending on the intake depth. The local thermal structure of the sea water can be substantially influenced by OTEC plant operation. In case of very large Gulf Stream OTEC power plants the climate of Northern Europe may also be affected.

Other details about OTEC may be found in chapter 5 of this book.

3. MANAGING THE EARTH'S ENERGY BALANCE

At least in part, Geo-engineering is the use of perfected and available technology to affect the Earth's radiation balance in order to stabilize or alter global and/or regional climate regimes. The geophysical scope and economic value of "Geo-engineering" is defined in standard references (See: Encyclopedia of Climate and Weather (Oxford University Press, 1996) and Encyclopedia of Global Change: Environmental Change and Human Society (Oxford University Press, 2002)).

3.1 Space Mirrors

By 1923, Herman Oberth had envisioned the application of planet-orbiting solar collectors and reflectors as a means to illuminate parts of the nighttime Earth-surface. More detailed calculations have since been performed for the illumination of large urban areas (Rush 1977). Also, frost control and fog removal could be considered. A proposed project would place a series of large, flat, light-weight mirrors into geostationary orbit to provide directly reflected sunlight for ground illumination at night. The geostationary satellite mirror has an orbital period equal the 24 hr period of the Earth's rotation, a condition satisfied by a circular orbit about 36000 km above the surface. A second motion of rotation of the satellite about its own axis every 48 hours is required in order that the sunlight be directed toward a fixed point on Earth. The two motions serve to deliver the reflected solar light on a ground area at the Equator. Latitudes not on the Equator and different longitudes can be reached by motions which leave the revolution and rotation rates unaltered. The geometry of the Earth-Sun-satellite system dictates that the satellite will be eclipsed (moved into the shadow of the Earth). The longest duration eclipse is about 71 minutes. During such eclipses, the Earth surface target can be illuminated by the use of a series of mirrors distributed along the Equator. Occasional adjustments to the orbit can be provided by addition of ion thruster. Energy for the ion thruster is supplied by a bank of conventional solar cells.

Any geostationary mirror, independent of shape, will spread its light over a circular region of about 300 km in diameter. This is because the Sun has an apparent angular diameter of 0.5°, rather than being a point source. If F_0 is the direct solar flux and F_r is the reflected flux reaching the ground, then

$$F_i = \frac{A_m}{A_i} F_0, \tag{3}$$

where A_m and A_i are mirror surface area and the illuminated area, respectively. Use of Eq. (3) shows that the illuminating flux is about $4.8 \cdot 10^{-4}$ times the solar flux under ideal conditions with no reflection losses and clear sky and assuming a 1.6 km diameter mirror. Moderate cloud cover would reduce the light level by a factor of 2–4 (which is still acceptable) but heavy cloud cover reduces light levels by a factor of 8–10 (which requires use of supplementary illumination of the type presently in use). Eye safety is not a concern because human eyesight has a solar Wien peak sensitivity that just barely protects people from very bright unnatural nighttime lighting (Laframboise and Chou, 2000).

Assuming a system of 15 mirrors, each of 1.6 km in diameter, the estimated lifetime is 60 years and the payback is about 10 years.

The first operational tests of a sunlight-reflector were conducted during 1993 when the Russian Znamya-2 experiment was successfully completed.

3.2 Satellite Parasols

Outer Space-based mechanical satellite parasols have been proposed to shade the Earth-surface, thereby reducing by a significant amount the surface heating. An artificial, unstructured, Outer Space debris ring encircling the Equator, a super-structure thereby dimming the sunlight impinging on that belt-shaped region, could eventually be deployed to cause an anthropogenic global cooling—if other, totally ground-based infrastructure is ineffectual or absent (Pearson, 2002).

3.3 Artificial Cloud Generation

Above the Earth-surface, the air's water vapor is the most important greenhouse gas; unlike carbon dioxide gas, it is not produced directly in large amounts by human activity—so far; most water vapor in the atmosphere results from evaporation at the ocean surface.

Clouds, which cover ~60–65% of the Earth-surface continuously, are ill-defined, yet discrete volumes of liquid water droplets; water vapor aloft is ~0.3% of the atmosphere by mass and 0.5% by volume. Luke Howard developed a scientific classification of Earth's clouds during 1803–04 (Cumulus, Cirrus, Nimbus and Stratus). UK researchers John Latham (2002) and Stephen Salter (2002) used in addition a two-fold classification (natural and Anthropic clouds, respectively). Earth's cloud cover was imaged first by TIROS on 1 April 1960; natural and artificial clouds will be monitored beginning in 2006 by CloudSat, an experimental orbiting satellite that employs millimeter radar to measure the vertical structure of ice and water clouds to better understand climate change (Belmiloud, 2000).

3.3.1 Cloud manufacture

Bai Ying was awarded international Patent Number CN1335054 on 13 February 2002 for devising an "Artificial Cloud Making Method" that replicates, in a regional setting, Earth's Hydrologic Cycle; using a pipeline facility constantly conveying air from a lower altitude to a higher altitude, water vapor is condensed to form a pervasive artificial cloud. Ying hopes such an installation will "... regulate local weather conditions, improve ecological environment and enlarge human living space". Latham (2002) and Salter (2002) may realize some of the major goals of the World Water Forum held in Kyoto, Japan, during March 2003; like Bai Ying both researchers hope to ultimately control the rate of global warming by enhancing the earth system albedo by increasing cloud cover and cloud longevity. Anthropic clouds would be generated using installed ocean wave-powered buoy machines, seawater pumps and specially designed aerial sprayers. The power sources for the anthropic cloud generators are similar to the machines that have used ocean wave energy to pump upwards deep-ocean seawater to the ocean's surface (Chen, 1995).

Long-lived, large-scale Anthropic Clouds, which will be mainly of the low-lying Marine Stratus type—or various types of mist and sea fogs—have already been successfully created industrially: one above the Pepsi-Cola Pavilion during Expo '70 in Osaka, Japan (Kluver et al., 1972) and the other over a part of a large freshwater lake in Switzerland during Expo.O2 (Diller and Scofidio, 2002). The second of these, named "Blur" by its architect-manufacturers, was a very short open-structure (unclad) skyscraper supporting pedestrian walkways inside a "permanent" artificial fog bank above the southernmost section of Lac de Neuchatel near Yverdon-les-Bains.

Latham (2002) and Salter (2002) hope to demonstrate a much larger version of Blur to mitigate regional aridity (Saudi Arabia, Iran, Australia) by manufacturing vast expanses of Anthropic Clouds in the marine boundary layer (below an altitude of ~1 km). By greatly increasing the seawater droplet concentration in Marine Stratus, this macro-project will cause an increase in regional albedo and perpetuation of the Anthropic Cloud for a long period. The local Earth-surface would be cooled significantly, potentially helping to mitigate future global warming. Their technique involves seawater droplet sprays originating at the ocean's surface of such quantity and uniformity of salt mass that the resulting Anthropic Cloud will have more drops and more uniform drops so that precipitation will be reduced and albedo increased relative to that in natural Marine Stratus. The identical (industrially mass-produced) seawater sprayers probably will have insignificant ecosystem impact—chiefly, reduction of sunlight's penetration into the top-most 200 m-thick ocean layer—and use only seawater and air as construction materials.

A second type of Anthropic Cloud has been proposed as a cure for global warming and as a boon for populous water-short regions. Imagined by Josh Storrs Hall, such an Anthropic Cloud could consist entirely of invisible mechanical "foglets"—a swarm of tiny motile robots—perfected by molecular nanotechnology (Mulhall, 2002; Hall, 2005). Instead of using self-replicating tiny robots to construct some required macro-object atom-by-atom, Hall foresees usage of sub-microscopic robotic "foglets" as linked human-controlled builders of macroscopic clouds with desired properties. Some speculators think humanity's 21st Century freshwater supply ought to become an industrially synthesized pure product, made exclusively by molecular nanotechnology's future safe-to-utilize constructors "condensing" water vapor directly from Earth's air and/or cheap mechanical combination of two hydrogen atoms from an inexpensive source (either abiogenic or biogenic) with one easily gathered oxygen atom to make water molecules *en masse* (Merkle, 1991). This macro-project proposal is, perhaps, one of the most imaginative contrived by EM proponents. Certainly, it far surpasses the well-documented idea for total management of the world's water cycle put forth by Jose Pinto Peixoto (1922–96) more than three decades ago (Peixoto and Kettani, 1973).

Human activities already produce a variety of water clouds: condensation from industrial smoke stacks; aircraft contrails; and cloud tracks from ocean or lake ships underway. Soon, Artificial Clouds will be deliberately manufactured to reflect sunlight to Outer Space and to increase total precipitation over now barren, but potentially habitable, landscapes that, afterwards, may become farms and ranches. It is suspected that a 15–20% increase in Earth's coverage by unbroken low cloud decks (such as Marine Stratus) could effectively offset a doubling of the atmosphere's carbon dioxide gas content. Monitoring by CloudSat will accurately measure the impact of any albedo change caused by EM.

3.3.2 *Effects on human habitability*

One could estimate the effect of artificial clouds on human habitability as follows. A rough relationship due to Hourwitz relates the monthly solar global irradiation for cloudy sky conditions on a horizontal surface H_C in terms of the same quantity for clear sky conditions H_{cs} (both in W/m²/month) as (Colacino and Giorgi, 1968):

$$H_C = H_{cs}(1-0.2C_h)(1-0.55C_m)(1-0.7C_l). \tag{4}$$

Here C_h, C_m and C_l are the amount of high, average height and low level clouds, respectively (all of them in tenths of the celestial vault). They obey the relationship $C = C_h + C_m + C_l$ with C $(0 \leq C \leq 1)$ the total cloud amount (sometimes called point cloudiness).

A number of statistically derived relationships exist between the monthly solar global irradiation on cloudy sky H_C and the monthly mean air temperature T_a and surface sea-water temperature T_{sw}. Here we shall use (Hatzikakidis and Sakas, 1973):

$$T_{sw} = 10.202 + 0.7313 H_C \tag{5}$$

$$T_a = -7.335 + 1.343 T_{sw}. \tag{6}$$

Here T_a and T_{sw} enter in deg Celsius while H_C enters in kcal/cm²/month. Equations (5) and (6) were derived for Athens (Greece) at latitude 38°N. In June the clear sky solar irradiation on a horizontal surface at the latitude of Athens is about $H_{cs} \equiv 28$ MJ/m²/month (see Fig. 3.4.1 of Duffie and Beckman, 1974).

Figure 1 shows results for a clear sky and three different cases, with artificial point cloudiness C created at high, average height and low altitude, respectively. This means $C = C_h$, $C_m = C_l = 0$ in the first case, and so on. Figure 1 is based on Eqs. (4)–(6) and the data for Athens in June. One can see that the most efficient way to decrease the ambient temperature is by creating low altitude artificial clouds, as expected. An artificial point cloudiness of 0.3 is enough to decrease the monthly mean air temperature by as much as four degrees Celsius.

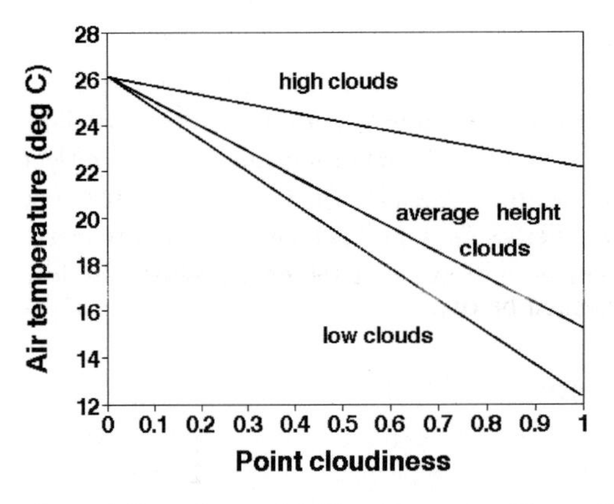

Figure 1. Dependence of ambient temperature (degree Celsius) on artificial point cloudiness. Clouds created at low, mid-level and high altitude, respectively. Data refers to Athens (Greece) in June

3.3.3 Secondary effects of artificial cloud generation

An engineered change of the Hydrologic Cycle links to major global cycles within the Earth-biosphere (Vorosmarty, 2002). If all the water vapor in the air ($\sim 13 \cdot 10^{15}$ kg) were to condense suddenly and fall as rain, on average the Earth-surface would receive a coating of freshwater ~25 mm deep. Deliberate weather modification will, inevitably, affect drainage basins on the downwind landscape. Water droplets of 1 mm fall with a velocity of 4.3 m/sec and 10 mm droplets fall at 13.6 m/sec. Solid materials eroded during channel enlargement ultimately will be deposited downstream, perhaps causing navigation problems at the coast and/or geochemical pollution from dissolved mineral deposition; on the very long time scale sedimentary rock lithologies will be initiated. Flood peak freshwater discharges, streambed changes and floodplain instability will have to be addressed. However, if increases in rainfall are gradual, then ecosystems/landscapes might evolve to reduce runoff. A speculated 15% increase of the annual rainfall on previously semi-arid lands adjacent to the Red Sea and the Persian Gulf could lead to changes in stream channel geometry and sediment yields that may result in sufficient economic damage to nullify many of the obvious benefits of such focused geo-engineering macro-projects.

Macro-engineers must assess the importance of biological elements affected by all cloud-making macro-projects. Low-level Marine Stratus reduces UV-A and UV-B Earth-surface dose rates by 60–75%, depending on the zenith angle and stratospheric ozone concentration. Solar radiation has a prominent ecological role on phytoplankton growth in the top, 100 m-thick layer of the ocean. Modification of the radiation budget of the ocean's surface will, obviously, have a remarkable oceanographic effect on the resident phytomass and may lead to a decrease of oceanic biomass, with associated effects on the productivity of the oceans and the CO_2 balance. However, this depends also on the limiting growth factors, which could be nutrients or Fe, and not necessarily the sunlight.

3.4 Changing the Earth's Surface Albedo

During 1957—the year humankind's Space Age began—J.G. Ballard, a UK science-fiction novelist postulated a massive metal and concrete building totally enclosing the Earth-surface (Cathcart, 2002). This sort of macro-project may significantly change the albedo of the planet. Badescu and Cathcart (2006) have calculated the thermodynamics of such an enormous planet-enveloping building. Similar (smaller scale) macro-projects are briefly presented below.

3.4.1 The insol concept

Fabric covered multi-story braced pneumatic greenhouses supercomputer modeled as well as actually built in Australia seem more appropriate for the 21st Century's global human infrastructure than other, more costly, types of big-footprint buildings and structures (Pohl and Montero, 1975).

Macro-engineers have calculated the probable climatic effect resulting from the physical emplacement of a Sahara-covering, distended tent-like building (called "the insol") intended to mitigate the harsh present-day aerial conditions confronting humans struggling to survive in that region (Cathcart and Badescu, 2004). Covering about $3.5 \cdot 10^6$ km^2 of North Africa can reduce that region's temperature by ~2 °C. Such a direct beneficial temperature reduction, plus further possible benefits accrued through irrigated vegetation inside the domes, could have a significant impact on the Sahara's future climate regime, cutting the UK's Tyndall Center's A.D. 2100 temperature projection by 33–50%, depending on the particular North Africa ecosystem-nation mapped by the Center (Pearce, 2000). This macro-project ought to be constantly monitored interactively *via* the World Wide Web and, in some instances, *via* directed roving macroscopic robots. The tent will have an immediate and dramatic effect on the region's albedo. Dust suppression is also anticipated and desired since particulates driven off the landscape of North Africa are proved detrimental, even harmful, to downwind life forms (Garrison, 2003). Overall, this computer-tested tenting may be seen as a realizable, and having a positive influence on the planetary biosphere's health under conditions of technology-accelerated human settlement and global warming.

What are the consequences of insol implementation on Sahara's climate? 110 North African places were selected to represent a region between 4.86°N and 36.86°N latitude and 17.05°W and 32.23°E longitude (Cathcart and Badescu, 2004). Figure 2a shows the geographical distribution of multi-year average values of solar global radiation G in July.

The insol installation was then simulated to cover almost all the Sahara's surface, for instance between 10–30 °N and between 5°W and 30°E longitude. This area is however smaller than the area with available meteorological data (see Fig. 2).

Generally, when a cloud cover exists, the radiation incident on a given location on the ground comes both directly from the sun and from other locations on the ground, after subsequent reflections on the ground and on the clouds base. Changing the natural surface albedo artificially changes the radiative transfer between the surface and the cloud base. In some special conditions, decreasing the ground surface albedo may lead to a decrease in the solar radiation incident on the ground (because the additional solar radiation flux due to multiple reflections decreases, too). This is accompanied by a decrease in the ambient temperature as the energy balance at ground level is altered. Decreasing the ambient temperature T is associated with an increase in the air's relative humidity, that finally leads to an increase in the cloud cover and, consequently, to a decrease in the relative sunshine σ (defined as the ratio between actual number of sunshine hours and the total number of daylight hours).

The available meteorological data (multi-year monthly means) allowed us to derive regression relationships of solar radiation versus other meteorological parameters. The following two linear regressions were used:

$$G/G_{ext} = \alpha + \beta\sigma \tag{7}$$

$$G = \alpha' + \beta' T_m, \tag{8}$$

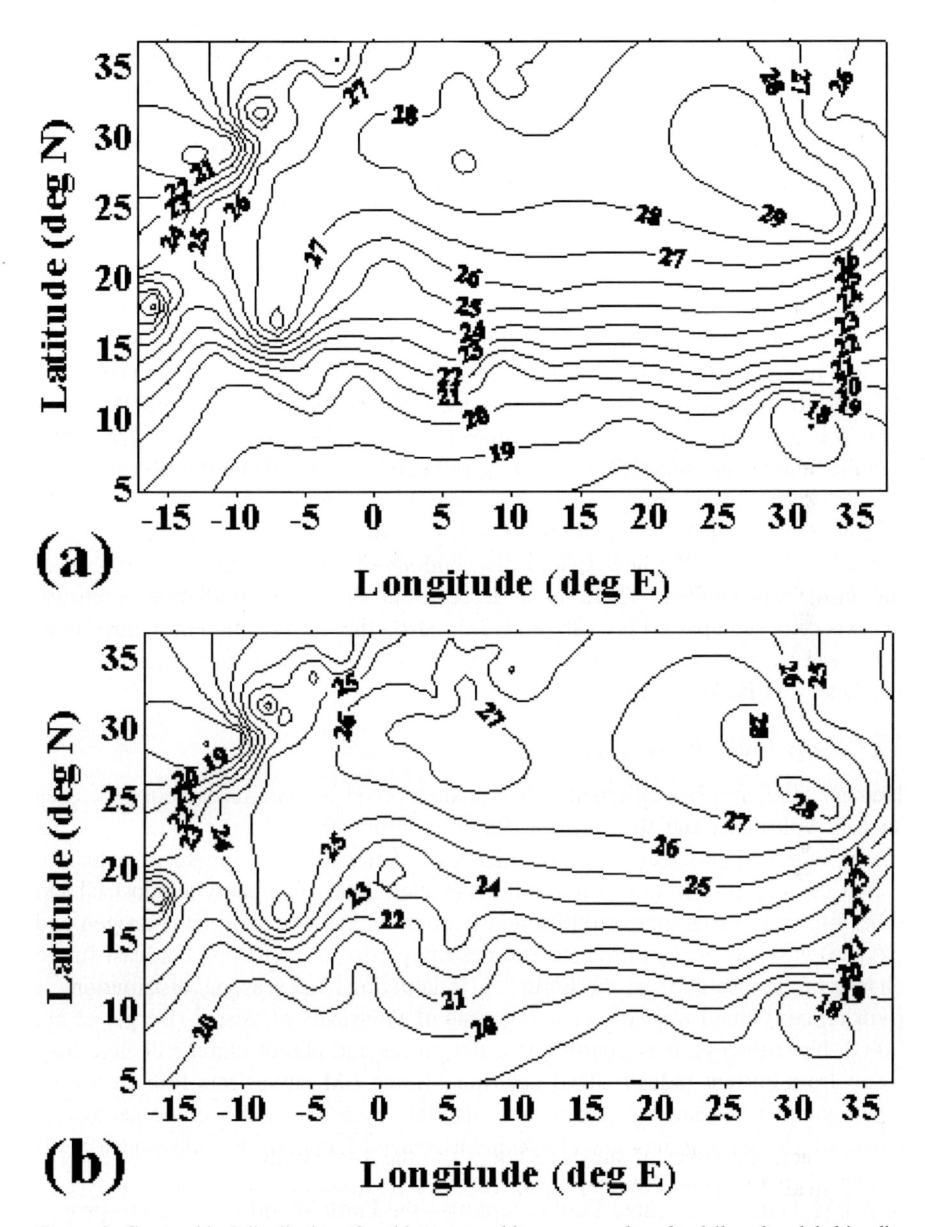

Figure 2. Geographical distribution of multi-year monthly average values for daily solar global irradiation on a horizontal plane ($MJ/m^2/day$) over Sahara in July. (a). Current conditions (the ground albedo was assumed $a_{soil} = 0.3$); (b) After the tent/insol was built (the tent/insol albedo was assumed $a_{insol} = 0.1$ (Cathcart and Badescu, 2004)

where T_m is the monthly mean ambient temperature. Equation (7) is the well known Angstrom-Prescott-Page regression (Cathcart and Badescu, 2004) where G_{ext} is solar irradiation on a horizontal surface at the top of the Earth-atmosphere. The regression coefficients α, β, α' and β were obtained by a usual least squares procedure, fitting Eqs. (7) and (8) to the available data from all the northern African sites.

The following relationship due to Averkiev (1951) relates G, σ, the ground albedo a and the solar irradiation incident on cloud tops, $G_\downarrow$

$$(9) \qquad G/G_\downarrow = 1 - a[0.2 + 0.5(1-\sigma)].$$

One can obtain from Eq. (9) the value $G_\downarrow$ of the solar irradiance incident on the top of the clouds, by using as known entries the measured data G and σ and the present-day soil albedo a_{soil} (sand's albedo was assumed $a_{soil} = 0.3$). The value $G_\downarrow$ just obtained is assumed to be the same after the "insol" is erected. The insol albedo is denoted a_{insol}. Then, new values can be obtained for the solar irradiation and the relative sunshine (G_{new} and σ_{new}, respectively) by solving the system of the Eqs. (7) and (9). Finally, by using Eq. (8) and the now known value G_{new} one could find the new calculated value of the monthly mean ambient temperature, $T_{m,new}$.

Figure 2b shows the new spatial distributions of solar irradiation in July after the insol was erected. There is a decrease in the solar irradiation intensity, as expected (compare Figs. 2b and 2a); also, the mean ambient temperature decreases by one or two centigrade degrees and the relative sunshine decreases (Cathcart and Badescu, 2004).

3.4.2 *The Worldhouse concept*

Humans often ascribe malignant or benignant motives to inanimate event-processes such as volcanoes and the weather. Rapid environmental change poses a special species survival problem—the potential loss of the Earth-biosphere as a habitat means the total loss of all possible refugia (Potter, 2000). Humanity's industrial and other life-style sustaining activities add (carbon dioxide) or subtract (nitrogen and oxygen) gases as well as produce and destroy particulate aerosols. Forecast future change of the insolation on the Earth "... in latitudinal and seasonal distribution ... [will remain] small over the next few tens of thousands of years" (Berger et al., 2003). Nevertheless, it is possible that dangerous and abrupt climate change may result from current industrialized activities. Some EM enthusiasts foresee a time when complete closure of energy and materials exchange will become necessary. Changed climate regimes may force industrialized humanity to construct a well-regulated ecosystem-world.

A EM devised finite-area Earthly tenting—the Earth Worldhouse macro-project described above—is a mirror of the proposed Mega-Float Giganticus, a mirrored or matte black collection of flat-decked barges anchored (or dynamically positioned) on the ocean precisely at the Equator covering $120 \cdot 10^3$ km^2 that is intended to alter the region's albedo for global climate control. Earth Worldhouse imitates, to a certain remarkable degree, the Mars terraforming prototype Worldhouse solution first advanced by Richard L.S. Taylor in 1992; Mars Worldhouse is a special structure, a long-span planet-enclosure structural system comprised of a relatively thin roof 3 km above the Mars-surface kept in place by fixed, ground-anchored towers (Taylor, 2001). A similar, secured special structure, if installed above the Earth, would indisputably transform its landscape from humankind's long-term bivouac and its barely explored seascape into a global anthropogenic civilization site separated from Outer Space by a new, human-made exterior.

So far, the world's largest greenhouse is the Eden Project in the UK's Cornwall region. The restrictive and stringent ethical considerations involved in the assessments for terraforming Mars do not logically apply to Earth-based EM (Pinson, 2002). Construction of an Earth Worldhouse can be much easier than building a Mars Worldhouse because humans are already settled here so that vital Worldhouse constructional aspects that are paramount to the ecologically established are not all that daunting to EM's experts.

Encasement of the Earth-surface—i.e., landscape and seascape—by a translucent and semi-permeable Earth Worldhouse roof will surely promote a simpler controlling mechanism for the weather contained therein than for the natural weather that now exists (Hoffman, 2002). See also chapter 7 of this book.

4. CONCLUSIONS

This chapter briefly reviewed a number of macro-projects related to power generation and to the deliberate management of the Earth's energy balance. As such these projects are examples of large-scale procedures of reducing un-wanted gas emissions and of engineering the climate.

REFERENCES

Assaf G (1976) The Dead Sea: a scheme for a solar lake. Solar Energy 18:291–299

Averkiev MS (1951) Meteorology. Moscow State University, Moscow

Badescu V, Cathcart RB (2006) Environmental thermodynamic limitations on global human population. Int J Global Energy Issues 25:129–140

Ball J (1933) The Qattara depression of the Libyan Desert and the possibility of its utilization for power production. Geogr J 82:289–314

Bassler F (1972) Solar depression power plant of Qattara in Egypt. Solar Energy 14:21–28

Belmiloud D (2000) Could water vapour be the culprit in global warming?. Geophys Res Lett 27:3703

Berger AL, Loutre MF, Crucifix M (2003) The Earth's climate in the next hundred thousand years (100 Kyr). Surv Geophys 24:123

Cathcart RB (2002) Unnatural envelopment: fieldwork on the active tectonics of J.G. Ballard's 'Build-Up'. J Geosci Educ 50:176–181

Cathcart RB, Badescu V (2004) Architectural ecology: a tentative Sahara restoration. Int J Environ Studies 61(2):145–160

Chen H-H (1995) Hydraulic modeling of wave-driven artificial upwelling. J Marine Environ Eng 1:263–277

Colacino M, Giorgi M (1968) Le rayonnement solaire et les echanges energetiques mer-air. Bulletin de la COMPLES 17:91–96

Defries RS (2002) Human modification of the landscape and surface climate in the next fifty years. Global Change Biol 8:438–458

Diller E, Scofidio R (2002) Blur: the making of nothing. Abrams, New York

Duffie JA, Beckman WA (1974) Solar energy thermal processes. Wiley, New York

Fogg MJ (1995) Terraforming: engineering planetary environments. SAE, Warrendale

Garrison VH (2003) African and Asian dust: from desert soils to coral reefs. Bioscience 53:469–480

Hall JS (2005) Nanofuture: What's next for nanotechnology. Prometheus, Amherst

Hatzikakidis AD, Sakas JG (1973) Factors affecting the thermic interaction between the sea and the atmosphere. Rev Internat Heliotechnique, Avril, p 6–13

Hawthorne C (2003) Turning down the global thermostat. Metropolis: Architecture Design 23:102

Hoffert MI (2002) Advanced technology paths to global climate stability: energy for a greenhouse planet. Science 298:981–987

Hoffman RN (2002) Controlling the global weather. Bull Am Meteorol Soc 83:241–248

Kettani MA, Scott RE (1974) A Red Sea heliohydroelectric (HHE) plant, Rev Internat Heliotechn, 2eme semestre, pp. 19–25

Kluver B, Martin J, Rose B (1972) Pavilion: by experiments in art and technology. Dutton, New York

Laframboise JG, Chou BR (2000) Space mirror experiments: a potential threat to human eyes. J Roy Astr Soc Can 94:237–240

Latham J (2002) Amelioration of global warming by controlled enhancement of the albedo and longevity of low-level maritime clouds. Atmospheric Sci Lett 3(2–4):52–58

Mangarella PA, Heronemus WE (1979) Thermal properties of the Florida current as related to ocean thermal energy conversion (OTEC). Solar Energy 22:527–534

Merkle R (1991) Theoretical studies of a hydrogen abstraction tool for nanotechnology. Nanotechnology 2:187–195

Mulhall D (2002) Our molecular future: How nanotechnology, robotics, genetics and artificial intelligence will transforms our world. Prometheus, Amherst

Pearce F (2000) A searing future. New Scientist 168:4

Pearson J (2002) Earth rings for planetary environment control, International Academy of Astronautics-02-U.1.01. 53rd International Astronautical Congress, Houston, Texas, USA, 15 pp 10–19 October 2002

Peixoto JP, Kettani MA (1973) The control of the water cycle. Scientific Am 228:46–61

Pinson RD (2002) Ethical considerations for terraforming Mars. Environ Law Reporter 32:11333–11341

Pohl JG, Montero J (1975) The multi-story air-supported greenhouse—a feasibility study. Architectural Sci Rev 18:50–59

Potter JF (2000) Seeking a new home: some thoughts on the longer term trends in planetary environmental engineering. The Environmentalist 20:191–194

Rush WF (1977) An orbiting mirror for solar illumination at night. Solar Energy 19:767–773

Salter S (2002) Spray turbines to increase rain by enhanced evaporation from the sea. Preprint, 10th Congress of International Maritime Association of the Mediterranean, Crete, May 2002

Schuiling RD, Badescu V, Cathcart RB, Seoud J, Hanekamp JC (2006) Power from closing the Red Sea. Int J Global Energy Issues (to be published)

Taylor RLS (2001) The Mars atmosphere problem: Paraterraforming—the Worldhouse solution. J Br Interplanet Soc 54:236–249

Vorosmarty CJ (2002) Global water assessment and potential contributions from Earth Systems Science. Aquatic Sci 64:328–352

CHAPTER 2

MINERAL SEQUESTRATION OF CO_2 AND RECOVERY OF THE HEAT OF REACTION

ROELOF DIRK SCHUILING

Institute of Geosciences, Utrecht University, PO Box 80021, 3508 TA Utrecht, The Netherlands

Abstract: Apart from saving energy, sequestration of CO_2 is the most direct way of combating the excessive greenhouse effect. Current approaches focus mainly on CO_2 storage in gaseous form in abandoned gas fields or aquifers. Sequestration in mineral form is still in its infancy, because the dry carbonation of common Mg- or Ca-silicates is unsuccessful. It can be deduced from natural examples that wet sequestration, combining hydration and carbonation is likely to be more successful. Several approaches are explored in this paper, either *in situ* in dunite massifs (olivine-rich rocks), or by reacting crushed olivine off-site in contained spaces with the off gases of thermal plants. The reaction produces a large amount of heat, which can be recovered as high enthalpy steam. In order to be effective, however, it should only be applied to large volumes of olivine, in a typical macro-engineering fashion, as the heat losses become unacceptably high in small systems with a high surface to volume ratio. One possibility would be to fill half of abandoned deep opencast mines with ground olivine and cover it by backfill. In the bottom part a mixture of hot CO_2 and steam is injected in order to set up a convective system similar to geothermal systems

Keywords: dunites; mineral carbonation; heat of carbonation and hydration

1. INTRODUCTION

Olivine-rich rocks known as peridotites or dunites are found in many locations in the world. The major olivine mines are in Norway, but olivine is also produced in Spain, Italy and Turkey. Olivine is a Mg-silicate with the formula Mg_2SiO_4. In natural olivines a small part of the magnesium is always replaced by divalent iron. These rocks originally stem from the mantle of the earth (a 2,900 km thick shell situated between the Earth's crust and the core) that consists mostly of the mineral olivine. In view of the availability of very large masses of olivine-rich rocks, many

V. Badescu et al. (eds.), Macro-Engineering: A Challenge for the Future, 21–29.

attempts have been made to sequester CO_2 by reacting it with olivine, but such attempts at dry reaction have failed (Lackner, 2002), or require a pre-treatment at high temperatures to make the MgO available for carbonation in a subsequent step (Kohlmann et al., 2002). A review of technologies for mineral sequestration of CO_2 has recently appeared (IEA Greenhouse Gas, 2005, and references therein). In this report it is confirmed that dry carbonation is not successful, and that so far all technologies for mineral sequestration of CO_2 are uneconomical, but that the direct aqueous carbonation route has the best prospects. Nature also suggests that a two-step reaction might be more favorable. When hot fluids pass through olivine-rich rocks at shallow depth, the rocks are transformed into a mixture of serpentine and magnesite (for the chemical formula of these minerals, see the reaction equation below).

The accompanying photograph (Fig. 1) of a hand specimen from a dunite at Orhaneli/Turkey demonstrates this phenomenon on a small scale. It shows a lens of chromite ore in a dunitic host rock. The chromite lens shows a number of parallel tension cracks filled with white magnesite. At the places where the cracks pinch out into the dunite rock, the dunite has reacted to form serpentine. Obviously, the fluids that cause these reactions are mainly mixtures of water and CO_2 in varying proportions:

$$(1) \qquad 2\underbrace{Mg_2SiO_4}_{\text{olivine}} + CO_2 + 2H_2O \rightarrow \underbrace{Mg_3Si_2O_5(OH)_4}_{\text{serpentine}} + \underbrace{MgCO_3}_{\text{magnesite}}.$$

The reaction has been experimentally verified in an autoclave, where fine-grained olivine was kept in a water-saturated steam mixture at 200 °C and a pressure of CO_2 of 36 bar for 24 hours. The X-ray pattern of the products showed clear reflections of serpentine and magnesite, and from the consumption of CO_2 during the experiment it can be judged that the reaction had proceeded approximately 10%.

Because the process offers a possibility to sequester CO_2 in mineral form, and recover the associated heat of reaction, a patent application has been filed at the patent office ("Octrooicentrum") of the Netherlands, entitled "Underground storage of CO_2 with geothermal effect" (Schuiling, 2005).

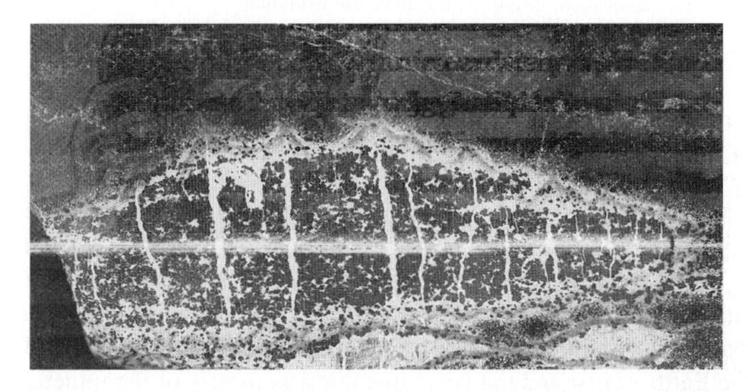

Figure 1. Hand specimen of olivine rock (dunite) near Orhaneli/Turkey, showing magnesite veinlets inside a chromite lense, topped with serpentine where the veinlets pinch out into the surrounding dunite. [Please see this figure in the color section at p. 309]

Table 1. Enthalpies at temperature 500 and 700 K of components participating in the serpentine/magnesite reaction Eq. (1)

Component	ΔH (kJ/kmol) (500 K)	ΔH (kJ/kmol) (700 K)
CO_2	−393.7	−394.0
H_2O	−243.8	−245.6
Mg_2SiO_4	−2170.2	−2168.1
$Mg_3Si_2O_5(OH)_4$	−4360.9	−4352.6
$MgCO_3$	−1112.0	−1112.0

2. THE REACTION RELEASES A LARGE AMOUNT OF HEAT

This can be easily understood from the fact that water in the crystalline lattice of serpentine and CO_2 in magnesite are in a well-ordered position, so their thermodynamic properties are similar to those of ice and solid CO_2. It is well known that the melting of ice followed by the evaporation of water, as well as the burning of limestone cost a large amount of energy. If the reaction is going in the opposite direction, forming hydrous silicates and magnesium-carbonate, this same amount of heat is set free. The importance of hydration energy as a possible contribution to heat flow was pointed out in 1964 (Schuiling, 1964a, 1964b). The thermodynamic data at 500 and 700 K are given in Table 1 (Robie et al., 1978).

This means that the reaction as written releases 250.3 kJ per 280 gram of olivine, or 890 Joules per gram at 500 K and a very similar value (240.4 kJ) at 700 K. As the specific heat of ultramafic rocks like dunites is of the order of $1250\,J.kg^{-1}.K^{-1}$ (Landolt-Bornstein, 1985), this means that a pure olivine rock would heat itself by 712 °C. In real life, however, the temperature will not rise above ~500 °C, because the equilibrium temperature of the reaction is around 500 °C, and part of the heat is required to raise the temperature of the introduced steam and CO_2.

If this man-made geothermal energy can be recovered, it qualifies as very green energy, because it is associated with a negative CO_2-emission!

3. CO_2 SEQUESTRATION

It can also be seen from the above reaction equation that we can sequester 44 gram of CO_2 with 280 gram of olivine (mol. wt olivine is 140). Even a not exceptionally large dunite massif like Orhaneli, NW Turkey (Orgun et al., 2004) contains already in its top kilometer several tens of billions ton of olivine, by which in the order of several billion ton of CO_2 can be sequestered in mineral form. This CO_2 is permanently fixed, in contrast to sequestration of gaseous CO_2 in aquifers or abandoned gas fields, for which there always remains a chance of escape.

It is interesting to note that one of the largest thermal power plants of Turkey, using lignite, is situated at a short distance from this dunite.

By capturing the off-gases of this plant, that consist mainly of steam and CO_2 (and nitrogen, of course), and injecting them down a borehole into the dunite at a depth of several hundred meters, one can start the reaction described above. While the CO_2 is fixed in mineral form, the temperature of the rock rises, and after some time it should be possible by using heat exchangers to tap this geothermal energy in the form of high enthalpy steam.

4. PROBLEMS

Although this scheme sounds attractive, it encounters a number of problems. The first one concerns the kinetics of the reaction. At ambient conditions, the reaction is essentially a weathering reaction. Olivine weathers relatively fast as compared to other silicates (Schuiling and Krijgsman, 2005), but even so the reaction rates near room temperature are much too slow to be of interest from a technological point of view. Even at 200 °C with fine-grained olivine (120 μ) no more than 10 % of the olivine had reacted in 24 hours.

There are two, possibly three ways, however, to increase the rate of reaction. One should, first of all, raise the initial temperature of the system. As a rule of thumb, most reaction rates more or less double for each ten degree temperature rise. This means that a reaction that would take 100 years to complete at ambient temperatures would take only a month at 110 °C. Secondly, the available surface of the reactants should be increased as much as possible, which means that the rock should be crushed and ground as finely as possible, or that the slimes of olivine mining operations should be used. Thirdly, if the gas mixture of steam and CO_2 to be injected contains any sulfuric acid or SO_2, this could be an advantage as well, because olivine dissolves fairly rapidly under acidic conditions (Jonckbloedt, 1997). It has been observed that the dissolution rate of olivine is barely affected by the CO_2 pressure, but the major parameter governing basic silicates dissolution is the pH of soil solutions in contact with minerals (Golubev et al., 2005). A higher initial acidity of the fluid leads to a rise of the Mg^{2+}-concentration in the pore fluid. After this acid is neutralized by further reaction with olivine, the resulting high Mg-concentration in the fluid is expected to enhance magnesite precipitation.

Another type of problems concerns the logistics. It will not often be the case, such as in Orhaneli, that dunite massifs are found in close proximity to large thermal plants. This means that one either has to move the CO_2 to a location with olivine, or the olivine to a location close to a thermal plant. Although one could check the possibility of transporting CO_2 in empty LNG tankers on their return trip, it will usually be necessary to transport the olivine.

In cases where crushed olivine must be transported and used in other locations, this will require a good thermal isolation of the volume of olivine to be reacted. If abandoned deep mines are used, it will also require enough backfill on top of the olivine in order to withstand the pressure of the injected high-pressure steam. The pressure of saturated steam at 250 °C is about 41 bar, so the backfill should have a thickness of about 200 meter (at a specific mass of the rock of 2,000 kg/m^3) to withstand this.

5. CONSEQUENCES FOR IMPLEMENTATION

If a thermal plant (or any other point source of CO_2) is close to a dunite massif, the off gases should be transported without cooling through an isolated pipe and injected into the dunite. One can search for tectonically crushed zones in the dunite and inject the gases in such a crushed zone. Unfortunately, nature in most cases will already have preceded us and will have altered these crushed zones and precipitated the same mixtures of serpentine and magnesite. The other possibility is to artificially crack a volume of fresh dunite by hydrofracturing or explosives, or to fill an underground mined-out cavity in the dunite with olivine dust and rubble. The reaction products occupy a larger volume than the original dunite, so some additional cracking of the rock may take place as a consequence of the reaction itself.

If the olivine must be transported close to the site where the CO_2 is produced, the operation of the system becomes complicated. One either needs huge pressurized vessels (industrial autoclaves) or a large underground cavity, where the surrounding rock can act both as a thermal isolator and as a pressure vessel.

Such underground cavities may take many forms. It would be an advantage to have underground cavities that are no longer in use, and even may have a negative value. One can think of cavities produced by solution mining of salt. For safety reasons, to prevent subsidence or even caving in, these must remain filled with saturated brine and indefinitely pressurized. Filling them with fine-grained olivine will push out an equivalent volume of brine, which can be recovered. Hot CO_2 will then be injected into the mixture of olivine grains with interstitial brine. Because the volume of serpentine + magnesite is larger than the volume of the olivine, the pore space between the olivine grains begins to be gradually closed, and the solid reaction product can support the cavity, obviating the need for further pressurizing and monitoring. Once the reaction is underway, the produced heat can be tapped by exchanging it from the solid rock to water circulating in pipes through the system.

Other possibilities involve the use of abandoned deep opencast mines. In most countries regulation requires that these holes in the ground are filled again and the landscape is rehabilitated after the ore deposit is mined out. A deep opencast dunite mine would obviously be the most appropriate choice, but such mines don't exist yet, as far as known. Several examples of deep open cast mines are diamond, copper, molybdenum and lignite mines. In the area between Cologne and Aachen in Germany, for example, several open pit lignite mines reach depths of over 500 meter. After the lignite is mined, these holes have to be filled again. Most of the lignite here is burnt in huge thermal power plants nearby. If one fills the bottom one to two hundred meters of the open-cast pit with ground olivine, lays out a tubing system for heat exchange while dumping the olivine, and then covers the top part of the pit with the original mining waste, the whole system can act as a huge reactor vessel. The backfill, of which the pore spaces are filled for the most part with ground water provides the required containing pressure, and the rock material surrounding the olivine provides an excellent thermal isolation. Part of the cost of filling the lower half of the abandoned pit with olivine can be offset against the costs required to fill the mine anyhow. If an opencast lignite mine with a length of 2 kilometer and a width of 1 kilometer is filled with a layer of 200 meter of olivine, it can sequester approximately 160 million ton of CO_2, and produce in the order of 150 billion kWh (note, however, that this is a non-realistic maximum, as it supposes that the complete heat of reaction can be recovered and converted to electric power).

Using huge abandoned holes as natural reactor "vessels" is a typical macro-engineering approach to mineral sequestration of CO_2. A major advantage of using natural cavities instead of huge industrial autoclaves is the fact that reaction kinetics do not play a major role. It is irrelevant whether the reaction proceeds in a few months or a few minutes, whereas reaction times must be as short as possible (preferably an hour or less) if expensive equipment like an industrial autoclave has to be used. This means that reaction times of 1,000 to 10,000 times longer than in an industrial autoclave are still acceptable. If even those reaction times are too short, one can trigger the process by placing a material that reacts faster with CO_2 (like steel slags, slags from the phosphorus industry or incinerator ashes) and also produces heat at the center of the olivine. This will locally increase the temperature, which in turn will speed up the olivine reaction.

A major disadvantage of the proposed sequestration technology is the fact that only one out of four Mg-ions that are made available as olivine is used for CO_2 sequestration, whereas the other three are used to make serpentine. This places a heavy burden on the transport costs of olivine, when it has to be transported off-site to a thermal power plant. A second disadvantage is that the reaction products are mixed together, which makes it virtually impossible to use the products.

Evidently, a reaction like

$$(2) \qquad Mg_2SiO_4 + 2CO_2 + 2H_2O \rightarrow 2MgCO_3 + Si(OH)_4$$

would considerably improve the economy of mineral sequestration, as all Mg-atoms are converted into magnesite. One could think of the following process. A long pressure vessel is heated at one end, whereas the other end is kept at a lower temperature. The olivine is loaded at the hot end, and a mixture of CO_2 and steam in a 1:1 molar ratio is added to the charge. The solubility of SiO_2 in water of 300 °C is in the order of 1 wt% (10,000 ppm) (Fyfe et al., 1978), whereas amorphous silica has a solubility of about 100 ppm at room temperature. Although no data on the variation of the solubility of magnesite with temperature are available, by analogy with Ca-carbonate (Fyfe et al., 1978) it can be expected that magnesite also becomes less soluble at high temperatures. This would mean that SiO_2 and $MgCO_3$ might precipitate at different places in a pressurized system with an imposed thermal gradient. Due to this gradient, there will be convection in the fluid. At the cold end of the vessel the SiO_2 will precipitate when the fluid cools, so the return water is strongly undersaturated when it comes again into contact with olivine at the hot end. More olivine will dissolve, and the Mg-concentration of the fluid will increase up to the point where magnesite starts to form. It is expected that the precipitation of magnesite will proceed more easily when carbonate and magnesium ions are both present in solution than in the case where a magnesium oxide or hydroxide must be carbonated. This is the case for many of the proposed mineral sequestration schemes that also use olivine rich rocks, but where Mg-oxide or hydroxide is first extracted from the rock (e.g. Lackner et al., 1997). If one can make the magnesite to precipitate in a different part of the pressure vessel than the silica, which is likely if their behavior is as predicted, it is possible at the end of the run to recover magnesite and precipitated silica separately, and use them as valuable products. Even if this turns out to be impossible, the proposed method has the advantage that all the magnesium of the olivine is converted to magnesite, by which the economy of the sequestration is improved fourfold.

This process also has its counterpart in nature. In many magnesite deposits like the magnesite deposits in the ultramafic rocks on the Kassandra Peninsula, the westernmost peninsula of Chalkidiki, Greece, it can be seen that the magnesite veins have formed in fissures of the host dunite, while at the same time extensive silicification has taken place, with precipitation of opal-like phases. The difference between the two processes in nature is probably that the reaction leading to the association of serpentine + magnesite takes place in a more or less closed system, whereas the combination magnesite + opal forms in open systems close to the earth's surface, which is also suggested by the observation that the magnesite veins become rarer and thinner towards depth. It would be worthwhile to see if we can create similar process conditions as in these magnesite deposits.

It is conceivable that a half open system as described for the abandoned lignite mines, in which the fluid is permitted to convect through the olivine grains between the hot bottom part and the cold top part could produce a similar result, and will possibly result in a silicified caprock. This in turn makes it possible to recover the heat of carbonation more completely. In fact, the reaction in which only magnesite and precipitated silica are formed seems to produce even more heat than the combined serpentine + magnesite reaction, as can be seen from Table 2.

For reaction Eq. (1) the enthalpy of reaction is calculated as −257.1 kJ (compare to −250.3 at 500 K and −240.4 at 700 K), whereas the enthalpy of reaction Eq. (2) after it is normalized to two olivine in order to make it comparable, comes out as −490.2 kJ, almost double the heat of reaction of the reaction leading to

Table 2. Enthalpies of components participating in reactions Eqs. (1) and (2) at 298 K

Component	ΔH (kJ/mol)
CO_2	−303.5
H_2O (ideal gas)	−241.8
Mg_2SiO_4	−2170.4
$Mg_3Si_2O_5(OH)_4$	−4361.7
$MgCO_3$	−1113.3
$Si(OH)_4$	−1460.0

Table 3. Energy production and CO_2 impact of raw materials discussed in this paper

	energy produced (MJ)	CO_2 emission/ immission (kg)
1 ton of coal	~29,300	~+2,830
1 ton of lignite	~17,000	~+1,200
1 ton of olivine (reaction Eq. 1)	917	−157
1 ton of olivine (reaction Eq. 2)	1,750	−628

magnesite + serpentine. Many of the parameters to operate such a system (rates of reaction, olivine grain size, optimum temperature and optimal temperature gradient) are not, or only poorly known, and will require extensive experimental study.

The energy production, and the CO_2 emitted or immitted per ton of raw material are summarized in Table 3. Although the energy produced by the mineral sequestration of CO_2 per ton of olivine is modest, although not insignificant compared to coal, it becomes quite important if the comparison is based on balancing the CO_2 emitted by coal combustion with an equal amount of CO_2 sequestered by olivine.

If all the CO_2 produced by the combustion of olivine is sequestered by reaction Eq. (1), it would require 18 tons of olivine, and produce 17,500 MJ. If the sequestration is done by reaction Eq. (2), it would require only 4.5 tons of olivine, but produce less energy, namely 7,900 MJ. This means that at locations where olivine resources are nearby and abundant, reaction Eq. (1) is to be preferred, whereas in situations where the transport costs of olivine dominate, reaction Eq. (2) would be more economical.

6. CONCLUSIONS

Dunitic rocks, or the olivine contained in them offer the possibility to sequester large volumes of CO_2 by applying a combined hydration/carbonation approach. The heat of reaction from the reaction by which olivine is converted to serpentine + magnesite is substantial and can be recovered. It is worthwhile to study a non-equilibrium process, where olivine is subjected to a thermal gradient in a convecting mixture of steam and carbon dioxide. This permits a more efficient use of the olivine for the mineral sequestration of CO_2, because all the magnesium is converted to $MgCO_3$. Many of the parameters governing the reactions as described are not, or poorly known and will require extensive experimental studies.

The macro-engineering approach outlined in this paper makes use of large man-made holes as pressure vessels for mineral sequestration and bypasses the problem of poor kinetics that must be faced when huge and expensive industrial autoclaves are used. If the reaction leading to magnesite + amorphous silica can be successfully applied in deep opencast mines, it may represent a breakthrough to the solution of the greenhouse problem, as the recovered heat of reaction is substantial, compared to the power produced by the combustion of fossil fuels.

ACKNOWLEDGMENTS

I wish to thank Mr. Koral Korkut of SETAT Mining company for his kind hospitality at the olivine mine near Orhaneli. I also thank Prof. P. Krijgsman for carrying out the experiment leading to partial conversion of ground olivine to a mixture of serpentine and magnesite.

REFERENCES

Fyfe WS, Price NJ, Thompson AB (1978) Fluids in the Earth's Crust. Elsevier Scientific Publishing Company, New York, p. 383

Golubev SV, Pokrovsky OS, Schott J (2005) Experimental determination of the effect of dissolved CO2 on the dissolution kinetics of Mg and Ca silicates at 25 °C. Chem Geol 217:227–238

IEA Greenhouse Gas (2005) Carbon dioxide storage by mineral carbonation. Report no 2005/11, prepared by Huijgen WJJ and Comans RNJ, 37 p

Jonckbloedt RCL (1997) The olivine process for the neutralization of waste acids. PhD-thesis, Utrecht University

Kohlmann J, Zevenhoven R, Mukherjee AB (2002) Carbon dioxide emission control by mineral carbonation: the option for Finland. INFUB, 6th European Conference Industrial Furnaces and Boilers, Lisbon, Portugal

Lackner KS (2002) Carbonate chemistry for sequestering fossil carbon. Annu Rev Energy Environ 27:193–197

Lackner KS, Butt DP, Wendt CH, Goff F, Guthrie G (1997) Carbon dioxide disposal in mineral form. Los Alamos National Laboratory, LA-UR-97–2094, 74 p

Landolt-Bornstein (1985) Geophysics of the solid earth, the moon and the planets, sub-vol B. Springer-Verlag, Berlin

Orgun Y, Gultekin AH, Cevik E, Copuroglu M (2004) Mineralogical and geochemical characteristics of the Orhaneli dunite and its importance in point of olivine, Orhaneli-Bursa, Western Anatolia, Turkey. In: Chatzipetros A, Pavlides S (ed) 5th Int Symp Eastern Medit Geol, Thessaloniki, Greece, vol 3, pp 1193–1196

Robie RA, Hemingway BS, Fisher JR (1978) Thermodynamic properties of minerals and related substances at 298.15 K and 1 bar (10^5 Pascal) pressure and at higher temperatures. Geol Surv Bull 1452:456

Schuiling RD (1964a) Serpentinization as a possible cause of high heat-flow values in and near oceanic ridges. Nature 201:807–808

Schuiling RD (1964b) Hydration energy – a possible contribution to crustal heat – a comment. Econ Geol 59:937–938

Schuiling RD (2005) Storage of CO_2 in the subsoil with geothermal effect. Patent Application No. 1028399 (in Dutch), Netherlands Patent Office ('Octrooicentrum')

Schuiling RD, Krijgsman P (2006) Enhanced weathering; an effective and cheap tool to sequester CO_2. Climatic Change, 74, 1–3, 349–354

CHAPTER 3

LARGE-SCALE CONCENTRATING SOLAR POWER (CSP) TECHNOLOGY

Solar electricity for the whole world

EVERT H. DU MARCHIE VAN VOORTHUYSEN

GEZEN Foundation for Massive-Scale Solar Energy, Groningen, The Netherlands

Abstract: Solar energy is the most abundant source of energy for mankind. Concentrating solar power (CSP) will become just as cheap as electricity from coal fired power stations or nuclear power plants. Operation S(un) is the macro-engineering approach for the implementation of solar energy on the globe. After an investment of 6300 billion dollar (which is 15% of the global Gross Domestic Product of one year) the solar thermal power stations produce just as much electricity as the global electricity consumption in 2003. One half of the plants is located at the coasts and their waste heat is applied for the desalination of seawater. The production of these stations satisfies the global municipal consumption of fresh water

Keywords: solar energy, solar electricity, solar thermal power, desalination, solar economy

1. THE HISTORY OF AN EXPLODING PLANET

The planet Earth, one of the smaller planets of the solar system, is experiencing a remarkable history. The conditions were favorable for the appearance of life. A tiny fraction of the energy of the solar rays was used to form complicated organic compounds together with oxygen from the carbon dioxide in the atmosphere and the water at the surface. A small fraction of these compounds did not end up as carbon dioxide, but was stored in the soil of the earth, and a small fraction of the oxygen remained in the atmosphere. Oxygen is a chemically active element, nearly all chemical elements and all compounds of carbon with hydrogen burn readily. During the billion years of photosynthesis nearly all carbon dioxide in the atmosphere was replaced by oxygen, and a large store of chemical potential energy was created at the surface of the earth in the form of the plants in the biosphere.

V. Badescu et al. (eds.), Macro-Engineering: A Challenge for the Future, 31–44.

Nearly all this potential energy was stored in the earths crust, in layers of coal, lignite, peat, oil, and gas. They are safely prevented from burning by layers of sediments and rocks.

After millions of years of storage of peat, lignite, coal, etc. the oxygen concentration in the atmosphere became so high that the potential energy of the plants became available. The second stage of life started. Animals do not obtain their energy from the sun, but from plants burnt inside their bodies, using the oxygen that is available now. A kind of equilibrium developed at an oxygen content of 20% and a carbon dioxide content of 0.2% in the atmosphere. The oxygen can do no harm, the fossil fuels are safely protected from the oxygen by thick layers of ground and rock, mainly consisting of oxides. The plants are protected from burning by their high water content.

Then, suddenly, enter mankind. The natural environment of man is the tropical climate. But after the discovery of fire, and after the discovery of the fossil fuels, mankind created a tropical climate in his houses, offices and public buildings over the entire globe. The natural time for resting is the night. But mankind created daylight in the streets, and in the houses and buildings. The natural limit of travel is one day on foot, but mankind created railways, cars, and airplanes, and the limit of traveling nowadays is the opposite side of the globe. The natural number of persons was limited by the yield from hunting, fishing and agriculture. But mankind has increased the output of agriculture tremendously. The population is growing exponentially, and there seems to be no limit to this growth. Right now the animal called homo sapiens is the most abundant species on earth, when measured in kilograms of living flesh.

Looking on a geochronological time scale, the proliferation of this single species appears as an explosion and it was made possible by a chemical explosion. The fossil fuels are returning to their ground states at an increasing rate, delivering their chemical energy to a mankind that is addicted to energy. The safe barrier between the components of the explosion oxygen and fuel is being removed by mankind at an accelerating rate. The explosion is now propagating at a rate of oxygen consumption equivalent to 2 ppm per year. At this rate the oxygen will have been removed from the atmosphere within 100,000 years, and all animal life, including homo sapiens, will have disappeared by then.

Long before that, other major changes of the global environment will have extinguished human civilization, and the global chemical explosion will come to an end. The most apparent physical effect is becoming visible: the rising average temperature on earth. Most scientists agree that this change of climate is the result of the anthropogenic greenhouse effect. The increased amounts of carbon dioxide (35%) and other greenhouse gases absorb an increasing fraction of the long-wavelength heat radiation originating from the surface of the earth that normally disappears into outer space.

1.1 Operation S

Let us now seek the best macro-engineering approach to stop the chemical explosion. Mankind is unable to stop the consumption of energy. We can reorganize our economies such that the efficiency of using energy is increased, but this cannot compensate the accelerating consumption of energy in fast growing countries like China, India and Brazil.

The best macro-engineering approach is to harvest the energy of the sun, solar energy, and to stop burning fossil fuels. In this chapter we will describe a project on a really global scale, that is the construction of sufficiently large numbers of solar thermal power stations in order to generate the total global electricity consumption.

We will give this macro-engineering project of global dimension a name: Operation S that is Operation Sun.

2. GENERATION OF SOLAR ELECTRICITY TO FEED THE WORLD

Most people are familiar with Photo-Voltaics (PV) as the technology for harvesting solar energy. However, Concentrating Solar Power (CSP) is a technology which produces solar electricity at much lower cost than PV (IEA, 2004; Trieb, 2005). In a solar thermal electric power station the rays from the sun are focused by mirrors onto (*i*) the boiler of a conventional thermal power station, (*ii*) a Stirling motor, (*iii*) (in the future) a gas turbine.

There exist two different types of mirrors, those with a line focus and those with a point focus. Most experience up-till-now is achieved with parabolic trough mirrors. A variant is the linear Fresnel reflector. Both systems have a line focus in which oil is heated, or water is boiled. Systems with a point focus are the solar tower with a field of mirrors having their focal point at the top of the tower, and the solar dish mirror. The linear Fresnel reflector will probably become the most economic CSP configuration (Mills and Morrison, 2000), see Fig. 1. The receiver is given by

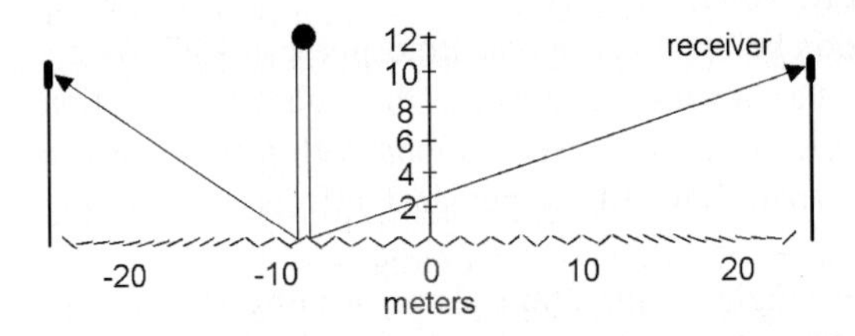

Figure 1. Schematic overview of the compact linear Fresnel mirror field of a solar thermal power station (Mills and Morrison, 2000). The long, flat, North-South oriented mirrors reflect the rays towards common receivers, which become extremely hot. Water which is pumped through the receiver is turned into high-pressure steam. During sandstorms/hailstorms the mirrors are inverted

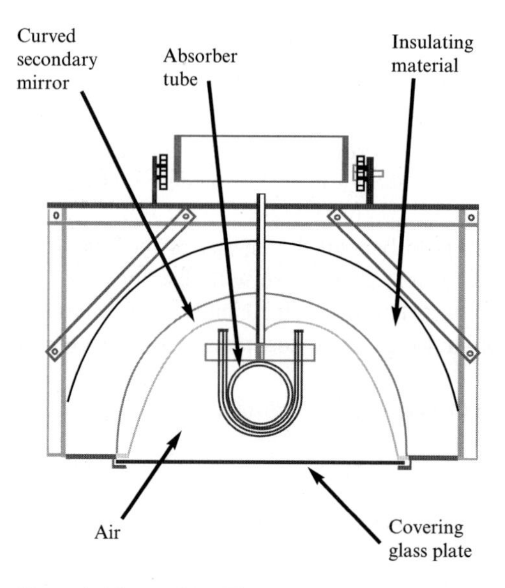

Figure 2. The receiver. The central part is the high-pressure boiler tube. The tube is coated with a spectral-selective layer which absorbs the short-wave solar radiation and inhibits the emission of the long-wave infrared radiation due to its own high temperature, typically 400 °C. The tube is surrounded by a curved secondary mirror and a glass entrance window (Morin, 2005)

Fig. 2 (Morin, 2005). In this chapter we calculate the capital investments that are necessary for the implementation of Operation S. All solar thermal power stations are of the type that is shown in Fig. 1.

2.1 Calculation of the Required Area of Mirrors

In the year 2003 electricity was consumed at an average rate 1686 GW (NEIC, 2005) which is 266 Watt per capita. The total amount of electric energy consumed in 2003 was $1.477 \cdot 10^{13}$ kWh (NEIC 2005).

The intensity of the solar radiation before entering the atmosphere is 1367 W/m^2. So an area of 1233 km^2 receives just as much solar energy as the global consumption of electricity. Unfortunately a cluster of solar power stations with a total surface of 1233 km^2 will never produce 1686 GW. Many physical effects are spoiling the game:

a) Absorption and scattering of sunlight in the atmosphere, leading to a reduction with absorption factor τ. The Direct Normal Irradiation, DNI, is equal to 1367 W/m^2.
b) The horizontal Fresnel mirror field is generally illuminated by oblique solar rays, leading to a reduced perpendicular irradiation $I_{hor} = \text{DNI} \cdot \cos\theta_z$ (the zenith angle θ_Z is the angle between the solar rays and the perpendicular).

c) Water droplet clouds and dust clouds. The thousands of solar power stations are expected to be located in climates where clouds are virtually absent.
d) Optical loss in concentrating the solar heat to high-temperature heat, we estimate this loss factor to be 50%.
e) Loss in generating electricity from high-temperature heat, including the parasitic losses of the pumps; we estimate the net electric production to be 30% of the high-temperature heat.

The efficiency of the solar thermal power plant is the product of the effects d) and e): 15%. Photovoltaic cells have the same efficiency, which is a remarkable coincidence.

The reduction effects a) and b) depend on the geographical location of the plant. We choose a representative latitude $\phi = 28°$, which corresponds to optimum locations for CSP plants such as Northern Mexico, Southern Morocco, Central Egypt, Kuwait, New Delhi (India), South-West China, Northern Chili, Central South Africa, and Brisbane (Australia). The absorption factor τ depends on the path length of the sunlight through the atmosphere, which in turn depends on the zenith angle θ_Z. As θ_Z depends on the time of day and the time of year, the irradiation on the horizontal mirrors I_{hor} must be integrated over the day in order to obtain the average irradiation on the mirror field as a function of date. Such calculations were performed by Du Marchie (2005). Table 1 gives the results for the daily DNI and the daily I_{hor} for 4 different days of the year for a location at latitude $\phi = 28°$ North.

The average daily Direct Normal Irradiation DNI is $7.46\,\text{kWh}/(\text{m}^2\text{day}) = 2723\,\text{kWh}/(\text{m}^2\text{year})$. The average daily irradiation on the mirrors of the solar power stations is $4.83\,\text{kWh}/(\text{m}^2\text{day})$ or $201\,\text{W}_{\text{th}}/\text{m}^2$ continuously. The efficiency of the solar thermal power plant is 15%, so $30.2\,\text{W}_{\text{e}}/\text{m}^2$ is available in the form of electricity.

The average citizen in the world consumes 266 Watt/person of electricity, which could be harvested from an area of $9\,\text{m}^2$ of mirrors. However, (*i*) these mirrors cannot be situated on the roofs of private houses like solar panels, but in large solar thermal power plants only, and (*ii*) a large part of the population is not living in the sunny climate which is needed for an economical operation of CSP plants. Their solar electricity has to be transported by means of long-distance transmission lines.

The total area of mirrors that is needed for producing the solar 1686 GW of Operation S is $1686 \cdot 10^9/30.2 = 55900\,\text{km}^2$, or a square area of 236 km in side. The total space to be reserved in the deserts of the world for the solar power stations of Operation S is less than Holland and Belgium added together.

Table 1. Daily irradiation data for latitude = 28° North

	December 21	March 21 and September 21	June 21
zenith angle θ_Z at noon	51.5°	28.0°	4.5°
Daily DNI (kWh/(m^{2}day))	5.67	7.53	9.09
Daily I_{hor} (kWh/(m^{2}day))	2.59	5.02	6.70

The electricity production per km^2 of mirror surface is $30.2\,MW/km^2$. An agglomeration of solar thermal power stations with the capacity of an average nuclear or coal-fired power station (about 1 GW), occupies an area of $33\,km^2$.

During the day the mirrors receive much more solar heat than the steam turbines can handle. This extra heat is stored in insulated tanks containing molten salts. When the sun is low, hidden by clouds or absent at night, this heat is used for the uninterrupted production of 1686 GW of electricity. The storage losses are negligible.

2.2 Cooling and Sea Water Desalination

Any thermal electric power station needs cooling. In the arid climate where the solar thermal power stations will be exploited cooling water is scarce, except in coastal regions, where sea water is available. The waste heat of these coastal CSP stations is extremely useful for the desalination of sea water. The product is distilled water, which can be used for producing drinking water, for irrigation and for industry. Figures 3 to 5 give the simplified block diagram of a complete solar thermal power station with sea water desalination. The plant is equipped with an auxiliary gas or oil fired boiler that provides heat when the heat storage tanks are empty. Gradually, during the transition of the fossil-fuel based economy towards the solar economy, natural gas and oil will be replaced by solar-generated hydrogen. The

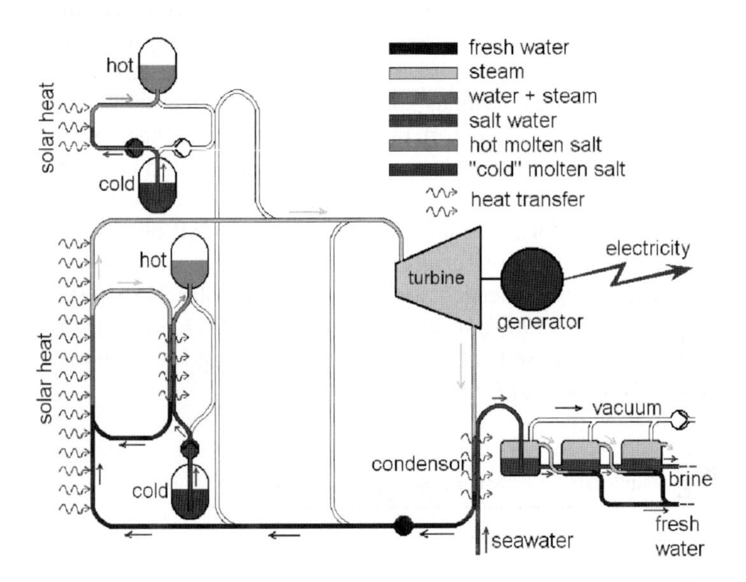

Figure 3. Block diagram of a solar thermal power station equipped with heat storage, auxiliary gas heating and sea water desalination in the three modes of operation. Solar operation at daytime, two thirds of the solar heat is stored in tanks containing molten salts. The upper tanks are destined for superheating the steam. [Please see this figure in the color section at p. 309]

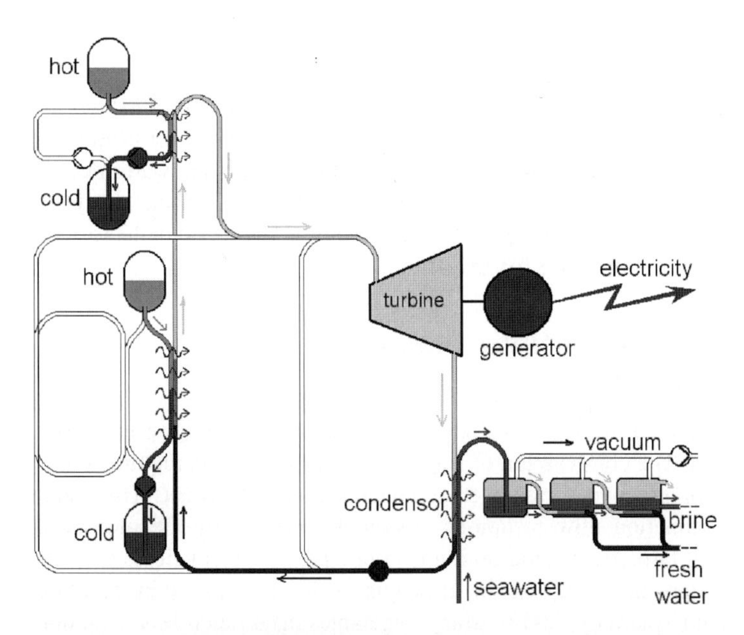

Figure 4. Same as Fig. 3. Operation at night, steam is produced out of the heat in the storage tanks. [Please see this figure in the color section at p. 310]

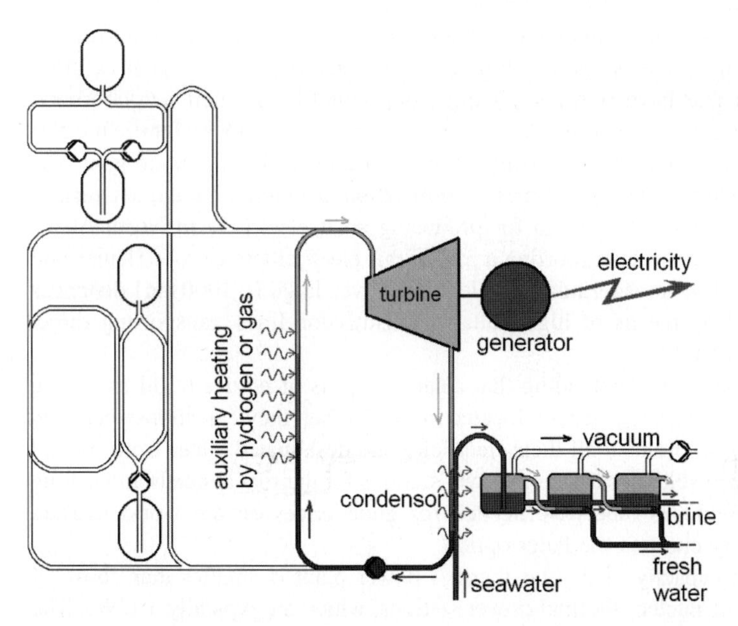

Figure 5. Same as Fig. 3. Operation on fossil fuel or hydrogen during cloudy days, when the heat storage tanks are empty. [Please see this figure in the color section at p. 310]

Multiple Effect Desalination (MED) factory consists of some 15 distillation vessels in series, operating at decreasing temperatures and pressures.

The production capacity per MW of electric capacity is about 1000 m^3/day of distilled water. If one half of the solar thermal power stations needed for producing all the electricity in the world (Operation S) were equipped with sea water desalination, the water production would be 843 million m^3 per day or 308 km^3 per year. This is equal to the total municipal consumption of water in the world or 12 to 15% of the world consumption of water for irrigation.

2.3 Economic Integration of Solar Thermal Power Stations in Developing Countries

Many deserts are situated in poor developing countries with growing populations and weak economies. The construction of solar thermal power plants requires much local labor and is of great importance for combating unemployment. The water from the solar desalination power stations is urgently needed by the municipal water utilities and by farmers. A good option is to operate the mirror fields at some elevation, 4 or 5 meters, and to utilize the underlying soil for advanced horticulture using drip irrigation (Nokrachy, 2004). Many vegetables need much less irrigation water when they grow in the semi-shade of the Fresnel mirrors (Bassam, 2004). A part of the irrigation water can be used beforehand for cleaning the mirrors.

The electricity is initially fed into the local grid, replacing oil, coal, or gas, and thereby relieving the national financial burden. Following the increase of the oil prices in 2004/5 the vulnerable economies of the developing countries are heavily burdened by the necessary imports of oil. Desert countries such as those in North-Africa that have rich energy importing countries in their neighborhood have the opportunity to export solar electricity. They can earn a significant part of their national income by exporting solar electricity and solar hydrogen. The Trans-Mediterranean Renewable Energy Cooperation is an international network of scientists that promotes a scenario for producing solar electricity in North-Africa and the Middle-East and for exporting a part of this electricity to Europe (Knies and Kabariti, 2004). The transportation of electricity over large (>1000 km) distances can be achieved by means of high voltage transmission lines transporting direct current, see Section 2.5.

It is a common misunderstanding that solar energy is of no use to oil-exporting countries. When these countries are located in the sun belt they can choose between either continuing to produce all their electricity and desalinated water from oil and gas, or investing in solar desalination power stations for their local needs and selling the unused oil and gas otherwise needed for good prices on the world market. A sensible society chooses the latter option.

The optimum capacity of a solar thermal power plant is smaller than those of fossil-fuel fired or nuclear thermal power stations, which are typically 1 GW_e. The area of the collector field of a 1 GW_e solar power station would be 33 km^2, leading to an average transport distance for the steam from the receiver to the turbine of more than 2 km. The inevitable losses of this transport make a solar thermal power plant of this size unattractive. The optimum size is the result of a trade-off between these losses and the optimum size of the power block (turbine+generator). It will probably be around 100 MW_e. Many of these units will probably be combined and will together form one large solar thermal power station.

The transition of the world economy towards a solar society in a short time, say 25 years, requires an enormous stream of investments in solar thermal power stations in the sunny countries all over the world. The world market for solar electricity to be delivered to the (inter)continental grid will be highly competitive, and only large internationally operating solar exploitation companies will be able to negotiate the most attractive power purchase agreements with the buyers of electricity and to obtain the best concessions in the desert countries. The management of the CSP stations has to be perfect and the working mentality has to be according to international standards. In many developing countries the culture of such efficient organizations will be regarded as rather alien and many large-scale projects fail. In many developing countries the culture and the existing economy favors artisans, small shops and farms, and disfavors paid employment. Solar thermal power is able to meet this preference of the local people for economic independence, at least for a fraction of the working people.

Figure 6 gives a schematic bird's-eye view of a coastal CSP-desalination plant of 100 MW. The produced electricity is delivered to the grid and most of the

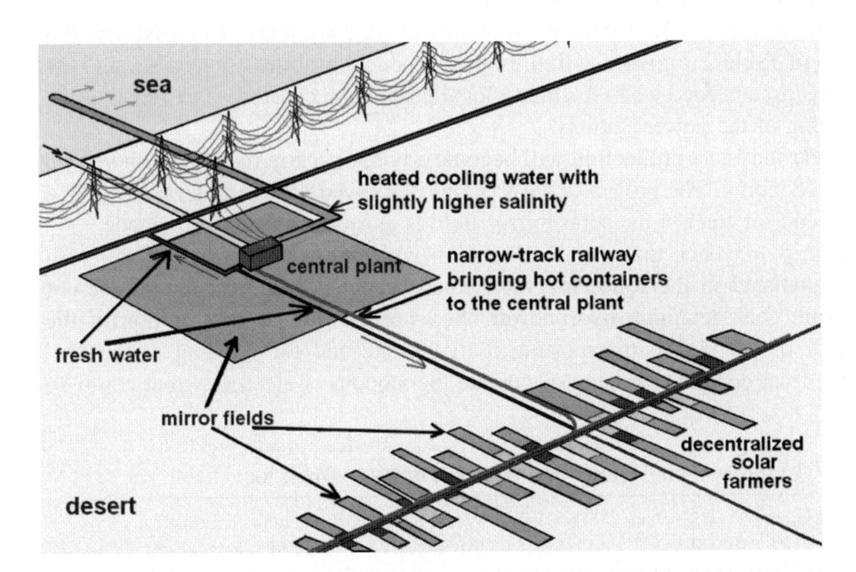

Figure 6. Birds-eye view of a coastal CSP-desalination plant of 100 MW surrounded by solar farmers. At daytime the independent solar farmer pump heat into containers. Hot containers are transported to the central plant and used for electricity generation at night. The solar farmers are partially paid by means of fresh water

produced water is pumped to the city nearby. By day the steam for the power block is produced in the 1.1 km^2 central mirror field that surrounds the central building.

At the same time some hundred independent "solar farmers" operate their own mirror fields and gather solar heat at high temperature in specially designed, well insulated heat containers. At the end of the day the hot containers at the farms are replaced by cold ones, and the hot containers are transported to the central building over a narrow-track railway. At night the turbine consumes steam generated with heat from the containers.

The solar farmers sell solar heat to the solar exploitation company and are paid pro rata in cash and in fresh water. The water is sufficient to produce food for their own families.

2.4 Total Cost of the Solar Thermal Power Stations Needed for Operation S

There are three ways to quantify the costs of an energy source: (i) the inventory of materials that are needed, (*ii*) the energy payback time, and (*iii*) the financial cost: dollars and euros, specified for a certain reference year because of the inevitable inflation.

At the solar energy department of the German institute for Aviation and Space, DLR, a thorough inventory has been made of the consumption of materials and energy during the construction and the exploitation of various types of solar thermal power stations (Viebahn, 2004). For CSP stations equipped with Fresnel mirrors, having a total capacity of 1686 GW_e (continuous), the amounts of needed material are given in Table 2, together with the world production of these materials, and the production time needed when the total global production capacity is assigned to the construction of the power stations.

It is clear that steel production will become a bottle-neck in the transition toward the solar economy. The project is feasible, but improved engineering resulting in a smaller need for steel in the solar mirror field is an essential step to be made.

The energy payback time of a CSP station with Fresnel mirrors is 6.7 months. It was calculated in the following way. Keep account of all primary sources of energy (oil, coal, uranium, hydropower, etc.) needed for the construction of the equipment of the plants, transporting it to the site and for building at the site, including final decommissioning. Calculate the amount of electricity that could be

Table 2. Need of materials for producing all electricity with CSP

Material	CSP plants (1686 GW) (Mton)	World production (Mton/year)	Production time for CSP plants (years)
Steel	3370	1050	3.2
bauxite	3.05	144	0.02
copper	2.17	15	0.14

generated from these amounts of primary energy. The solar power plant needs to operate a certain time to produce the same amount of electricity. This time is the energy payback time.

In the following paragraph, dealing with financial costs, euros and dollars are quoted directly from the literature, without any attempt to correct for inflation and without calculating any change of currency.

No new solar thermal power plants have been constructed during the last fifteen years. Therefore it is not easy to calculate the financial cost. Fortunately contracts for two major projects have been signed recently, the 100 MW Andasol Project in Spain and the 500 MW Solar Dish Project in California.

The Andasol Project comprises two 50 MW CSP stations with 6 hours of thermal storage. The solar field consists of parabolic trough mirrors produced by the German firm Schott. The investment is 520 M€ all told (Flagsol GmbH, 2005) for a solar power station that produces electricity 14 hours a day. If these stations were to be upgraded to 24 hours production, the mirror field would have to be expanded by a factor $24/14 = 1.7$ and the heat storage capacity by a factor $16/6 = 2.7$. The cost of the mirror field dominates the total cost, so we estimate a factor 2 increase of investment costs, making 1040 M€ for a 100 MW_e plant of base-load solar electricity, or €10.40 per Watt of installed base-load generation capacity.

A Californian grid operating company has recently signed the power purchase contract with the Stirling Energy Systems company (PESNetwork, 2005) for 20,000 solar dishes of 25 kW_e each. The total investment will be 600 M$ for a 500 MW_e plant running at daytime only. If this project were to be upgraded to a 500 MW_e base-load station, the number of parabolic dish mirrors would have to be tripled and some kind of heat storage would have to be added. We estimate the total investment of a dish-based 500 MW_e base-load solar plant to be 2100 M$, or $4.20 per Watt.

A very competitive solar power station using Fresnel mirrors has been proposed by the Australian company Solar Heat & Power (SHP, 2005) using a low-temperature 240 MW steam turbine and heat storage by means of underground pressurized hot water. A 240 MW_e solar power station with a capacity factor of 68% would cost 496 M$. According to their numbers upgrading to a base-load plant would increase the total investment to 708 M$, or $2.95 per Watt.

The mirror field needs more than half of the investment costs for a base-load solar thermal power plant. The main emphasis for reducing the costs should therefore be given to the mirror field. At the moment these costs are 206 €/m^2 for parabolic trough mirrors (Pitz-Paal et al., 2005). Fresnel mirrors are cheaper, 120–150 €/m^2 (Bockamp et al., 2003) and 103 $/m^2 (SHP, 2005) due to their more simple and more robust design. Therefore we have chosen to equip the thousands of solar thermal power stations that are needed for Operation S with linear Fresnel mirrors, see Fig. 1.

As with any emerging technology the costs of CSP technology components will decline, following a so-called learning curve. In the scenario of TREC-DLR (Trieb and Knies, 2004; Du Marchie, 2004) the costs of parabolic trough mirrors drop from 320 $/m^2 in 2006 to 124 $/m^2 in 2020 and ultimately to 106 $/m^2 in 2040. In the report by Sargent & Lundy (2003), these costs decline from 234 $/m^2 in 2004 to 181 $/m^2 in 2020. There are no learning curve estimates available for Fresnel mirror fields, but starting from the current situation, 103 $/m^2 to 150 €/m^2 it is not unreasonable to expect a reduction to 50 $/m^2 in the future for the total cost of a linear Fresnel field.

For the investment costs of the power block and the desalination equipment we conform to the TREC-DLR scenario. We apply heat storage in molten salts for the future costs as quoted in the TREC-DLR scenario (Trieb and Knies, 2004). We assume that half of the solar power stations will be built at coastal sites, and that these stations will be equipped with Multiple Effect Desalination (MED) facilities. For the total investment see Table 3.

A solar thermal power station with desalination produces 1000 m^3/day per MW$_e$ of installed continuous power.

In the start-up phase of Operation S the investment per GW$_e$ solar capacity will be higher. These costs will decrease towards their saturation value because of learning effects well before 10% of all investments are made. Therefore we hardly make an error in assigning this saturation value to all investments. So the total investment that is needed for the solar thermal power stations is about 5500 G$.

2.5 Total Cost of the Power Lines from the Solar Power Stations to the Consumers

Large sections of the population live in regions where solar thermal power stations cannot be exploited because of excessive cloud coverage or geographical latitude. This is the case for most of the Europeans, the Russians, the Canadians, the Japanese, many Chinese, many Americans, and most people living in a tropical monsoon climate. The solar electricity has to be brought to them by means of High Voltage Direct Current (HVDC) transmission lines and sea cables. Europe for instance has to be supplied from North Africa by means of 3500 km long power lines and 50 km of submarine cables.

We make the rough estimate that one half of the population of the world can obtain electricity from CSP stations that are located within 500 km, and for the

Table 3. Overview of the investments for the solar thermal power stations

	Unit costs	Needed for 1686 GW	Investment(G$)
Solar field	50 $/m^2	55900 km^2	2793
Power block etc	800 $/kW		1350
Storage cost	9 $/kWh$_{th}$	77560 GWh$_{th}$ (15.3 hours)	698
Subtotal			4841
Desalination	800 $/m^3/day	½ : 0.843 km^3/day)	675
Total			5516

other half the solar electricity has to be transported over 3500 km. In the latter case the resistive loss of electricity together with the loss at the AC-DC conversion stations is 14.5% (Trieb 2004). This loss has to be compensated for by adding 7.25% to the world capacity of solar power stations. The additional investment is $0.0725 \cdot 4841 = 351\,\text{G\$}$. The investment for 3500 km of power line, 50 km of submarine cable, and the conversion stations is 430 M$/GW (Trieb 2004). The total investment for transporting the electricity is $0.5 \cdot 1686 \cdot 1.0725 \cdot 0.43 = 390\,\text{G\$}$. Together with the extra stations that are needed for compensating the resistive loss we arrive at 741 G$.

3. CONCLUSIONS

The total investment that is needed for Operation S is $5516 + 741 = 6257\,\text{G\$}$. This is the investment for producing $1.477 \cdot 10^{13}\,\text{kWh/year}$ of electricity and 308 km^3/year of fresh water. This electricity and water is for free, because solar energy is free. The only costs are capital costs and operational costs. The capital costs depend on the interest rate and the period of repayment. When the power stations and the power lines are financed at a rate of 6% with a repayment period of 30 years, the capital costs are $0.07265 \cdot 6257 = 454.6\,\text{G\$/year}$. The operational, maintenance, and insurance costs are estimated to be 3% of the investment, leading to 187.7 G$/year. So the total cost is 642.3 G$/year.

The annual production of solar electricity is $1.477 \cdot 10^{13}\,\text{kWh/year}$ so the cost per kWh becomes $0.0435, which will certainly be cheaper than all other methods for producing electricity.

Besides electricity the power stations also produce 308 km^3 of fresh water annually. If this water is sold for $0.50 per m^3, the price of the electricity is reduced by one cent per kWh.

The total investment is just 15% of the Gross Domestic Product (GDP) of the whole world. If all countries invest 1% of their GDP in the solar infrastructure for the next 15 years, Operation S would become reality. The benefits are obvious. The consumption of coal would be drastically reduced and the consumption of oil would go down too, but at a lower rate. The production of radioactive waste would stop. The emission of carbon dioxide would decrease by about 30%. The most important benefit will be the proliferation of solar energy to other branches of energy. Solar hydrogen will be produced that will replace the oil in the transport sector. Additional solar power stations will produce extra electricity for heating houses and buildings by means of heat pumps. The global economy will be powered by renewable energy sources, and solar energy will provide the largest contribution.

The transition of electricity production from fossil fuel and uranium towards solar energy is a major operation, to be compared with the American space program in the sixties. When President Kennedy proclaimed: “Before the end of the decennium an American will walk on the Moon” a wave of enthusiasm swept through the nation and the objective was achieved. The world is now waiting new leaders who proclaim: “All electricity will become solar within a quarter of a century.”

REFERENCES

Bassam N (2004) Private communication. http://www.ifeed.org/

Bockamp S, Griestop T, Fruth M, Ewert M, Lerchenmüller H, Mertins M, Morin G, Häberle, A.Dersch (2003) Solar Thermal Power Generation. Presented at the International Conference Power-General Europe, Track 5, Paper 4, Düsseldorf, Germany

Flagsol GmbH (2005) http://www.flagsol.com/andasol_projects.htm

Freedman DH (2005) Looking Into the Sun, Inc., Vol 27, Issue 7, p 76. http://www.stirlingenergy.com/news/Inc_Article.pdf. Cited July 2005

Häberle A, Zahler C, Lerchenmüller H, Mertins M, Wittwer C, Trieb F, Dersch J (2002) The Solarmundo line focussing Fresnel collector. Optical and thermal performance and cost calculations. Presented at SolarPaces Conference, Zürich, Switzerland. http://www.solar-power-group.de/cms/upload/pdf/Fresnel_trough_cost_compare.pdf

IEA (2004) World energy outlook 2004, International Energy Agency (IEA), Paris, page 233

Knies G, Kabariti M (2004) Memorandum, how a trans-mediterranean renewable energy co-operation between Europe, the Middle east and North Africa (EU-ME-NA) can contribute to common water, energy and climate security. TREC, http://www.trec-eumena.org

Marchie van Voorthuysen EH du, (2004) Economic prospects of CSP, explanation of the TREC-DLR scenario of Trieb and Knies, 2004. http://www.gezen.nl/index.php?option=com_content&task=view&id=26&Itemid=53

Marchie van Voorthuysen EH du, (2005) The promising perspective of Concentrating Solar Power (CSP). Presented at the International Conference on Future Power Systems, Amsterdam, November 16–18

Mills DR, Morrison GL (2000) Compact Linear Fresnel Reflector solar thermal power plants. Solar Energy 68:263–283. http://solar1.mech.unsw.edu.au/glm/papers/CLFR–Geelong99V6.PDF

Morin G (2005) Presented at the Groningen Energy Convention, Groningen, The Netherlands, October 26–28

NEIC (2005) National Energy Information Center. http://www.eia.doe.gov/emeu/ international/electric.html#Consumption

Nokraschy HE (2004) Private communication. http://www.nokraschy.net

PESNetwork (2005) Worlds largest solar installation to use Stirling engine technology. http://pesn.com/2005/08/11/9600147_Edison_Stirling_largest_solar/

Pitz-Paal R, Dersch J, Milow B (2005) European Concentrated Solar Thermal Road-Mapping, DLR, Stuttgart, 2005, page 38. ftp: //ftp.dlr.de/ecostar/

Sargent & Lundy Consulting Group (2003) Assesment of Parabolic Trough and Power Tower Solar Technology Cost and Performance Forecasts Chicago.http://www.osti.gov/ dublincore/ gpo/servlets/purl/15005520-kLbVbt/native/

SHP (2005) Solar Heat and Power Ltd. http://www.solarheatpower.com

Trieb F (2005) Concentrating Solar Power Now, brochure of DLR, Stuttgart. http: //www.dlr.de/tt/institut/abteilungen/system/publications/Vortrag_CSP_07Druckversion.pdf

Trieb F, Knies G (2004) A Renewable Energy and Development Partnership EU-ME-NA for Large Scale Solar Thermal Power & Desalination in the Middle East and in North Africa. MENAREC Conference Sana'a, April 21/22. http://www.trec-eumena.org/documents/sanaa_paper_and_annex_2004_04_15.pdf

Viebahn P (2004) Ökobilanzen von SEGS-, FRESNEL- und DSG-Kollektoren, part of the SOKRATES-Projekt, Solarthermische Kraftwerktechnologie für den Schutz des Erdklimas, DLR internal report, Stuttgart, Germany. http://www.dlr.de/ tt/institut/abteilungen/system/projects/Stk/Sokrates/Technologien/AP%202-%20Oekobilanzen.pdf

CHAPTER 4

WIND PARKS, MARICULTURE, NUTRIENTS FROM ORGANIC WASTE STREAMS, CO_2 SEQUESTRATION: A FRUITFUL COMBINATION?

ROELOF DIRK SCHUILING AND GERRIT OUDAKKER
Inst.Geosciences, P.O.Box 80021, 3508 TA Utrecht, The Netherlands

Abstract: Major applications of macro-engineering are in the field of energy production. Wind energy, in the form of offshore wind farms is becoming a popular concept. Yet, some of the technical problems surrounding the maintenance and repair of wind turbines under harsh conditions make the economic feasibility of such wind farms doubtful. This is aggravated by the inherent weaknesses of large-scale energy production by wind, namely the unpredictability of the wind, and the low energy density of flowing air. Some possibilities to overcome some of these weaknesses are discussed in this paper, leading to the proposal to place the wind turbines on a ring dike. The enclosed basin has some potential for additional uses. Part of it can serve as a pump accumulator basin, where water is pumped in during periods of low power demand, when the power price is low, and passed through a turbine at periods of high demand, when the price is high

Keywords: offshore wind farms; mariculture; struvite and silica fertilizer; power optimization

1. INTRODUCTION

In the North Sea, off the coast of Denmark, a large wind farm has been constructed, and plans for the construction of more wind farms in the North Sea are in an advanced state. It is possible to adapt the infrastructure of those farms in such a way, that the system is optimized and can serve several different goals at the same time. A major problem with wind turbines at sea is that their accessibility for maintenance and repair is limited in periods of strong winds and high waves (Gore and Danby, 1999). This has become evident in the case of the Danish Horns Rev Park, 20 km offshore, where most of the wind turbines were out of commission after 2 years of operation due to break downs which could not be remedied in a timely fashion on

V. Badescu et al. (eds.), Macro-Engineering: A Challenge for the Future, 45–51.

account of the inaccessibility of the stand-alone turbines. The Danish government has decided to abandon its plans for the construction of additional offshore wind farms. It is possible, however, that accessibility to offshore structures may become easier in the future thanks to a recent invention (Tempel, 2004). Moreover, attempts are being made to reduce maintenance costs, e.g. by corrosion protection from sea spray (Gore and Danby, 1999).

By constructing the turbines on a ring dike (Fig. 1), their accessibility can be improved and the downtime of the system can be reduced. One can think of a basin with a diameter of 5 km. This will permit the emplacement of ~30 large 5 MW turbines approximately 1 km apart in order to reduce aerodynamic interaction. This number can be increased to 37 if the basin is subdivided into a power accumulator basin and two subbasins as in Fig. 2. The costs of construction can probably be considerably reduced if a high concrete ring is used instead of a dike. The concrete elements must be sunk in the seafloor, and on top of it a transport system can be constructed for the servicing of the wind turbines. The columns of the wind turbines can serve as anchoring points for the concrete elements. This is similar to the construction that is used to protect the central operating and living units of the Ekofisk oil field. The Ekofisk oil field, situated about halfway between Norway and Scotland in the transition between the North Sea and the Atlantic Ocean, often experiences very harsh conditions and extreme waves.

The optimal shape of such a structure depends on currents and frequent wind directions, and must be determined by a modeling study. Spare parts for turbine repair can be stored in close proximity in a shed on the enclosing dike, and can easily be transported to the required site if a railway track is constructed on top of

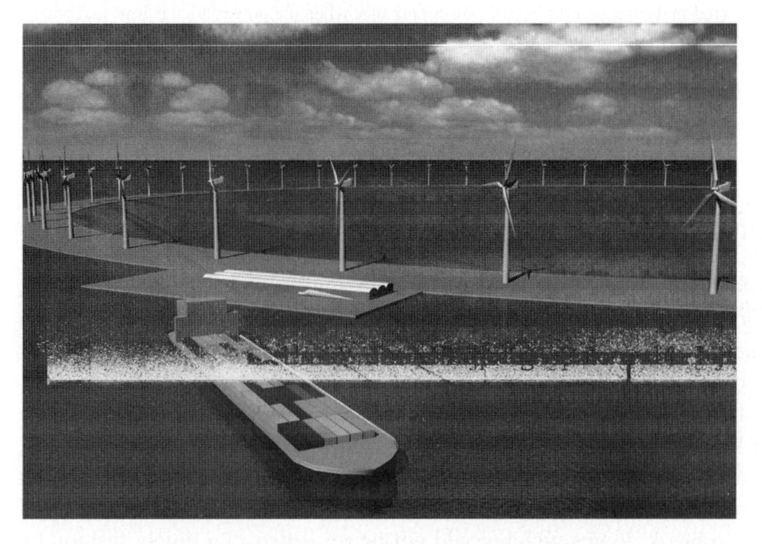

Figure 1. Artist view of an offshore wind farm on a ring dike for better accessibility of wind turbines for maintenance and repair. [Please see this figure in the color section at p. 311]

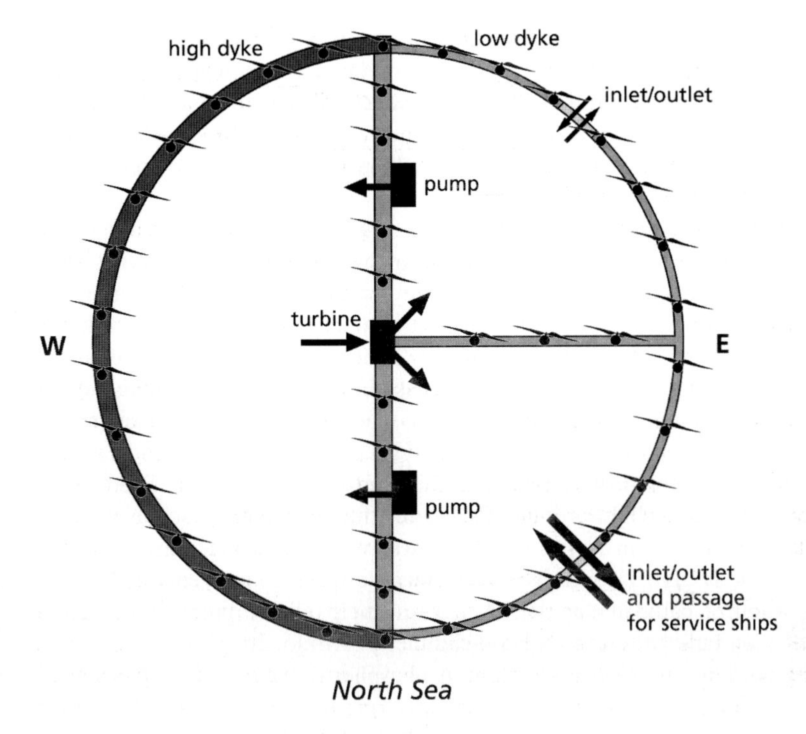

Figure 2. Lay-out for pump accumulator basin, turbine, pumps and seawater inlets/outlets. Wind turbines are constructed on top of the ring dike, as well as on the containing dike for the pump accumulator basin and the dike that separates the two subbasins

the dike or concrete ring. Moreover the enclosed basin can be used for mariculture, or more generally for biomass production and/or for storage of energy by pumping water to a high level when power demand is low, and releasing the water through turbines when power demand is high.

2. FISH CULTIVATION OR BIOMASS PRODUCTION

In most current hatchery systems, the food is provided as fishmeal. A more natural approach would be to arrange the system in such a way that the fish food is produced in the system itself. This can be achieved by adding the required nutrients in the form of fertilizer. The added nutrients will lead to an increased production of phytoplankton in the water, followed by an increase of zooplankton, which in turn will lead to a larger fish population, a system comparable to the Norwegian Maricult project (Olsen et al., 2001). In short, supplying the required nutrients will lead to the formation of a complete marine food chain in the system.

Instead of using commercial fertilizer one can also add the nutrients recovered from excess manure or other phosphate-rich organic waste streams. These must first be anaerobically digested at their place of origin, in order to liberate organically bound phosphate. The digestion produces biogas. After the digestion, the phosphate is recovered in the form of struvite, an ammonium magnesium phosphate with excellent properties as a slow-release fertilizer (Schuiling and Andrade, 1999; Ganrot, 2005), although it has not yet been tested as a fertilizer for mariculture. Struvite is a powdery finely crystalline material with the composition $(NH_4, K) MgPO_4.6H_2O$.

This struvite is transported to the wind farm, and spread over the waters of the enclosed basin. From some preliminary experiments that were carried out as part of a high school curriculum it appears that the productivity of the system can be increased threefold by adding a limited amount of amorphous silica to the fertilizer (Schuiling en van Haaften, 2001). This probably indicates a rapid growth of diatoms, which have a siliceous exoskeleton. The increase in biomass also accounts for the sequestration of a significant quantity of CO_2. The mean composition of marine biomass is characterized by the Redfield ratio of the components C:N:P in an atomic ratio of 116:16:1, although it is clear that this Redfield ratio has no fixed value. Most notably, resource (light or nutrients) acquisition machinery, such as proteins and chlorophyll, is high in N but low in P, whereas growth machinery, such as ribosomal RNA, is high in both N and P. Because these components make up a large proportion of cellular material, changes in their relative proportions have a marked effect on bulk cellular C:N:P stoichiometry (Arrigo, 2005). Even so, if P is the limiting nutrient, by adding one atom of phosphorus, we can fix 116 atoms of C as biomass. In terms of weight this means that 1gram of P fixes about 165 grams of CO_2 that it effectively withdraws from the atmosphere. It seems reasonable that the operators of the system are compensated by their governments under the rules of the Kyoto agreement. The governments can then claim this reduction of their CO_2 emissions as a partial fulfillment of their Kyoto obligations.

Instead of using the basin as a fish hatchery, there is also the possibility to use it for the cultivation of biomass for energy production. This will modify the hatchery concept in certain ways, because it opens the possibility to adapt the fertilization to obtain maximum productivity, probably in the form of algae or diatoms, instead of aiming to set up a complete food chain with optimum conditions for (commercial) fish cultivation. The use of the basin for biomass production implies that a simple system for harvesting the biomass is available. Production of algae in ponds and harvesting them for cosmetic use or as animal fodder is already operational on land.

3. ALTERNATIVE POWER USE

In itself, the production of electricity for the power net by wind turbines is a costly operation. When the wind speeds are too high, the turbines must be stopped, and when the wind speed drops by a factor n, the power production drops by a factor of n^3. In the future this situation may improve with the development of turbines that can operate even at force 12 winds (Hondebrink, 2005). The unpredictability of the wind requires that an approximately equal capacity of power production by conventional means must always be available on standby. This makes wind energy costly and unreliable, when used for power production. One may argue, however, that even if it is impossible to say whether next Thursday there will be sufficient wind to supply the electric grid, one can make a reasonable guess regarding the *average* power production of one turbine during a whole year, if the Weibull distribution, hat gives the statistical distribution of the wind speed for that particular site is known (e.g. Fernandez et al., 2005).

If wind farms can be spread over a geographically large area, this will also to some extent level out the variability of the wind (the wind will blow somewhere provided the area is sufficiently large). A case in point is provided by the wind farms on the south coast of Australia. They are sited on hills around 4 hundred meters up right on the coast over a wide area to obtain a more uniform average output. The distance covered for such a system to be effective probably needs to be at least 1000 km (Whisson, 2005). For the North Sea system such an approach is out of question.

It seems logical, therefore, to match wind energy with purposes where the exact moment at which the power is produced is not of prime importance. In view of the likelihood that, pushed by the exhaustion before long of fossil fuels (e.g. ASPO, 2004), our society will switch to hydrogen as the main energy carrier, one possibility would be to use wind turbines, and particularly wind farms at sea primarily for the production of hydrogen.

Another possibility to improve the economy of offshore wind parks on ring dikes would be to provide power to the net only during (expensive and profitable) peak hours, and to divert the remaining power to underwater lighting during the cheap night hours in order to increase the photosynthesis and thus the biomass production, particularly in the deeper part of the basin. Similar lighting systems to increase crop growth are widely used in greenhouses in the Netherlands.

The most practical approach to energy storage is probably the following. One can surround half of the basin with a high seawall, and use the turbines outside the peak hours to pump water into this basin. For improved protection the high part of the dike should face the major storm directions, which are southwest to northwest in the North Sea (Fig. 2). During peak hours the difference in water levels can be used for power generation. This is essentially a revival of the “plan Lievense”. Lievense proposed in 1979 to use part of the Ijsselmeer in the same fashion by surrounding it with a high ring dike on which wind turbines could be placed. At times of low power demand and high winds, the energy could be stored by pumping water into the basin, and this energy would be released again during peak hours. The plan was never executed, because it was alleged to be too costly, and it would lead to “visual pollution”. Part of the problem may have been that government and industry in the Netherlands are traditionally wary of innovations, because even at the time of conception it was calculated to be already economically feasible. If the plan Lievense is revived in the basin of an offshore wind farm, the cost aspect may be more favorable than in the proposed location in the early eighties, because part of the surrounding wall of the accumulator basin must be built anyhow as part of the dike surrounding the wind farm, and the reference level for competing power production, the price of oil, has risen dramatically in recent years.

Biomass, probably as phytoplankton, can be harvested more easily from flowing water while it is being pumped, or while it flows back during hydropower generation. In ponds on land, algae are already harvested for use in cosmetics or as animal fodder.

If the ring dike is constructed in a sea or an estuary with an appreciable tidal range, this can be used to advantage to increase the power production. The lower basin in the description above must be divided into two sub-basins A and B, which can be opened or closed to the sea. Additional wind turbines can be placed on the dikes separating these lower level basins from the accumulator basin and from each other. By pumping the water from a basin filled at high tide one spends less energy to fill the accumulator basin, and one produces more energy by releasing the water through the turbines into the sub-basin with its water level at the low tide level, thus providing the largest hydraulic head.

Instead of building the dike around the accumulator basin to its full height of 50 to 60 meter, it is probably cheaper to construct a concrete or steel seawall on top of a considerably lower dike, and still obtain the same hydraulic head. This is similar to the protective seawall that was built around the central service unit and the living quarters of the Ekofisk oil field in the North Sea, between Norway and Scotland, where weather conditions are fierce.

It is clear that even if parts of the ring dike are substituted by such concrete walls, the building of a ring dike to support wind turbines is more costly than constructing stand alone wind turbines each on their own column. It should be calculated if the considerable reduction in downtime of the turbines, the on-site production of hydrogen or the selling of power during peak demand, and its use for energy storage during low demand, the proceeds from the mariculture (either in the form of fish or biomass) and the associated CO_2 sequestration, justify the costs of building such structures.

4. CONCLUSIONS

Stand alone wind turbines in offshore wind farms suffer from their inaccessibility during periods with high winds or high waves, which makes timely repair and maintenance impossible, and causes a high percentage of down time.

The wind turbines become more accessible if they are placed on a ring dike, surrounding a basin. Spare parts can be stored on location.

The basin can be used for fish farming or biomass production.

The wind is erratic, it may blow when energy demand is low, and there may be calm when demand is high. By subdividing the enclosed basin into several parts, it is possible to store energy during periods of low demand by pumping water into an accumulator basin, and release it as hydropower during periods of low demand.

ACKNOWLEDGMENTS

I wish to thank Max Whisson for reviewing the paper and for his remarks that permitted me to widen its scope. Ir. D. Zwemmer is thanked for the information he provided on the "plan Lievense", the Pump Accumulator Power Plant. Fred Trappenburg is thanked for his artist impression of the ring dike with wind turbines. Frits Verbruggen kindly provided some information on the concrete protective ring that was built around the central servicing unit of the Ekofisk oil field.

REFERENCES

Arrigo KR (2005) Marine microorganisms and global nutrient cycles. Nature 437:349–355

ASPO (2004) The Association for the Study of Peak Oil & Gas. http://www.peakoil.net

Fernandez P, Gomez-Aleixandre A, Zubillaga S (2005) Wind energy on the shore coast: study case on Galicia's coast, Spain. In: Ozhan E (ed) Proceedings Seventh Intern Conf Mediterranean Coastal Environment. Medcoast, vol 1. Middle East Technical University, Ankara, pp 337–344

Ganrot Z (2005) Urine processing for efficient nutrient recovery and reuse in agriculture. PhD Thesis, Goteborg University

Gore D, Danby G (1999) Wind Power, Res. Paper 99/55, Library House of Commons, pp 1–31

Hondebrink JP (2005) Personal communication

Lievense (1979) Voorstel brandstofbesparing door toepassing van geaccumuleerde windenergie, te zamen met piekegalisatie van het landelijk electriciteitsnet. (Proposal for fuel saving by applying accumulated wind energy in conjunction with peak shaving of the national power net). Note presented to the Ministry of Science Policy in the Netherlands

Olsen Y, Bockmann T, Bokn S, Bremdal E, Oiestad V, Skjoldal E, Svendsen, Vadstein O (2001) Maricult Research Programme, 1996–2000, Final Scientific and Management Report from the Steering Committee, Publication of the Board of the Maricult Programme, printed by Tapir, Trondheim, 32 p

Schuiling RD, Andrade A (1999) Recovery of struvite from calf manure. Environ Technol 20:765–768

Schuiling MP, en van Haaften RM (2001) Meer biomassa in de oceaan, minder CO_2 in de lucht (More biomass in the ocean, less CO_2 in the air). Final year's essay, Christelijk Gymnasium Utrecht, 18 pp

van der Tempel J, (2004) Der Ampelmann. Safe and easy access to offshore Wind-Turbines. Poster presented at the European Wind Energy Conf., London, 22–25 November 2004

Whisson M (2005) Personal communication

Windenergie en Waterkracht (Wind energy and hydropower) (1981) Report of the Commission for the study of the 'plan Lievense'. Nertherlands Government Printing Office, the Hague, May 1981

CHAPTER 5

A MACRO-PROJECT TO REDUCE HURRICANE INTENSITY AND SLOW GLOBAL SEA LEVEL RISE

RICHARD LAROSA
sealevelcontrol.com, 317 Oak Street, South Hempstead NY 11550-7713, USA

Abstract: This chapter describes a macro-project to reduce hurricane intensity by cooling the tropical sea surface with cold water pumped up from the ocean bottom. The pumps are powered by heat engines that exploit the temperature difference between the surface and bottom water. If implemented in the Caribbean area, the cooler water carried northward by the Florida Current and Gulf Stream will reduce the loss of land-supported polar ice sheets. This will reduce the rate of sea level rise

Keywords: Hurricanes; ocean thermal energy conversion; effluent diffuser

1. INTRODUCTION

Hurricanes have devastated large portions of the U.S. Gulf States, Mexico, Central America, and the Caribbean Islands. The critical sea surface temperature required to form hurricanes is about 27–29 °C (Colling, 2001, p. 32). Available data show that the temperature of the sea surface in this area has been in this range during the hurricane season.

Meanwhile, northern glaciers are melting or sliding into the sea. This, plus thermal expansion of the water, causes the sea level to rise. The trend appears to be accelerating, and there are warnings of great future troubles (Barnett et al., 2005; Curry and Mauritzen, 2005; Guisan et al., 1995; Hansen et al., 2005). The world's population is concentrated along the coasts. Refugees from inundated coasts are hardly ever welcomed by residents of inland regions which, for the most part, already have strained economies. Flood control macro-projects are possible only in limited areas whose wealth can support their great cost.

The Earth is warming because the incoming short wave solar energy exceeds the outgoing long wave infrared radiation. The excess solar energy is mainly stored in the oceans because water has high solar absorption and oceans cover much of

V. Badescu et al. (eds.), Macro-Engineering: A Challenge for the Future, 53–63.

the Earth's surface. Greenhouse gases, such as carbon dioxide, water vapor, and methane, limit the rate at which infrared energy is radiated to space. The Kyoto Treaty and conservation measures attempt to reduce the rate at which greenhouse gases are being added to the atmosphere. Even if these measures are fully implemented, the injection rate of carbon dioxide will exceed its very slow decay rate. [Carbon dioxide concentration in the atmosphere would decay to 37% of its initial value in 100 years if no additional carbon dioxide were injected.] We have no direct control over water vapor. Methane concentration may increase due to emissions from thawing tundra and methane hydrates on the continental shelves.

Therefore, the radiation imbalance will continue and land and sea temperatures will rise. We have no adequate plan in place to combat this warming, the resulting sea level rise, and the increase in hurricane intensity and frequency. The author advocated the extraction of power by turbines and other devices from the Florida Current and Gulf Stream in order to slow them down (LaRosa, 2005). This was for the purpose of reducing the heat transport from the tropics to the Arctic, thereby slowing the destruction of ice sheets and the resulting sea level rise.

It was estimated that a power decrease of 28.6 GW would result in a 10% slowing of the North Atlantic Ocean current system. This might supply 14.3 GW to the electric power grid with 14.3 GW being dissipated in system losses, especially turbulence caused by deep-sea mooring cables. This delivered power is considerably less than the peak power supplied by Consolidated Edison and the Long Island Power Authority to their combined service area on a hot summer day.

After the 2005 hurricanes in the Gulf of Mexico, it became clear that slowing the Florida Current and Gulf Stream would decrease the heat transport from the Caribbean Sea and the Gulf of Mexico. This would raise the sea-surface temperature in these areas and increase the frequency and severity of hurricanes. This observation should serve as a warning to all those who might want to extract more than a small amount of power from these ocean currents. It also suggests that the extraction of power or slowing by other means should be prohibited by law.

Finding a method to cool the tropical sea surface is highly desirable. Previous attempts to devise ways to stir the tropical thermocline layer and transfer heat from the surface to the cold bottom water had not been encouraging. The only practical solution was to pump the cold water up from the bottom and mix it with the warm surface water. An estimate was made of the rate at which bottom water must be pumped to the surface and the pumping power required. It was evident that the power must come from converting the thermal energy stored in the ocean into mechanical power. Some partial solutions were found to the serious problem of preventing the cold water from sinking before it has a chance to cool the surface water.

This chapter offers a macro-engineering solution to lower the tropical sea-surface temperature, thereby reducing hurricane activity and exporting cooler water to the north via the Florida Current. This should reduce the heating of the Arctic area somewhat and increase the thermohaline circulation. If the rate of ice sheet destruction is not sufficiently reduced, other macro-projects beyond the scope of this chapter must be devised.

2. REDUCING SEA SURFACE TEMPERATURE

There is a vast reservoir of very cold water supplied by the Deep Western Boundary Current (Siedler et al., 2001) which flows in the deep abyss adjacent to the North American continent. The warm water at the surface is insulated from the cold water by the thermocline layer (Thurman and Burton, 2001). A stable temperature gradient is maintained in the thermocline because the warm water is less dense and stays on top. Buoyancy suppresses vertical motion, so there is little mixing between the different-temperature layers. Still water is a poor heat conductor.

The cold water can be brought up to the surface through long large-diameter vertical pipes that extend down 1000 m or more from floating pumping stations. These would be located in the Caribbean Sea and Gulf of Mexico. Bathymetric contours (Leier, 2001) suggest that the flow of large volumes of cold water into the Caribbean Sea basins and the Gulf of Mexico is sufficient.

To estimate the size of the cooling system involved, note that the water that enters the Caribbean Sea through the passages between the islands of the Antilles chain passes through the Yucatan Channel into the Gulf of Mexico, and then through the straits of Florida. At 27 °N the Florida Current contains water whose temperature ranges from 27.5 °C at the surface down to 7 °C at the bottom (Rosenstiel School, 2002). This data reflects the temperatures present in the Gulf of Mexico and the Caribbean Sea where hurricanes form. The temperatures at the high end of the range will move up or down seasonally and yearly, but the distribution of velocity over the temperature range should stay relatively constant. The volume transport rate of water at a particular temperature at the 27 °N location in the Florida Current should be the same as the transport rate for the corresponding temperature water going through the Antilles island passages.

The data referred to above is in the form of temperature and velocity profiles plotted on the 27 °N cross section of the Florida Current. The cross section was partitioned into small areas and the average velocity and average temperature of each area element were noted. For this estimate, a decision was made to cool all water above 24 °C down to 24 °C. The temperatures listed in Column 1 of Table 1

Table 1. Calculation of required cooling rate

T (°C)	Transport (Sv)	(T – 24) (°C)	Cooling rate (Sv °C)
27.5	0.086	3.5	0.30
27.3	0.089	3.3	0.29
27.0	4.103	3.0	12.31
26.8	0.193	2.8	0.54
26.5	1.764	2.5	4.41
26.0	1.346	2.0	2.69
25.0	0.986	1.0	0.99
24.5	0.132	0.5	0.07
Total	8.70		21.60

are the temperature readings that applied to the centers of differential areas included in the greater-than-24 °C category. The velocity times the differential cross section area is the volume transport of each area element. The elemental volume transport rates were summed for each temperature. Column 2 lists the volume transport in Sv (Sverdrups, equivalent to $10^6\,m^3\,s^{-1}$) of water at each temperature T.

The 8.7 Sv sum at the bottom of Column 2 in Table 1 is the total volume transport of all water whose temperature is greater than 24 °C. This agrees with the 8.9 Sv total Caribbean inflow (Schmitz, 1991) of water warmer than 24 °C.

The third column in Table 1 lists the temperature drops required to bring water at each temperature down to 24 °C. The product of the temperature drop times the volume transport rate is the cooling rate required. It is expressed in units of Sv °C. The cooling rates are listed in Column 4 for the respective water temperatures. Summing over this column gives a total cooling requirement of 21.60 Sv °C.

The somewhat arbitrary graphical partitioning of the 27° N cross section into area elements resulted in an uneven distribution of area elements and velocities among the different temperature bins. If the 27.5, 27.3, and 27.0 °C bins are grouped together, the required cooling rate is 12.9 Sv °C. The 26.0, 26.5, 26.8 °C group requires 7.64 Sv °C, and the 24.5, 25.0 °C group requires 1.06 Sv °C cooling rate. These smoothed results show the expected trend: the warmest water requires the greatest cooling rate.

The 21.6 Sv °C cooling requirement is to be satisfied by water pumped up from great depth. The pumping stations are in a location where the water at 1000 m depth is at approximately 6 °C (Colling, 2001, p. 207). This water is to be brought to the surface and mixed with the water described by the first two columns of Table 1. The result should be a mixed layer with a uniform temperature of 24 °C. The temperature of the cold water will be raised 18 °C and it will require a pumping rate of 21.6/18 = 1.2 Sv. This is 4% of the approximately 30 Sv total transport (Colling, 2001, p. 109) of the Florida Current. It will take many vertical pipes distributed over the area to carry water up to the surface at the required rate.

Now it is necessary to calculate the pump power required to bring the cooling water up from 1000 m depth. Pipe friction is a major factor that determines the pumping power required. Pipe friction is minimized by choosing large inside diameter, smooth inside walls, and low velocity. However, with the pump at the top, pulling up the water, the interior of the pipe is at a pressure lower than the outside water pressure. A large diameter pipe is more prone to collapse. Diameters up to 9.14 m were considered practical in the OTEC (Ocean Thermal Energy Conversion) program (Avery and Wu, 2004, p. 18). The present design is based on sections of fiberglass-reinforced plastic pipe 10 m long with flanges at each end and intermediate rings and ribs to resist buckling. The pipe is surrounded with foam to provide neutral buoyancy, and it is believed that a 10 m inside diameter is possible. The water velocity in the pipe is assumed to be $2.36\,m\,s^{-1}$, the velocity chosen by Vega (2002) for the cold-water pipe of an OTEC plant. The water in the pipe is at 6 °C and its kinematic viscosity (White, 1979, p. 676) is $1.45 \cdot 10^{-6}\,m^2s^{-1}$. The Reynolds number Re is given by

$$\text{(1)} \qquad \mathrm{Re} = \frac{VD}{\nu},$$

where V is the velocity, D is the pipe diameter and ν is the kinematic viscosity. The result predicted by Eq. (1) is $Re = 1.63 \cdot 10^7$.

Assume the inside surface roughness is 3.29 cm which, when divided by the pipe diameter, gives a roughness factor of $3.29 \cdot 10^{-3}$. Entering the roughness factor and the Reynolds number into a Moody chart (White, 1979, p. 333) gives a friction factor f = 0.0265. The suction pressure p at the top of the pipe required to overcome pipe friction is given by

$$\text{(2)} \qquad p = f \frac{L}{D} \frac{\rho V^2}{2},$$

where L is the pipe length and ρ is the density of the water in the pipe, obtained from Table 2.

Readily available (Colling, 2001, p. 113) temperature and salinity measurements at a point just east of the Gulf Stream on a line between Chesapeake Bay and Bermuda were used for Table 2. The data would be slightly different for the assumed pump locations, but the accuracy should suffice for this initial investigation. Table 2 lists the temperature T, salinity S and density ρ at depths from 0–1600 meters. The density values were obtained from the temperature and salinity values with the aid of Figure 2.1 of Pond and Pickard (1983).

The water in the pipe is taken from 1000 m depth, so its density is $1027.7\ \mathrm{kg/m^3}$. Using this value in Eq. (2), the suction pressure required to overcome friction is $7.6 \cdot 10^3\ \mathrm{N/m^2}$. Additional suction is required to lift the water against gravity because the water in the pipe has a density slightly greater than the average density of the water outside the pipe, which is $1026.925\ \mathrm{kg/m^3}$. The difference in density is $0.775\ \mathrm{kg/m^3}$ which, when multiplied by the pipe length and the acceleration of gravity, gives an additional suction requirement of $7.60 \cdot 10^3\ \mathrm{N/m^2}$. The total suction is therefore $1.52 \cdot 10^4\ \mathrm{N/m^2}$.

Table 2. Temperature, salinity, and density vs. depth

Depth (m)	T (°C)	S	P (kg m^{-3})
0	22	36.5	1025.4
200	19	36.6	1026.3
400	17	36.4	1026.6
600	15	36.0	1026.7
800	10	35.4	1027.4
1000	6	35.2	1027.7
1200	4.7	35.0	1027.7
1400	4.2	34.99	1027.8
1600	3.8	34.97	1027.8

The required pump power is obtained by multiplying the flow rate by the pump pressure. The pump power is 18.24 GW. It is prudent to almost double this value to 36 GW to allow for bends, constrictions, and the effluent diffuser to be described later.

3. OCEAN THERMAL ENERGY CONVERSION

The pumping power estimated in Section 2 must be derived from the thermal energy stored in the ocean. Ocean thermal energy conversion (OTEC) has a long history (Avery and Wu, 1994; Meyer, 2004; Vega, 2002). OTEC uses the temperature difference between the warm surface water and the cold bottom water to supply power to a heat engine. Cold water is pumped up through a pipe like the many pipes that would be needed to bring up the cooling water, except that the OTEC cold-water pipe will be longer in order to reach cooler water. This increases the efficiency of the OTEC thermal cycle. The cold water for the OTEC plant is pumped through a heat exchanger to condense a working fluid vapor that is exhausted from a turbine. The working fluid condensate is pumped into another heat exchanger. Warm ocean-surface water on the other side of this heat exchanger vaporizes the working fluid. The vapor enters the turbine and does mechanical work on the rotor, making it turn. The rotor shaft drives an electrical alternator, and in the present application, one or more pumps.

A detailed OTEC plant design (Vega, 2002) is used as a model because it incorporates much practical experience. The model OTEC plant produces 5.26 MW net electrical output using 26 °C surface water and 4.5 °C bottom water. The flow of top water is about twice the flow of bottom water. The model uses $13.9\,\mathrm{m^3 s^{-1}}$ of cold water brought up from a depth of 1000 meters. Our proposed system will bring the 4.5 °C OTEC water up from 1400 meters, but we can use a pipe with a diameter larger than the 2.74 meters of the Vega design in order to lower the parasitic pumping loss.

The warm water intake for the OTEC plant will face upstream. The surface-cooling water and the warm- and cold-water discharge from the OTEC plant will be pumped downstream in long perforated fabric hoses. These effluents are cooler and, therefore, more dense than the surface water coming from upstream. If the entire effluent is released in the neighborhood of the pumping station, it will form a plume that sinks without cooling the surface water. Therefore it must be released over a large area by the perforated fabric hoses, which are inflated by the pressure of the discharged water. The hoses will be kept horizontal and near the surface by floats so that the surface water will be cooled. The hoses present an impediment to navigation and fishing, so careful design will be required to properly disperse the cooling water with minimum hose length. OTEC plume studies (Avery and Wu, 1994, pp. 413–417) may be useful. Many OTEC floating plants discharged vertically downward (Avery and Wu, 1994, pp. 341–342, for example) and the sinking of the discharge plume was desired to avoid cooling the warm water intake.

Assume that pumps operating at 72% efficiency are substituted for the 95%-efficient alternator of the Vega OTEC model plant. To determine the OTEC cooling water flow rate required to supply the 36 GW of surface-cooling pump power, multiply the 13.9 m^3s^{-1} cold water input of the model by the power ratio $(36 \cdot 10^3/5.26)$ times the efficiency ratio (95/72). The result is $1.255 \cdot 10^5$ m^3/s, which is 10.5% of the cold water to be pumped up for surface cooling.

The OTEC cold water is passed through the condenser heat exchanger and is discharged at 10.3 °C so it adds to the cooling of the sea surface. The OTEC warm water is passed through the evaporator heat exchanger and is discharged at 3.1 °C below the surface temperature, so it adds a bit more to the sea surface cooling capacity. The overall plant could be scaled down to account for this added cooling capacity, but it is better to have some extra capacity to compensate for all the losses that are not being addressed in this initial design.

4. SYSTEM CONFIGURATION

The total surface-cooling water flow, excluding the OTEC plant discharge, is $1.2 \cdot 10^6$ m^3/s. Using 2.36 ms^{-1} velocity in 10 m diameter pipes, 6474 pipes are required. Consider placing two of these pipes and their associated OTEC plant on a pumping station. There would be about 3240 pumping stations.

Even with long perforated discharge hoses to distribute the cold water, it is easy to see that much warm surface water will pass around these stations without being cooled. The cold water near the discharge hoses may sink to a level where it may not mix with the warm water passing between the stations. Therefore it will be necessary to concentrate the stations in hurricane-prone areas of the Caribbean Sea and the Gulf of Mexico. The plants are intended to operate throughout the year with, perhaps, reduced output when the surface water is cooler in winter. The result will be that the water exported by the Florida Current and transported by the Gulf Stream and the North Atlantic Subtropical Gyre will cool the surface of the entire Atlantic Ocean slightly. The water transported to the Arctic by the North Atlantic Current will also become cooler. The Caribbean and Gulf surfaces might eventually need less cooling, and some pumping stations can be moved further out into the Atlantic to suppress hurricanes over a greater area.

The word "barge" has been avoided in describing the pumping station. Barges are flat-bottomed and their large area at the waterline is well suited for transporting heavy cargo or large structures. Because they float high in the water, they bob up and down on the waves, and are subjected to lateral forces in storms. This presents a serious problem when a barge has two 1000-meter-long pipes with 10-meter inside diameter and a 1400-meter pipe of smaller diameter suspended from its bottom. The pipes have just the limited flexibility allowed by play in the water-tight couplings joining the 10-meter-long sections. The couplings consist of grooved bands that are strapped around the perimeters of mating flanges. The stress at the barge attachment is relieved by ball-and-socket joints (Avery and Wu, 1994, pp. 193–199) or equivalent structures.

If the buoyancy is provided by a submerged hull with only columns of small cross section protruding through the sea surface, the waves can ride up and down the columns with only small changes in the buoyant force. This reduces the vertical motion of the hull due to waves. Also, this structure is subjected to less impact from breaking waves and horizontal surges. The pumping station for this project might use a submerged hull with one or two columns connected to hatches above the waterline. The hatches would have to be large enough to pass the largest components that are to be removed for servicing. The three cold-water pipes (two surface-cooling and one OTEC) would be on the center line of the hull. Allowing 12 m for each of the larger pipes and another 12 m for the smaller OTEC pipe, we have a length of 36 m plus the space required for the heat exchangers, which are quite large. Aluminum cold-water pipes would be rigidly attached to the bottom of the hull, connecting to the ball-and-socket connections to the plastic cold-water pipes. This would seem to lower the center of gravity and provide a righting moment to improve the pitch and roll stability of the hull. All of this is quite speculative, and much design work lies ahead.

The pipes would have a 90° elbow inside the hull so that the exit could be through the side of the hull, which would face downstream. The pumps would be axial flow with rotors in the horizontal exit of the elbow. This allows the shaft to protrude through the outside of the elbow. The two surface-cooling pumps would be driven by turbines. The OTEC cold- and warm-water pumps, as well as the working fluid pump and other auxiliary pumps, would be driven by electric motors so that the OTEC plant can be started from electric power provided by a service boat. After startup, the electric power can be provided from an alternator driven by the OTEC turbine.

5. MODIFIED CONFIGURATION

To improve the mixing of the cold effluent with the warm surface current, it may be better to use pumping stations which have only one cold water pipe. The pipe diameter would be 10 m and its length would be 1400 m to bring up the 4.5 °C water required for maximum efficiency of the OTEC thermodynamic cycle. Bringing all the water up at 4.5 °C will result in a lower quantity being required to cool the surface, which may offset the effect of the greater required suction on the total pumping power.

Reducing the size of the pumping station will result in more stations and closer spacing. Long perforated effluent diffusion hoses extended downstream may cool the adjacent surface water only to the point where it spreads out in a thin layer underneath the warm current that passes between the pumping stations. This close proximity will result in complete mixing and uniform surface temperature. Theoretical studies and experiments are needed to determine whether the foregoing desired result is obtained as opposed to the cooled water sinking to a lower depth. In the latter case, all is not lost because the cooled water should stay in the mixed layer and fully influence the surface temperature further downstream. Once again, theoretical analysis and tests are required.

The 10-m-diameter cold-water pipe will pass through the bottom of the hull, turn through a 90° elbow and exit the downstream side of the hull. The main cold-water pump is axial flow with its rotor in the downstream leg of the elbow. Its horizontal shaft comes through the elbow wall and is driven by the OTEC turbine, which also drives an alternator.

The cold water for the OTEC condenser is taken from the main pipe downstream from the main pump rotor. An electric-motor-driven pump will supply the pressure needed to force cold water through the condenser heat exchanger. The water leaving this heat exchanger will be about 10.3 °C (Vega, 2002) and will be discharged into a second perforated hose.

The warm surface water for the OTEC evaporator will be taken from the upstream side of the hull. An electric motor will drive a pump that forces the water through the heat exchanger and discharges it from a third perforated hose.

The ammonia (or other working fluid) pump and other auxiliary pumps will also be driven by electric motors. This enables the OTEC plant to be started from electric power supplied by a service vessel. Once the system is running, the electric power can be supplied by the alternator driven by the OTEC turbine. Electric-motor drive of the OTEC pumps also allows flexibility in the placement of these components within the confines of the semi-submersible hull. Many design iterations will be required to adjust the weight, hull volume, buoyancy, center of buoyancy, center of gravity, righting moments, hatch access, etc. to achieve the desired stability and storm survival. Progress in the design of this system will be reported in future publications.

6. ENVIRONMENTAL IMPACT AND FEASIBILITY

The intended impacts are the elimination of hurricanes, reduction of ice sheet destruction, and slowing of sea level rise. This is to be accomplished by lowering the sea-surface temperature in the Caribbean and Gulf of Mexico areas.

A theoretical study of the effect of 1000 OTEC plants in the Gulf of Mexico, each supplying 200 MW of electrical power, showed (Avery and Wu, 1994, p. 421) that after 10 years of operation the surface temperature would drop 1 °C and the temperature at the cold-water source depth would increase 5 °C. This system pumps up cold water at about half the rate of the hurricane suppression system proposed in this chapter.

The increase in bottom temperature reported above is not relevant to the surface cooling system of this chapter. The referenced OTEC plants discharged their effluent into the thermocline. The proposed surface cooling system will confine the effluent to the surface until it is mixed with the surface water and exported through the Caribbean Sea-Gulf of Mexico-Florida Current system.

OTEC has generally been regarded as a feasible technology by Avery and Wu (1994), Vega (2002), Meyer (2004), and many others. The proposed system is basically a large number of OTEC plants that pump water instead of making electric power or ammonia. By this reasoning, the proposed system is technically feasible. The only difference is a very long perforated hose required to disperse the effluent over the downstream surface.

There are many biological and environmental effects that must be investigated. Given that the pumping stations would be away from shore, the environmental problems should be less severe than those faced by OTEC plants located on or near the shore. Upwelled nutrients will be dispersed evenly over the surface to augment the nutrients upwelled by "natural" processes. This could be a benefit rather than a problem.

Because of its size, this macro-project will create a forced tropical upwelling of about $1.33 \cdot 10^6$ m^3/s. The water brought up from the bottom must be replaced, probably by water which flows along the bottom. This would strengthen the thermohaline circulation coming from surface water that sinks in the Arctic. Cooler water would be carried northward by the Florida Current and the Gulf Stream. This would make it easier for the water reaching the Arctic to sink. Studies are needed to quantify these effects, which appear offhand to be favorable.

To keep a proper perspective, it is also necessary to balance any negative environmental impact of this macro-project against the pollution, destruction of fisheries, health hazards, human suffering, and economic losses caused by hurricanes. The prevailing winds (Colling, 2001, pp. 20, 21) would carry slightly cooler air to Norway and Finland, as well as the British Isles and other parts of Europe. Considerable alarm (Guisan et al., 1995) has been expressed over the global warming effects on the Alps and Fennoscandian Mountains. People in these areas would be expected to welcome this macro-project.

The author has received some negative comments from people in the British Isles about a different macro-project (LaRosa, 2005) which might lower temperatures in their region. That particular project involved slowing the Florida Current-Gulf Stream system, and it has been abandoned as a result of the realization that these currents are carrying heat away from the tropical ocean. Whether these British people will have a negative reaction to the present macro-project proposal remains to be seen.

ACKNOWLEDGMENTS

The author thanks the referees who, based on experience with the OTEC program, emphasized the importance of designing the effluent dispersion system to spread the cold water evenly at the surface in order to avoid concentrated plumes that sink before they can mix with the warm surface water.

REFERENCES

Avery WH, Wu C (1994) Renewable energy from the ocean. A guide to OTEC. Oxford University Press, New York

Barnett TP, Pierce DW, Achuta Rao KM, Gleckler PJ, Santer BD, Gregory JM, Washington WM (2005) Penetration of human-induced warming into the world's oceans. Science 309(5732):254–255

Colling A (2001) Ocean circulation, 2nd edn. Butterworth–Heinemann, Oxford
Curry R, Mauritzen C (2005) Dilution of the northern North Atlantic ocean in recent decades. Science 308(5729):1772–1774
Guisan A, Holten JI, Spichiger R, Tessier L (1995) Potential Ecological Impacts of Climate Change in the Alps and Fennoscandian Mountains. Geneva Department of Cultural Affairs, Geneva, pp 15–184
Hansen J et al. (2005) Earth's energy imbalance: confirmation and implications. Science 308(5727): 1431–1435
LaRosa R (2005) http://www.sealevelcontrol.com
Leier M (2001) World atlas of the oceans. Firefly Books, Buffalo, pp 134–135
Meyer R (2004) OTEC–where is this promising technology today? Ocean News & Technol 2004: pp. 58–60 May/June
Pond S, Pickard GL (1983) Introductory dynamical oceanography, 2nd edn. Butterworth–Heinemann, Oxford, p 6
Rosenstiel School (2002) http://oceancurrents.rsmas.miami.edu/ atlantic/ img_florida/ Fl_trans.jpg
Siedler G, Church J, Gould J (2001) Ocean circulation and climate. Academic Press, San Diego, p 264
Thurman HV, Burton EA (2001) Introductory oceanography, 9th edn. Prentice Hall, Upper Saddle River, p 543
Vega LA (2002) Ocean thermal energy conversion primer. Marine Technol Soc. J 6(4):25–35
White FM (1979) Fluid mechanics. McGraw-Hill, New York

CHAPTER 6

MITIGATION OF ANTHROPOGENIC CLIMATE CHANGE VIA A MACRO-ENGINEERING SCHEME: CLIMATE MODELING RESULTS

GOVINDASAMY BALA[1] AND KEN CALDEIRA[2]

[1] *L-103 Energy and Environment Directorate, Lawrence Livermore National Laboratory, Livermore, CA 94550*

[2] *Department of Global Ecology, Carnegie Institution, 260 Panama Street, Stanford, CA 94305*

Abstract: It has been suggested that climate change induced by anthropogenic CO_2 could be cost-effectively counteracted with macro-engineering schemes designed to diminish the solar radiation incident on Earth's surface. It is clear that such schemes could counteract global and annual mean global warming. However, the spatial and temporal pattern of radiative forcing from greenhouse gases such as CO_2 differs from that of sunlight, therefore it is uncertain to what extent these macro-engineering schemes would mitigate regional or seasonal climate change. The NCAR atmospheric general circulation model, CCM3, has been used to study this issue; in these simulations, the solar radiation incident on the Earth was diminished to balance the increased radiative forcing from a doubling and quadrupling of atmospheric CO_2 content. The results indicate that, despite differences in radiative forcing patterns, large-scale macro-engineering schemes could markedly diminish regional and seasonal climate change from anthropogenic CO_2 emissions. However, there are some residual climate changes in the Macro-engineered $4xCO_2$ climate: a significant decrease in surface temperature and net water flux occurs in the tropics; warming in the high latitudes is not completely compensated; the cooling effect of greenhouse gases in the stratosphere increases and sea ice is not fully restored. The stratospheric cooling becomes larger also in the Macro-engineered $2xCO_2$ climate, and the additional cooling due to macro-engineering could enhance stratospheric ozone depletion. The impact of these climate stabilization schemes on terrestrial biosphere is also investigated using the same climate model. Results indicate that climate stabilization would tend to limit changes in vegetation distribution brought on by climate change, but would not prevent CO_2-induced changes in Net Primary Productivity (NPP) or biomass; indeed, if CO_2 fertilization is an important factor, then a CO_2-rich world with compensating reductions in solar radiation could have higher net primary productivity than our current world. However, CO_2 effects on ocean chemistry could have deleterious consequences for marine biota. Caution should be exercised in interpretation because these results are from a single model with many simplifying assumptions. The most prudent and least risky option to mitigate global warming may be to curtail emissions of greenhouse

V. Badescu et al. (eds.), Macro-Engineering: A Challenge for the Future, 65–86.

gases. Nevertheless, studying macro-engineering will provide us the scientific basis to understand the possibility of rapidly counteracting catastrophic global warming without inadvertently creating a bigger problem

Keywords: macro-engineering; solar radiation; mitigation; climate model; terrestrial biosphere

1. INTRODUCTION

Several schemes have been proposed to counteract the warming influence of increasing atmospheric CO_2 content with intentional manipulation of Earth's radiation balance (Budyko, 1977; Early, 1989; Seifritz, 1989; NAS, 1992; Watson et al., 1995; Flannery et al., 1997; Teller et al., 1997). These 'macro-engineering' schemes typically involve placing reflectors or scatterers in the stratosphere or in orbit between the Earth and Sun, diminishing the amount of solar radiation incident on the Earth. However, the radiative forcing from increased atmospheric carbon dioxide (Kiehl and Briegleb, 1993) differs significantly from that of a change in effective solar radiative flux (List, 1951). For example, carbon dioxide traps heat in both day and night over the entire globe, whereas diminished solar radiation would be experienced exclusively in daytime, and on the annual mean most strongly at the equator, and seasonally in the high-latitude summers (Fig. 1). Hence, there is little *a priori* reason to think that a reduction in the solar flux incident on the Earth would effectively cancel CO_2-induced climate change (Schneider, 1996). One might expect, on the basis of the considerations above, that a macro-engineered CO_2-laden world would have less of a diurnal cycle, less of a seasonal cycle, and less of

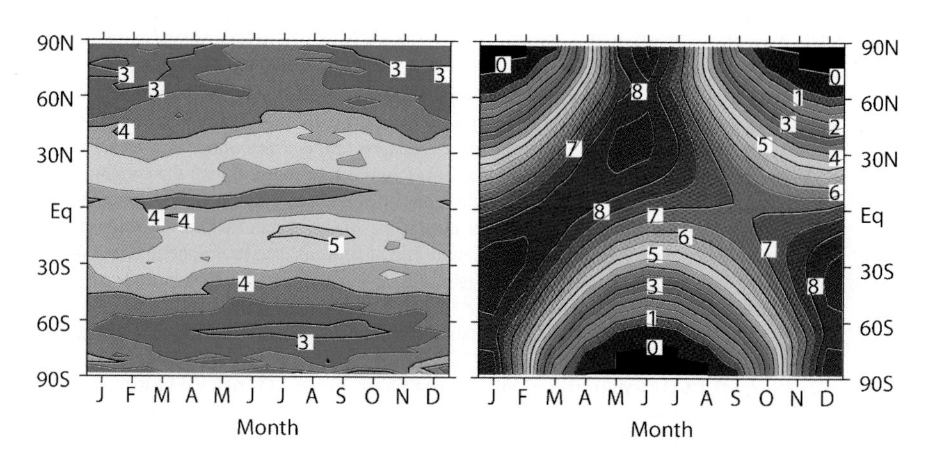

Figure 1. Change in net long wave radiative flux at the tropopause when CO_2 is doubled (left panel) and the reduction in incoming solar radiation (right panel) needed to compensate the global-mean of this forcing. Both values (Wm^{-2}) are zonally averaged as a function of time of year. Change in solar radiation has a latitudinal and seasonal pattern markedly different from the radiative forcing of CO_2

an equator-to-pole temperature gradient than would have existed in the absence of human interference in the climate system. Such changes, even in the absence of globally and annually averaged warming, could produce damaging regional and seasonal climate change.

This chapter will focus on the effects of macro-engineering schemes that diminish the amount of solar radiation incident on the Earth. Modeling results on residual regional and seasonal climate change are discussed. The term 'geoengineering' has been also used in the literature to refer to macro-engineering. In Section 2, the General Circulation Model (GCM) that has been primarily used to investigate this problem is described. Sections 3 and 4 discuss the mitigation and residual climate change in macro-engineered $2 \cdot CO_2$ and $4 \cdot CO_2$ worlds respectively. Section 5 discusses the effects of these schemes on the terrestrial biosphere. The concluding remarks are given in Section 6.

2. THE GENERAL CIRCULATION MODEL

To investigate the mitigation of climate change using macro-engineering schemes that reduce solar radiation, the Community Climate Model (CCM3) developed at the National Center for Atmospheric Research (Kiehl et al., 1996) has been used primarily (Govindasamy and Caldeira, 2000; Govindasamy et al., 2002; Govindasamy et al., 2003). This is a spectral model with 42 surface spherical harmonics to represent the horizontal structure of prognostic variables in the atmosphere: the horizontal resolution is approximately 2.8° in latitude and 2.8° in longitude. The model has 18 levels in the vertical, and the upper lid of the model is at 2.9 mb. An important aspect of CCM3 is that it has very little systematic bias in the top-of-atmosphere and surface energy budgets. A version of the model with a simple slab ocean-thermodynamic sea ice model, which allows for a simple interactive surface for the ocean and sea ice components of the climate system is used. The slab ocean model employs a spatially and temporally prescribed ocean heat flux and mixed layer-depth, which ensures replication of realistic sea surface temperatures and ice distributions for the present climate. For the climate-change simulations that represent deviations from the present day climate, we use the same prescribed ocean heat flux, and mixed layer depth. Therefore, this approach does not include the feedbacks from a dynamic ocean. Typically, the slab-ocean model needs to run for ~20 years to reach equilibrium.

3. MITIGATION OF CLIMATE CHANGE FOR A DOUBLING OF CO_2

3.1 Experiments

To evaluate the extent to which an effective reduction in solar radiation might mitigate the climate impacts of doubled atmospheric CO_2 content, Govindasamy and Caldeira (2000) performed three model simulations using the model described in Section 2: (i) "Control" or pre-industrial, with a CO_2 content of 280 ppm and a solar "constant" of 1367 W/m^2; (ii) "2 · CO_2", with doubled atmospheric CO_2 content (560 ppm), but the same solar "constant" as the Control simulation; and (iii) "Macro-2 · CO_2", with doubled atmospheric CO_2 content and the solar constant reduced by 1.8%. This reduction in solar radiation was chosen by them to approximately offset the radiative forcing from a CO_2 doubling in this model (4.17 W/m^2), taking into consideration the model's planetary albedo. In practice, this reduction in solar radiation incident on the Earth would be effected through the placement of reflecting or scattering devices between the Earth and Sun (Early, 1989; Flannery et al., 1997; Teller et al., 1997). The model was run for 40 years and the climate statistics are computed over the last 15 years of model simulations. Chaotic fluctuations in the simulated climate would introduce some year-to-year variation in the simulations, therefore the statistical significance of the difference in the means at each model-grid point are assessed using the Student-t test (Chervin and Schneider, 1976a, 1976b; Press et al., 1989), corrected for the influence of serial correlation (Zwiers and Storch, 1995).

3.2 Global Climate Change

Comparison of annual and global mean results (Table 1) suggests that the reduction in solar radiation largely compensates for the climatic impacts of increased CO_2 concentrations on surface temperature, precipitation, precipitable water vapor, and sea-ice volume. In the "2 · CO_2" simulation, the planet warms 1.75 K, leading to a reduction in sea-ice volume and an increase in precipitable water vapor. The global mean warming is in the lower end of the 1.5 to 4.5 K range estimated by IPCC (Houghton et al., 1995). The 1.8% reduction in solar radiative flux cools the Earth 1.88 K from its 2 · CO_2 state, slightly overcompensating the CO_2 doubling. It is estimated that a shielding of ~1.7% of solar radiation incident on the Earth, in this model, would more exactly compensate the effect of

Table 1. Annual- and global-means of surface temperature, total absorbed longwave and shortwave fluxes at the surface, precipitation, precipitable water, and sea-ice volume for the control simulation and changes in these variables for the other simulations as described in Sections 3.1 and 4.1

Case	Surface temperature (K)	Absorbed radiative flux (W/m^2)	Precipitation (mm/day)	Precipitable water vapor (mm)	Sea ice volume (10^{12} m^3)
Control	285.50	492.85	2.98	24.9	51.2
2 · CO_2 – Control	1.75	10.57	0.09	3.1	−12.5
4 · CO_2 – Control	2.27	14.04	0.14	4.4	−18.1
Macro-2 · CO_2 – Control	−0.13	−1.93	−0.06	−0.7	0.3
Macro-4 · CO_2 – Control	−0.07	−4.15	−0.10	−1.2	−2.4

a CO_2 doubling. Results from a highly idealized annual-mean general circulation model (Manabe and Wetherald, 1980) indicated that the model's annual mean tropospheric response to a 2% change in solar radiation was of the same magnitude and qualitatively similar to a doubling in atmospheric CO_2 content.

3.3 Surface Temperature Change

Comparison of results for annual mean surface temperature (Fig. 2) indicates that macro-engineering may largely compensate for impact of increased CO_2 concentrations, despite the differences in the spatial pattern of radiative forcing between changes in CO_2 and changes in solar flux (Fig. 1). The warming in the $2 \cdot CO_2$ climate is statistically significant at the 5% level over 97.4% of the globe, and is most pronounced in high latitudes where the warming is > 4 K. In general, land areas show more warming than adjacent oceans. In sharp contrast, the Macro-$2 \cdot CO_2$ simulation shows relatively little surface temperature change. There is a detectable difference (at the 5% significance level) in simulated annual mean temperature between the Macro-$2 \cdot CO_2$ and Control simulations over only 15.1% of Earth's surface; most of these significant differences are in areas (the tropics) with little change but low variability. The fact that ocean heat fluxes are prescribed is partly responsible for the low variability.

Comparison of surface temperature results by latitude band and season (Table 2) indicates that a reduction in solar radiation may largely compensate for the impact of increased atmospheric CO_2, despite the differences in the latitudinal and seasonal pattern of these radiative forcings (Fig. 1). The Macro-$2 \cdot CO_2$ simulation cools most

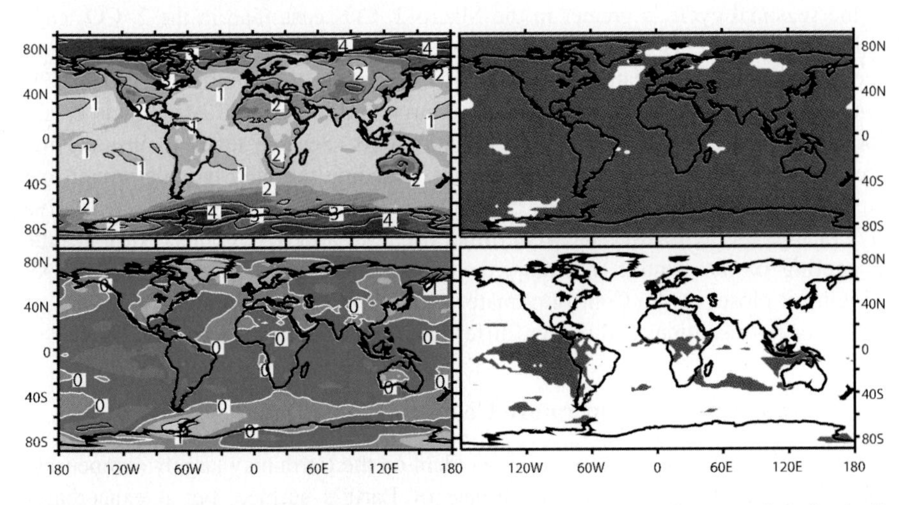

Figure 2. Surface temperature changes (left panels) and areas with changes that are statistically significant at the 5% level (right panels) for the $2 \cdot CO_2$ (top panels) and the Macro-$2 \cdot CO_2$ (bottom panels) simulations described in Section 3.1. Solar radiation has a spatial pattern that differs greatly from that of radiative forcing due to doubling atmospheric CO_2 content, yet a reduction in solar forcing largely compensates the temperature response to CO_2-doubling

Table 2. Changes in simulated mean surface temperature (K) in three latitude bands for the $2 \cdot CO_2$ and Macro-$2 \cdot CO_2$ cases relative to the Control case, for December, January and February (DJF), and June, July and August (JJA)

	$2 \cdot CO_2$			Macro-$2 \cdot CO_2$		
Latitude Belt	DJF	JJA	Change in seasonal amplitude	DJF	JJA	Change in seasonal amplitude
90 °N to 20 °N	+2.33	+1.67	−0.66	+0.15	−0.06	−0.21
20 °N to 20 °S	+1.31	+1.36		−0.31	−0.27	
20 °S to 90 °S	+1.70	+2.01	−0.31	−0.15	−0.08	−0.07

in equatorial regions, because in this region the reduction in radiative forcing from diminished solar flux is greater than the increase in radiative forcing from doubled atmospheric CO_2 content.

The $2 \cdot CO_2$ simulation warms more in the winters than summers at high latitudes (Table 2) in both the hemispheres, reducing the amplitude of the seasonal cycle. Macro-engineering this $2 \cdot CO_2$ world might be expected to diminish this amplitude further, because the reduction in solar radiation preferentially reduces solar insolation in the high latitude summers (Fig. 1), which would tend to preferentially cool. However, poleward of 20°N, the Macro-$2 \cdot CO_2$ case has average wintertime temperatures reduced by 2.18 K relative to the $2 \cdot CO_2$ case, but summer-time temperatures reduced by only 1.73 K, despite the fact that a reduction in solar flux decreases the insolation in summer more than in the winter. Hence, the amplitude of the seasonal cycle is greater in the Macro-$2 \cdot CO_2$ case than in the $2 \cdot CO_2$ case (Table 2). This occurs because there is more sea ice in the Macro-$2 \cdot CO_2$ simulation than in the $2 \cdot CO_2$ simulation (Table 1). The climate is colder in the Macro-$2 \cdot CO_2$ simulation and the ice-albedo feedback (Curry and Webster, 1999; Lewis et al., 2000) leads to more sea ice in the polar regions. Sea ice tends to insulate the ocean waters from the colder overlying air, reducing the high-latitude wintertime flux of heat from the ocean to the atmosphere. In the Macro-$2 \cdot CO_2$ case, relative to $2 \cdot CO_2$, the reduction in wintertime ocean-to-atmosphere heat flux results in cooling of the winters and amplification of the high-latitude seasonal cycle, bringing it closer to the Control climate. Macro-$2 \cdot CO_2$ temperatures poleward of 20°N, for both summer and winter, differ from the Control case by <0.2 K.

3.4 Stratospheric Temperature Change

Macro-engineering the solar radiation incident on the Earth may largely compensate for CO_2-induced changes on the climate of Earth's surface, but it exacerbates the impact of CO_2 on stratospheric temperature (Fig. 3). The addition of CO_2 to the atmosphere tends to warm the surface but cool the stratosphere (Manabe and Wetherald, 1975, 1980; Manabe and Stouffer, 1993, 1994; Washington and Meehl, 1989; Murphy and Mitchell, 1995), whereas a reduction in solar radiation tends to cool the atmosphere everywhere.

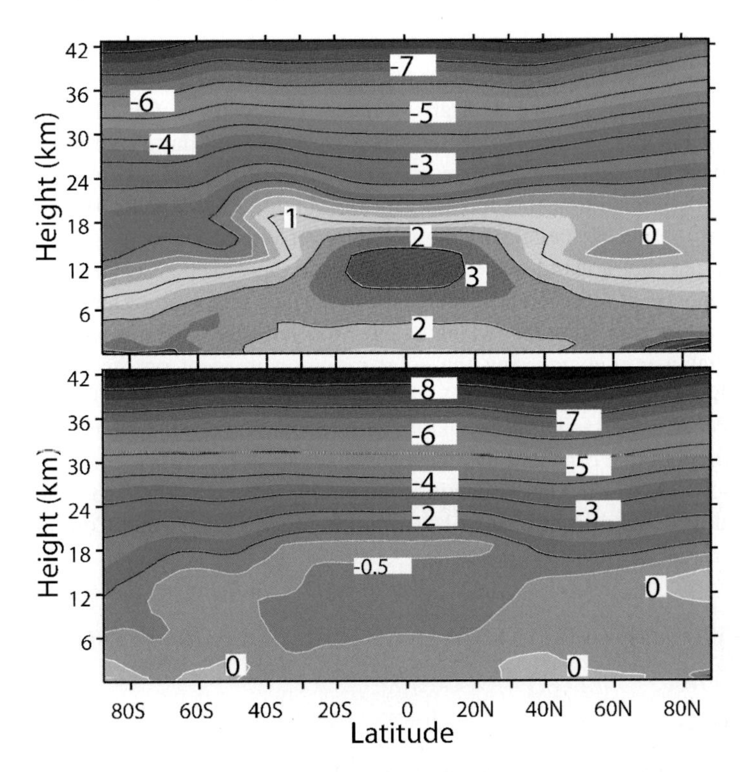

Figure 3. Zonal mean temperature changes for the $2 \cdot CO_2$ (top panel) and Macro-$2 \cdot CO_2$ (bottom panel) simulations as a function of latitude and height above Earth's surface. Diminishing the solar radiation incident on the Earth largely compensates the CO_2-induced warming in the troposphere, but cools the stratosphere by an additional ~1 K

In the $2 \cdot CO_2$ simulation, the equatorial tropopause warms over 3 K whereas the upper stratosphere cools up to ~8 K. The Macro-$2 \cdot CO_2$ simulation largely compensates for the tropospheric warming but cools the stratosphere by an additional ~1.0 K. Zonal mean temperature changes (Fig. 3) are generally significant at the 5% level when the change is >0.5 K. The cooling of the stratosphere could enhance the formation of polar stratospheric clouds, which could in turn contribute to the destruction of stratospheric ozone (Houghton et al., 1990). Macro-engineering approaches involving placing aerosols in the stratosphere (Flannery et al., 1997; Teller et al., 1997), could have additional adverse impacts on stratospheric chemistry (Kinnison et al., 1994).

3.5 Hydrological Changes

In general, the model's hydrological cycle (e.g., precipitation) does not show a strong sensitivity to doubling CO_2 (Table 1). Changes in the annual mean net fresh water flux (precipitation minus evaporation) were statistically significant at the 5% level over only 13.9% and 3.9% of Earth's surface, for the $2 \cdot CO_2$ and

Macro-2 · CO_2 simulations, respectively. As found in other studies (Murphy and Mitchell, 1995; Manabe and Stouffer, 1993), there was significant increase in the net fresh water flux to the surface at high latitudes. Poleward of 60°, the net fresh water flux in the 2 · CO_2 simulation increases by 0.130 mm day^{-1}, with the change in this flux significant at the 5% level over 51.7% of this area. However, in the Macro-2 · CO_2 simulation, this increase in high-latitude fresh-water flux is only 0.008 mm day^{-1}, and is statistically significant over only 1.8% of this area. It has been suggested that a shutdown of North Atlantic thermohaline circulation could be a consequence of CO_2-induced increases in surface temperature and net-fresh-water flux in the high latitudes (Manabe and Stouffer, 1993). The modeling results indicate that macro-engineering the solar radiation incident on the Earth might diminish the impact of increased CO_2 on both of these quantities, making a shutdown of the ocean's thermohaline circulation less likely. Further, the melting of Greenland and Antarctic ice caps and the consequent sea level rise is less likely to occur in a Macro-2 · CO_2 world. Other quantities, including zonal winds and specific humidity, showed little significant change between the Macro-2 · CO_2 and Control simulations.

4. MITIGATION OF CLIMATE CHANGE FOR A QUADRUPLING OF CO_2

4.1 Experiments

In this section, modeling results of the residual climate change in a quadrupled CO_2 world are presented. In addition to the experiments discussed in the previous section, two more experiments from Govindasamy et al. (2003) will be discussed: (i) "4 · CO_2", with quadrupled atmospheric CO_2 content (1120 ppm) from the pre-industrial levels, but the same solar constant as the Control simulation; and (ii) "Macro-4 · CO_2", with 4 times atmospheric CO_2 content and the solar "constant" reduced by 3.6%. The reduction in solar intensity in (ii) was chosen to approximately offset the global- and annual-mean radiative forcing from a CO_2 quadrupling in this model (8.34 W/m^2), taking into consideration the model's planetary albedo. At high latitudes, the resulting change in seasonal amplitude of insolation is about 5 times smaller than that associated with Milankovitch cycles (Imbrie et al., 1984).

4.2 Global Climate Change

Comparison of annual- and global-mean results (Table 1) suggests that the reduction in solar radiation in "Macro-4 · CO_2" largely also compensates for the climatic impacts of increased CO_2 concentrations on surface temperature, absorbed surface radiative flux, precipitation, precipitable water vapor, and sea-ice volume. In the "4 · CO_2" simulation, the planet warms 4.02 K, leading to a reduction in sea-ice volume and an increase in precipitation and precipitable water vapor. Other models with different climate sensitivity and a dynamic ocean model may not necessarily reproduce this compensation. In the "$2 \cdot CO_2$" simulation, the surface temperature increased by 1.75 K from the Control simulation (Table 1) whereas it increases by 2.27 K from "$2 \cdot CO_2$" simulation to "$4 \cdot CO_2$" simulation. The last two rows in Table 1 indicate larger changes in global-mean quantities for the climate change from "$2 \cdot CO_2$" to "$4 \cdot CO_2$" than for the change from Control to "$2 \cdot CO_2$". Radiative forcing is the same for a doubling of CO_2 ("$1 \cdot CO_2$" to "$2 \cdot CO_2$" or "$2 \cdot CO_2$" to "$4 \cdot CO_2$") because of its logarithmic dependence on CO_2 concentration. This suggests that the climate sensitivity may increase as the climate warms in this model presumably due to some positive feedbacks. The annual- and global-mean surface temperature shows a similar increase in climate sensitivity at higher concentrations of CO_2 (and hence warmer climates) in another modeling study using CCM3 (Kothavala et al., 1999).

The 3.6% reduction in solar radiation cools the Earth 4.09 K from its $4 \cdot CO_2$ state, slightly over-compensating the change due to CO_2 quadrupling (Table 1). Though the global mean surface temperature rise in "$4 \cdot CO_2$" is almost exactly balanced in "Macro-$4 \cdot CO_2$", there are some residual changes in other quantities like absorbed surface radiative flux, precipitation, precipitable water and sea-ice volume (Table 1). These global mean quantities except sea-ice suggest a slightly colder "Macro-$4 \cdot CO_2$" planet than the control. The decrease in sea-ice is due to a slight residual warming in high latitudes in the "Macro-$4 \cdot CO_2$". This result is expected, since the radiative forcing due to quadrupling of CO_2 in winter exceeds that due to the reduction in solar flux at high latitudes.

4.3 Temperature Change

The "Macro-$4 \cdot CO_2$" simulation cools most in equatorial regions (Fig. 4), because in this region the reduction in radiative forcing from diminished solar flux is greater than the increase in radiative forcing from increased atmospheric CO_2 content (Fig. 1). In the tropical latitude band (10°S to 10°N), the annual mean temperature decreases by 0.56 K, with this change significant at 5 % level over 76.8 % of this area. Therefore, small residual surface temperature change does occur in the tropics in "Macro-$4 \cdot CO_2$" simulation. The small decrease in global- and annual-mean precipitation and precipitable water (Table 1) are associated with this decrease in surface temperature in low latitudes.

Poleward of 60°, the annual mean temperature increases by 0.56 K, with the change significant at 5% level over 23.5% of the area. This increase in surface temperature in high latitudes is consistent with the decrease in sea-ice volume in the "Macro-$4 \cdot CO_2$" simulation compared to the control (Table 1). As in the "Macro-$2 \cdot CO_2$" simulation, the reduction in solar flux in the "Macro-$4 \cdot CO_2$" simulation largely compensates for the change in seasonal cycle associated with quadrupled atmospheric CO_2 content. The diminished solar flux largely compensates for the tropospheric warming, but it cools the stratosphere by an additional $\sim$1 K as discussed in Section 3. Therefore, macro-engineering schemes involving placing

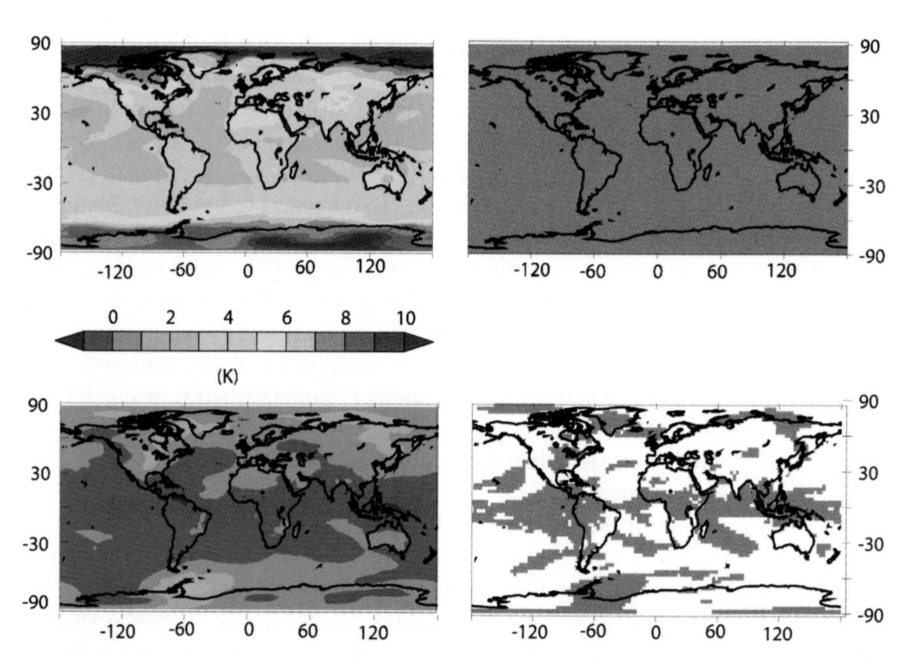

Figure 4. Surface temperature changes (left panels) and areas with changes that are statistically significant at the 5% level (right panels) for the $4 \cdot CO_2$ (top panels) and the Macro-$4 \cdot CO_2$ (bottom panels) simulations. A reduction in solar forcing via macro-engineering largely compensates the temperature response to CO_2-quadrupling

reflectors outside Earth's atmosphere do not mitigate the cooling effect of the greenhouse gases in the stratosphere while counteracting the warming effect in the troposphere. In the "Macro-$4 \cdot CO_2$" simulation, the changes in net water fluxes relative to the control are not significant except in tropics where there is a reduction in net water flux.

4.4 Sea Ice Change

In the pre-industrial control simulation, the simulated maximum annual mean thickness of sea ice is about 5 meters in Arctic and 1.44 meters in Antarctic (Fig. 5). In the model, as in real world, the sea ice in the Southern Hemisphere is seasonal; it almost vanishes in Southern Hemisphere summer. However, sea ice in the arctic is permanent. In the "$4 \cdot CO_2$" simulation, annual mean sea ice thickness decreases drastically at all sea ice points. The maximum thickness decreases to 1.9 meters in Arctic and 1 meter in Antarctic. The sea ice area coverage also shrinks, most notably in the Southern Hemisphere. In the "Macro-$4 \cdot CO_2$" simulation, sea ice thickness and area coverage are almost recovered to the levels in the control simulation, though the annual- and global-mean sea ice volume is slightly lower than in the control simulation (Table 1).

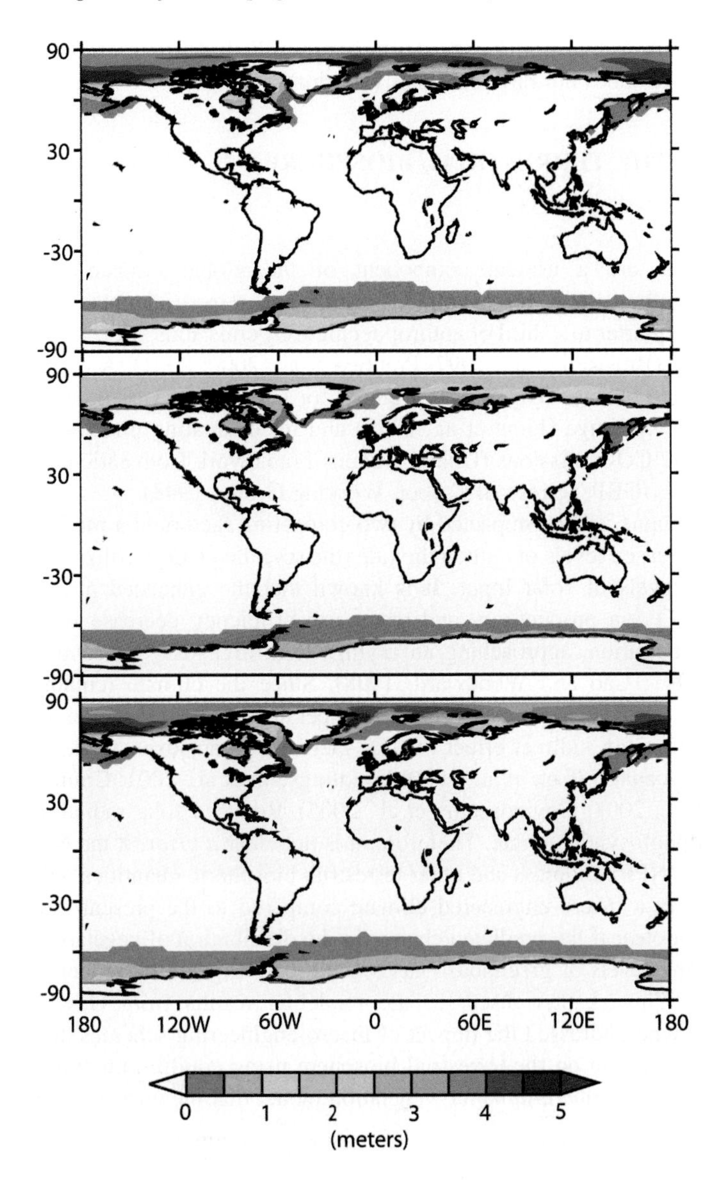

Figure 5. Annual mean sea ice thickness in the Control (top panel), $4 \cdot CO_2$ (middle panel), and Macro-$4 \cdot CO_2$ (bottom panel) simulations. The reduction in solar forcing in Macro-$4 \cdot CO_2$ simulation largely compensates the decrease in sea ice thickness and area coverage in the $4 \cdot CO_2$ simulation

It should be noted that the sea ice in this modeling study (Govindasamy et al., 2003) does have seasonal variations. However, the Arctic and Antarctic processes presented here have uncertainties due to the following limitations: 1) Sea ice dynamics is neglected; 2) the effect of fresh water flux and salinity on sea ice are not considered; 3) the heat transport by the ocean is prescribed; 4) the prescribed depth of the mixed layer does not have seasonal variations.

5. IMPACT ON THE TERRESTRIAL BIOSPHERE

5.1 Introduction

Terrestrial ecosystems are a critical component of the global climate and carbon cycle. During the 1980s, for example, oceanic and terrestrial uptake of carbon amounted to a quarter to a third of anthropogenic CO_2 emissions with strong interannual variability (Braswell et al., 1997; Prentice et al., 2000; 2001). A better understanding of carbon balance dynamics is required for interpreting variations in atmosphere-biosphere exchange (Fung et al., 1997) and for evaluating policies to mitigate anthropogenic CO_2 emissions (United Nations Framework Convention on Climate Change, 1997; IGBP Terrestrial Carbon Working Group, 1998).

Photosynthesis of plants will be impacted by two competing factors in a macro-engineered world: increased levels of carbon dioxide (the so-called CO_2 fertilization effect) and reduced levels of solar input. It is known that the enhanced physiological effects of CO_2 on productivity and water use efficiency decrease with increasing CO_2 concentration, approaching an asymptote at high CO_2 concentration (King et al., 1997; Cao and Woodward, 1998). Since the climate remains nearly constant in a macro-engineered world, a temperature related increase in heterotrophic respiration (an indirect effect of CO_2 increase) is not expected as it is in global warming scenario (Cox et al., 2000; Friedlingstein et al., 2001; Cramer et al., 2001; Joos et al., 2001; Govindasamy et al., 2005). Reduced solar radiation will lead to reduced photosynthetic rate. Therefore, it is not clear *a priori* if the Net Primary Productivity (NPP), biomass and other terrestrial biospheric quantities will increase or decrease in a macro-engineered climate compared to the present day climate. Also, it is not clear if the predicted change in the distribution of vegetation types due to increasing levels of greenhouse gases could be mitigated by reducing the solar insolation. This chapter discusses the modeling results from Govindasamy et al. (2002) who addressed the impact of macro-engineering schemes that reduce surface solar radiation on the terrestrial biosphere using equilibrium simulations from a coupled atmosphere-dynamic vegetation model that includes climate feedbacks.

5.2 Model

The slab ocean modeling strategy as discussed in Section 2 is used to investigate the impacts of macro-engineering on terrestrial biosphere (Govindasamy et al., 2002). The atmosphere model (Section 2) is now coupled to a terrestrial biosphere model, Integrated Biosphere Simulator or IBIS (Foley et al., 1996; Kucharik et al., 2000). IBIS was also used by Naik et al. (2003) to investigate the influence of macro-engineering schemes on the Net Primary Productivity (NPP) of the terrestrial biosphere. Land surface biophysics, terrestrial carbon flux and global vegetation dynamics are represented in a single, physically consistent modeling framework within IBIS. IBIS performs a coupled simulation of surface water, energy and carbon fluxes on hourly timesteps and integrates them over the year to estimate annual water and carbon balance. The annual carbon balance of vegetation is used to predict changes in the leaf area index and biomass for each of 12 plant functional types, which compete for light and water using different ecological strategies. IBIS also simulates carbon cycling through litter and soil organic matter.

When driven by observed climatological datasets, the model's near-equilibrium runoff, Net Primary Productivity (NPP), and vegetation categories show a fair degree of agreement with observations (Foley et al., 1996; Kucharik et al., 2000).

5.3 Experiments

To evaluate the extent to which an effective reduction in solar flux might mitigate the biospheric impacts of increased atmospheric CO_2 content, Govindasamy et al. (2002) performed four model simulations: (i) "Control", with a CO_2 content of 355 ppm and incoming solar flux of 1367 W/m^2; (ii) "$2 \cdot CO_2$bio", with doubled atmospheric CO_2 content (710 ppm), but the same incoming solar flux as the Control simulation; (iii) "Solar" with a CO_2 content same as control, but solar flux reduced by 1.8 %; and (iv) "Macro-$2 \cdot CO_2$bio", with doubled atmospheric CO_2 content and the solar flux reduced by 1.8%. The Control experiment discussed in this section is a present-day control whereas we discussed a pre-industrial control in Sections 3 and 4.

For all the experiments, the biosphere model was initialized with a state corresponding to present day conditions. From this initial state, the coupled atmosphere-biosphere model typically needs to run for at least $\sim$100 simulated years so that NPP, NEP, biomass and soil carbon approach quasi-equilibrium.

The climate statistics presented below are the averaged values over the last 25 years of model simulations.

5.4 Global Climate and Carbon Cycle Changes

The climate warms by 2.42 K in the $2 \cdot CO_2$bio experiment and cools by nearly identical amount in the Solar experiment (Table 3). Precipitation increases by 3.7% for $2 \cdot CO_2$bio and decreases by 5.8% in the Solar experiment. In $2 \cdot CO_2$bio experiment, the biospheric quantities like NPP, and biomass almost double in response to increased availability of CO_2 suggesting a strong response of the biosphere to CO_2. That this response is largely due to CO_2 is confirmed by comparing NPP in the Control to Macro-$2 \cdot CO_2$bio simulation. Heterotrophic respiration also increases because of the increase in the size of the soil carbon pool. Terrestrial disturbances, such as fire, that are prescribed in this model account for the difference between NPP and heterotrophic respiration in IBIS. These results from $2 \cdot CO_2$bio experiment

Table 3. Global and annual mean model results from Control, $2 \cdot CO_2$bio, Solar and Macro-$2 \cdot CO_2$bio simulations described in Section 5.3. Temperature and precipitation changes are relative to the Control simulation

	Surface Temperature Change (K)	Precipitation Change (%)	NPP (Gt-C/yr)	Biomass (Gt-C)	Heterotrophic respiration (Gt- C/yr)
Control	0.0	0.0	56.4	631.6	50.1
$2 \cdot CO_2$bio	2.42	3.7	100.1	1212.9	87.5
Solar	−2.40	−5.8	55.1	619.5	49.9
Macro-$2 \cdot CO_2$bio	0.14	−1.7	97.1	1156.2	84.7

are in fairly close agreement with previous results from six dynamic vegetation models run offline (Cramer et al., 2001). Biospheric quantities in Solar experiment are very close to the control climate experiment suggesting that photosynthesis and respiration of plants are relatively insensitive to small reductions in sunlight in this model.

In contrast to the physical climate variables, the biospheric quantities in Macro-$2 \cdot CO_2$bio are much closer to $2 \cdot CO_2$bio than to the Control with the fertilization effect of the increased CO_2, rather than climate, dominating the biospheric response (Naik et al., 2003). NPP and Heterotrophic (soil microbial) respiration almost doubles in the $2 \cdot CO_2$bio and Macro-$2 \cdot CO_2$bio cases because increased atmospheric CO_2 permits plant stomatal openings to narrow, thereby diminishing water loss and increasing water use efficiency. As a consequence, the carbon pools (biomass and soil carbon) almost double in these experiments. The effect of CO_2 fertilization is probably exaggerated in these simulations because we do not consider factors other than limitation by sunlight, water, and carbon dioxide. Inclusion of other factors, such as nitrogen or phosphate limitation would diminish the magnitude, but not the sign, of these effects. The biospheric quantities in Macro-$2 \cdot CO_2$bio are only slightly smaller than in $2 \cdot CO_2$bio.

5.5 Vegetation Distribution Changes

The vegetation distribution in the Macro-2·CO2bio simulation is intermediate between the vegetation distributions in the Control and 2·CO2bio simulations. It is more similar to the Control simulation than are the distributions in the $2 \cdot CO_2$bio and Solar simulations.

The kappa statistics (Monsrrud, 1990) can be used to compare maps of vegetation distributions. Kappa takes on a value of 1 with perfect agreement. It has a value close to zero when the agreement is approximately the same as would be expected by chance. A kappa value of 0.47 (fair agreement) is obtained for a comparison of IBIS simulated vegetation and observations (Foley et al., 1996).

Macro-engineering tends to limit changes in vegetation distribution. Global comparison of vegetation distributions of Macro-$2 \cdot CO_2$bio and Control gives a kappa value of 0.66 (good agreement), a value higher than 0.52 (fair) obtained for a comparison of $2 \cdot CO_2$bio with Control (Table 4). The reduction of solar input has less impact on the vegetation distribution than an equivalent (in terms of climate forcing) increase in atmospheric CO_2. Global comparison of Solar and Control gives a kappa value of 0.62 (good agreement), again a value higher than 0.52 (fair) obtained for $2 \cdot CO_2$bio and Control.

The first column (or row) of Table 4 can be used to assess the relative sensitivity of vegetation distribution to changes in CO_2 versus changes in climate achieved via changing solar flux. The vegetation distribution is slightly more sensitive to changing climate alone than to changing CO_2 concentration alone (Cramer et al., 2001). The kappa value of 0.62 obtained from comparing the Control and Solar cases results from changing climate alone. A slightly larger kappa value of 0.66 obtained from comparing Control and Macro-$2 \cdot CO_2$bio cases measures the sensitivity of vegetation distribution to changing CO_2 while leaving the climate unchanged to first order. This contrasts with the findings for biomass and NPP, which are more sensitive to changes in CO_2than to changes in climate. These findings are consistent with the expectation that vegetation distribution is controlled nearly equally by climate and CO_2, while biomass is controlled mainly by CO_2 concentration. Kappa values are lower (0.52 and 0.51) when both climate and CO_2 have changed (comparison of Control and $2 \cdot CO_2$bio, and, Solar and Macro-$2 \cdot CO_2$bio). Kappa is only 0.40 (poor agreement) for larger climate change (comparison of $2 \cdot CO_2$bio and Solar) that confirms the dependence of vegetation distribution on climate.

5.6 Terrestrial Carbon Cycle Changes

Regions of highest biomass are associated with warm and moist climates; areas of low or zero biomass are in the extreme subtropical deserts, high-altitude areas and polar latitudes (Fig. 6). The spatial patterns are similar in all cases with correlation exceeding 90% between any two cases. With the exception of the boreal regions where light is a limiting factor for plant growth, the magnitude of biomass is similar in Control and Solar cases, indicating roughly proportional influence of sunlight on NPP and biomass (Naik et al., 2003). Similarly, the magnitudes are similar in $2 \cdot CO_2$bio and Macro-$2 \cdot CO_2$bio cases, confirming the relatively weak influence

Table 4. Kappa statistics: Values of kappa for comparison between vegetation distributions from Control, $2 \cdot CO_2$bio, Solar and Macro-$2 \cdot CO_2$bio simulations described in Section 5.3

Experiment	Control	$2 \cdot CO_2$bio	Solar	Macro-$2 \cdot CO_2$bio
Control	1	0.52	0.62	0.66
$2 \cdot CO_2$bio	0.52	1	0.40	0.66
Solar	0.62	0.40	1	0.51
Macro-$2 \cdot CO_2$bio	0.66	0.66	0.51	1

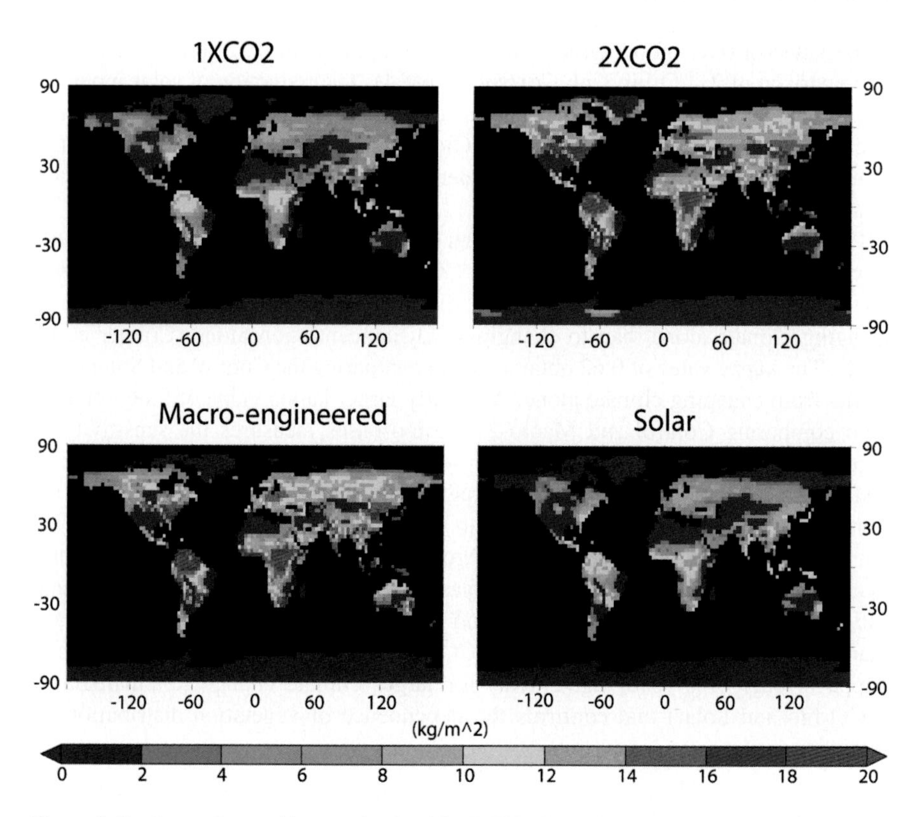

Figure 6. Total annual mean biomass simulated by IBIS in Control, $2 \cdot CO_2$bio, Macro-$2 \cdot CO_2$bio and Solar experiments. The spatial patterns are similar in all cases. The magnitudes in Macro-$2 \cdot CO_2$bio and Solar are closer to $2 \cdot CO_2$bio and Control respectively, suggesting strong influence of CO_2 and weak influence of sunlight. The model considers limitation by only light, water, and carbon dioxide. Consideration of nitrogen and phosphate limitation would diminish the magnitude, but not alter the sign, of the response of the model to increased atmospheric CO_2 content

of solar flux and strong influence of CO_2 on terrestrial biological activity in the simulations. Maps of other quantities such as NPP, leaf area index, heterotrophic respiration, and soil carbon exhibit similar characteristics.

Zonal mean NPP exhibit greater sensitivity in the summer hemisphere and tropics than in the winter hemisphere and high latitudes (Fig. 7). Latitudinal distribution of NPP is similar in $2 \cdot CO_2$bio and Macro-$2 \cdot CO_2$bio simulations except in Northern Hemisphere high latitudes during JJA (Fig. 7). In the Macro-$2 \cdot CO_2$bio case, the increase relative to Control is slightly less than in $2 \cdot CO_2$bio, because the climate is cooler and drier than the $2 \cdot CO_2$bio case. The latitudinal distribution is nearly the same in Control and Solar. In the northern high latitudes where light is a limiting factor for plant growth, the sunlight reduced Solar results in decreased NPP relative to Control.

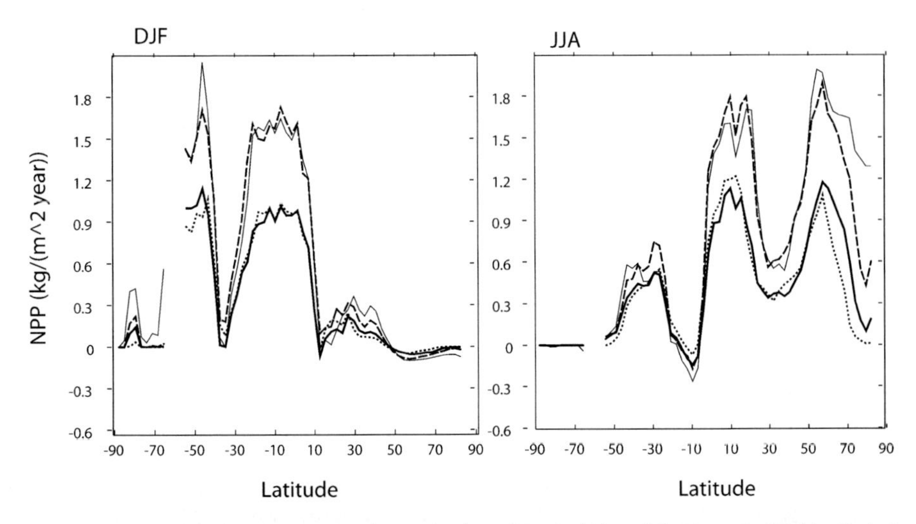

Figure 7. Zonal mean NPP for Control (thick solid), $2 \cdot CO_2$bio (thin solid), Macro-$2 \cdot CO_2$bio (dashed line), and Solar (dotted line) cases for December, January, and February (DJF), and June, July, and August (JJA). The magnitudes in Macro-$2 \cdot CO_2$bio and Solar are closer to $2 \cdot CO_2$bio and Control respectively, suggesting strong influence of CO_2 and weak influence of sunlight

6. DISCUSSION AND CONCLUSION

An analysis of several paleoclimates and paleo-radiative-forcing reconstructions indicated that the latitudinal structure of temperature response to climate forcing is insensitive to the details of the latitudinal structure of the radiative forcing (Covey and Thompson, 1989). In a set of atmospheric GCM simulations with specified ocean heat transport (Hoffert and Covey, 1992), total poleward heat transport was largely insensitive to the specified ocean heat transport, as changes in atmospheric heat transport largely compensated for changes in ocean heat transport. These findings suggest that the response of the climate system to external forcing is somewhat insensitive to the detailed spatial and/or temporal distribution of that forcing, which may complicate the attribution of climate changes to specific forcings. Of course, this relative independence of the geographic and seasonal climate response from the details of the climate forcing has its limits. For example, the climate-response "fingerprint" of sulfate aerosols is quite distinguishable from that of carbon dioxide (Taylor and Penner, 1994).

The modeling studies discussed in Sections 3 and 4 suggest that there may be effective engineering strategies to counteract most climate change, as it may not be necessary to replicate the exact radiative forcing patterns from greenhouse gases to largely negate their effects. However, subtle changes in the distribution of solar radiation associated with the Milankovitch cycles (Imbrie et al., 1984) may have produced large climate change on time scales $> 10^4$ yr, after ocean circulation and ice sheets adjusted to the slightly modified new climate. Even if macro-engineering schemes could largely compensate for the climate change induced by a CO_2 doubling on short time scales, there is no guarantee that long-term climate would remain relatively unaffected. For instance, as discussed in Section 5, the uptake of CO_2 by the biosphere will increase at elevated levels of atmospheric CO_2 irrespective of whether macro-engineering schemes are implemented or not. A climate-stabilized world could have higher NPP than the current world. The macro-engineering climate stabilization schemes might prevent some changes in vegetation distribution but would have little effectiveness in preventing changes in NPP and biomass. The macro-engineering schemes that diminish the surface solar radiation would not prevent ocean acidification from absorbed CO_2 (Caldeira and Wickett, 2003).

Caution should be exercised in interpretation because these results are from a single atmospheric model with many simplifying assumptions. So far, only equilibrium climate simulations using a single atmospheric general circulation model coupled to a slab-ocean and thermodynamic sea-ice model have been performed. Climate models exhibit a wide range of response for similar climate forcings (Hansen et al., 1997). It is possible that other atmospheric GCMs would yield quantitatively different results. Results may be highly sensitive to the formulation of the model and the parameterization of various physical processes (Hansen et al., 1999). For instance, in the GFDL coupled model simulations (Manabe and Stouffer, 1994), the global mean surface air temperature increases by 3.5 K and 7 K in the doubling and quadrupling experiments respectively. The GFDL model has a higher climate sensitivity and apparently constant climate sensitivity for increases in CO_2. In contrast, CCM3 has a much lower sensitivity (1.75 K and 4.02 K for doubling and quadrupling) that increases for increasing concentrations of CO_2 (Kothavala et al., 1999).

This chapter considered anthropogenic forcing from carbon dioxide. Results may differ for other radiatively active gases. The uncertainty of the radiative forcing of the aerosols, both direct and indirect, are large and could potentially compensate a large part of the radiative forcing of greenhouse gases (Houghton et al., 2001). Simulations using a coupled atmosphere, dynamic sea-ice and ocean general circulation models would include dynamical feedbacks that could amplify the regional or global climate change simulated for macro-engineering scenario (Manabe and Stouffer, 1993). Furthermore, the studies discussed here have considered only a steady-state doubled- and quadrupled CO_2 scenario, and the transient responses of the climate system need to be addressed.

Macro-engineering schemes impose a variety of technical, environmental and economic challenges (Early, 1989; NAS, 1992; Watson et al., 1995; Flannery et al., 1997; Teller et al., 1997). For instance, in the case of placing reflectors in space, since a doubling of CO_2 requires the interception of about 1.7% of the sunlight incident on the Earth, an interception area of $\sim 2 \cdot 10^6$ km^2 or a disk of roughly 800 km in radius has to be built. To counteract a transient warming, the solar input has to diminish over time as CO_2 increases. Since CO_2 is increasing at the rate of $\sim 0.4\%\,yr^{-1}$ (Houghton et al., 1995), to counteract this warming, $\sim 1.2 \cdot 10^4$ km^2 of interception area needs to be built each year. Other options also involve great difficulties. Placing small particles or aerosols in the stratosphere may not result in uniform diminution of radiation. The new particles will increase the cooling in the stratosphere and will lead to additional ozone depletion, as was the case after the eruption of Pinatubo volcano in 1991. Mirrors in low-earth orbit will lead to flickering of the sun ~2% of the time, and involves tracking problems so that mirrors don't collide with each other. Reflectors or scatterers at the Lagrange point between the Sun and Earth involve large costs. The failure of a macro-engineering system could subject the Earth to extremely rapid warming. Ethical and political concerns differ depending on whether global-scale climate modification is intentional or merely a predictable consequence of our actions.

Ecosystems would be impacted by changes in atmospheric CO_2 content and photosynthetically active radiation, even without climate change: the terrestrial biosphere is more sensitive to changes in CO_2 concentration than to changes in sunlight. Even with a stabilized climate, CO_2 fertilization could impact ecosystem goods and services such as species abundance and competition, habitat loss, biodiversity and other disturbances (Root and Schneider, 1993).

Many of the macro-engineering schemes are cooperative solutions that require continuous world management for multiple centuries. Given the history of non-cooperation at a global scale just in the 20th century, there is very high probability of the non-feasibility of macro-engineering of cooperative solutions (Schneider, 2001). Given these difficulties, the most prudent and least risky option to mitigate global warming may well be to curtail emissions of greenhouse gases (Hoffert et al., 1998). Nevertheless, the impact of macro-engineering options should be studied so that if greenhouse-gas emissions induce a truly catastrophic climate response, humanity will probably have some possibility of rapidly counteracting part of the problem, and the knowledge to assure that it will not inadvertently create a bigger problem.

REFERENCES

Braswell BH, Schimel DS, Linder E, Moore B (1997) The response of global terrestrial ecosystems to interannual temperature variability. Science 278(5339):870–872

Budyko MI (1977) Climate Changes. American Geophysical Union, Washington, DC, English translation of 1974 Russian volume, pp.244

Caldeira K, Wickett M (2003) Oceanography: Anthropogenic carbon and ocean pH. Nature 425:365

Cao M, Woodward FI (1998) Dynamic responses of terrestrial ecosystem carbon cycling to global climate change. Nature 393:249–252

Chervin RM, Schneider SH (1976) A study of the response of NCAR GCM climatological statistics to random perturbations: estimating noise levels. J Atmos Sci 33:392–404

Chervin RM, Schneider SH (1976) On determining the statistical significance of climate experiments with general circulation models. J Atmos Sci 33:405–412

Covey C, Thompson ST (1989) Testing the effects of ocean heat transport on climate. Paleogeogr Paleoclimat Paleoecol (Glob Planet Change Sec) 75:331–441

Cox PM, Betts RA, Jones CD, Spall SA, Totterdell IJ (2000) Acceleration of global warming due to carbon-cycle feedbacks in a coupled model. Nature 408:184–187

Cramer W, Bondeau A, Woodward FI et al (2001) Global response of terrestrial ecosystem and function to CO_2 and climate change: results from six dynamic global vegetation models. Global Change Biol 7:357–373

Curry JA, Webster PJ (1999) Thermodynamics and atmospheres and oceans. International Geophysics Series, Vol.65, Academic Press, San Diego, CA, p 465

Early JT (1989) The space based solar shield to offset greenhouse effect. J Br Interplanet Soc 42: 567–569

Flannery BP, Kheshgi H, Marland G, MacCracken MC (1997) Geoengineering climate. In: Watts R (ed) Engineering response to global climate change. Lewis Publishers, Boca Raton, FL, pp 403–421

Foley JA, Prentice IC, Ramankutty N, Levis S, Pollard D, Sitch S, Haxeltine A (1996) An integrated biosphere model of land surface processes, terrestrial carbon balance and vegetation dynamics. Global Biogeochem Cycles 10:603–628

Friedlingstein P, Bopp L, Clais P et al (2001) Positive feedback between future climate change and the carbon cycle. Geophys Res Lett 28:1543–1546

Fung I, Field CB, Berry JA et al (1997) Carbon 13 exchanges between the atmosphere and biosphere. Global Biogeochem Cycles 11:507–533

Govindasamy B, Caldeira K (2000) Geoengineering Earth's radiation balance to mitigate CO_2-induced climate change. Geophys Res Lett 27:2141–2144

Govindasamy B, Caldeira K, Duffy PB (2003) Geoengineering earth's radiation balance to mitigate climate change from a quadrupling of CO_2. Global Planet Change 37:157–168

Govindasamy B, Thompson S, Duffy PB, Caldeira K, Delire C (2002) Impact of geoengineering schemes on the terrestrial biosphere. Geophys Res Lett 29(22):2061. DOI 10.1029/2002GL015911

Govindasamy B, Thompson S, Mirin A, Wickett M, Caldeira K, Delire C (2005) Increase of carbon cycle feedback with climate sensitivity: results from a coupled climate and carbon cycle model. Tellus 57(B):153–163

Hansen JE, Sato M, Lacis A, Reudy R, Tegen I, Mathews E (1999) Climate forcings in the industrial era. Proc Natl Acad Sci USA 95:I. 12753–12758

Hansen JE, Sato M, Ruedy R (1997) Radiative forcing and climate response. J Geophy Res 102:6831–6863

Hoffert MI, Caldeira K, Jain AK, Haites EF et al (1998) Energy implications of future stabilization of atmospheric CO_2 content. Nature 395:881–884

Hoffert MI, Covey C (1992) Deriving global climate sensitivity from palaeoclimate reconstructions. Nature 360:573–576

Houghton JT, Ding Y, Griggs DJ, Noguer M, an der Linden PJ, Dai X, Maskel K, Johnson CA (2001) Climate change 2001: the science of climate change. Intergovernmental Panel on Climate Change, United Nations Environmental Program/World Meteorological Organization. Cambridge University Press, New York, USA, pp.881

Houghton JT, Filho LGM, Callander BA, Harris N, Kattenberg A, Maskell K (1995) Climate change 1995: the science of climate change. Intergovernmental Panel on Climate Change, United Nations Environmental Program/World Meteorological Organization. Cambridge University Press, New York, USA, pp.572

Houghton JT, Jenkins GJ, Ephraums JJ (1990) Climate change: the IPCC scientific assessment. Intergovernmental panel on climate change, United Nations Environmental Program/World Meteorological Organization. Cambridge University Press, New York, USA, pp.365

Imbrie J et al (1984) The orbital theory of Pleistocene climate: Support from a revised chronology of $\delta^{18}O$ record. In: Berger A et al (ed) Milankovitch and climate. D. Reidel, Dordrecht, Netherlands, pp 269–305

Joos F, Prentice IC, Sitch S et al. (2001) Global warming feedbacks on terrestrial carbon uptake under the Intergovernmental Panel on Climate Change (IPCC) emission scenarios. Global Biogeochem Cycles 15:891–907

Kiehl JT, Briegleb BP (1993) The relative roles of sulfate aerosols and greenhouse gases in climate forcing. Science 260:311
Kiehl JT, Hack JJ, Bonan GB, Boville BA, Briegleb BP, Williamson DL, Rasch PJ (1996) Description of the NCAR Community Climate Model (CCM3), NCAR Technical Note, NCAR/TN−420+STR. Boulder, CO, USA, pp.152
King AW, Post WM, Wullschleger SD (1997) The potential response of terrestrial carbon storage to changes in climate and atmospheric CO_2. Climate Change 35:199–227
Kinnison DE, Grant KK, Connell PS, Rotman DA, Wuebbles DJ (1994) The chemical and radiative effects of the mount pinatubo eruption. J Geophys Res–Atm 99:25705–25731
Kothavala Z, Oglesby RJ, Saltzman B (1999) Sensitivity of equilibrium surface temperature of CCM3 to systematic changes in atmospheric CO_2. Geophys Res Lett 26:209–212
Kucharik CJ et al (2000) Testing the performance of a dynamic global ecosystem model: Water balance, carbon balance, and vegetation structure. Global Biogeochem Cycles 14(3):795–825
Lewis EL, Jones EP, Lemke P, Prowse TD, Wadhams P (eds) (2000) The Fresh Water Budget of the Arctic Ocean, NATO Science Series 2-70, Kluwer Academic Publishers, Boston, MA, Dordrecht, pp. 625
List RJ (ed) (1951) Meteorological Table, 6th edn. Smithsonian Institute, Washington DC, pp.527
Manabe S, Stouffer RJ (1993) Century-scale effects of increased atmospheric CO_2 on the ocean atmosphere system. Nature 364:215–218
Manabe S, Stouffer RJ (1994) Multiple-century response of a coupled ocean-atmosphere model to an increase of atmospheric carbon dioxide. J Climate 7:5–23
Manabe S, Wetherald RT (1975) The effects of doubling the CO_2 concentration on the climate of a general circulation model. J Atmos Sci 32:3–15
Manabe S, Wetherald RT (1980) On the distribution of climate change resulting from an increase in CO_2 content of the atmosphere. J Atmos Sci 37:99–118
Monsrrud RA (1990) Method for comparing global vegetation maps, IIASA WP-90–40. Int. Inst. For Appl. Syst. Anal., Laxenburg, Austria, pp.31
Murphy JM, Mitchell JFB (1995) Transient response of the hadley center coupled ocean-atmosphere model to increasing carbon dioxide. J Climate 8:57–80
Naik V, Wuebbles DJ, DeLucia E, Foley JA (2003) Influence of geoengineered climate on the terrestrial biosphere, Environmental Management, DOI 10.1007/s00267–003–2993–7
National Academy of Sciences (1992) Policy Implications of Greenhouse Warming: Mitigation, Adaptation and the Science Base, Chap 28 (Geoengineering). National Academy Press, Washington DC, pp 433–464
Prentice IC et al. (2001) Climate Change 2001: The Scientific Basis: Contribution of Working Group I to the Third Assessment Report of the IPCC, Houghton JT et al (ed), Cambridge University Press, UK, pp 183–237
Prentice IC, Heimann M, Sitch S (2000) The carbon balance of the terrestrial biosphere; ecosystem models and atmospheric observations. Ecol Appl 10(6):1553–1573
Press WH, Flannery BP, Teukolsky SA (1989) Numerical recipes. Cambridge University press, Cambridge, UK, pp.702
Root TL, Schneider SH (1993) Can large scale climatic models be linked with multiscale ecological studies? Conservation Biol 7(2):256–270
Schneider SH (1996) Geoengineering—Could or should we do it? Climatic Change 33:291–302
Schneider SH (2001) Earth Systems engineering and management. Nature 409:417–421
Seifritz W (1989) Mirrors to halt global warming? Nature 340:603
Taylor KE, Penner JE (1994) Response of the climate system to atmospheric aerosols and greenhouse gases. Nature 369:734–737
Teller E, Wood L, Hyde R (1997) Global warming and ice ages: I. Prospects for physics based modulation of global change,UCRL-231636 / UCRL JC 128715. Lawrence Livermore National Laboratory, Livermore, California, USA
Washington WM, Meehl GA (1989) Climate sensitivity due to increased CO_2: experiments with a coupled model and ocean general circulation model. Climate Dyn 4:1–38

Watson RT, Zinyowera MC, Moses RH, Dokken DJ (1995) Climate Change 1995: Impacts, adaptations and mitigation of climate change: Scientific-technical analyses. Intergovernmental Panel on Climate Change, United Nations Environmental Program/World Meteorological Organization, Chap 25. Cambridge University Press, New York, USA, pp 799–822

Zwiers FW, Storch HV (1995) Taking serial correlation into account in tests of the mean. J Climate 8:336–351

CHAPTER 7

A DUAL USE FOR SPACE SOLAR POWER

The global weather control option

ROSS N. HOFFMAN, JOHN M. HENDERSON, GEORGE D. MODICA,
S. MARK LEIDNER, CHRISTOPHER GRASSOTTI
AND THOMAS NEHRKORN

Atmospheric and Environmental Research, Inc., Lexington, Massachusetts 02421, U.S.A

Abstract: One day mankind may be capable of controlling the weather on a global scale. The key factor enabling control of the weather is that the atmosphere appears to be chaotic and chaos implies sensitivity to small perturbations. Extreme sensitivity to initial conditions suggests that small perturbations to the atmosphere may effectively control the evolution of the atmosphere if the atmosphere is observed and modeled sufficiently well. It is shown that four-dimensional variational analysis (4d-VAR) is a data assimilation technique that has promise for calculating optimal perturbations for weather modification. Experiments described here demonstrate the ability of 4d-VAR to calculate perturbations to influence the evolution of a simulated tropical cyclone. In "damage cost function" experiments described here, 4d-VAR simultaneously minimizes the size of the initial perturbation and an estimate of property loss that depends on wind speed. In these experiments the hurricane surface winds decrease over the built-up area at landfall. It is as if the simulated hurricane "blinks its eye" at a precisely controlled time. The optimal perturbations usually include quasi-axisymmetric features centered on the hurricane. It appears that the perturbation evolves as a concentric wave disturbance that propagates to a focus at the hurricane center, and converts the kinetic energy of the hurricane into thermal potential energy at the appropriate time. The hurricane surface winds regenerate soon thereafter, so a continuous series of perturbations may be needed in practice. Experiments are described with different control vectors, including all prognostic variables, temperature only, and temperature only outside of the center core of the hurricane. The temperature only experiments suggest that precisely prescribed heating might serve to control hurricanes and other weather phenomena in the future. Microwave heating rate calculations are presented in support of the concept of dual use space solar power satellites for electric power and for weather control. These calculations show that by tuning within the 183 GHz water vapor absorption interval it would be possible to control the height of the maximum heating. The prototype experiments presented here suggest that global weather control will eventually become a reality especially since many of the supporting disciplines will naturally evolve at a rapid pace. The costs associated with recent damaging hurricanes should cause this pace to accelerate. It is plausible that two generations from now controlling the global weather

V. Badescu et al. (eds.), Macro-Engineering: A Challenge for the Future, 87–120.

may be within the capabilities and resources of several nations or groups of nations. In the future, NASA's mission may explicitly include mention of research to control the weather for the benefit of mankind

Keywords: weather modification, hurricanes, space solar power, 4d-VAR

1. INTRODUCTION

Our experiments to control computer simulations of hurricanes demonstrate the possibilities, as well as highlight the physical difficulties, involved in controlling hurricanes. It will likely be decades before controlling hurricanes becomes technically feasible. A prerequisite for responsible weather modification is the ability to predict the weather so accurately that the effect of small changes can be reliably forecast. Also, politically and socially it may never be acceptable to attempt to control a hurricane. Just think: A hurricane that devastates Florida may later provide much needed rain to Ohio. Who is to blame if it is not sent out to sea, or if it is sent out to sea, or if it is sent out to sea but by mistake strikes Bermuda, or if some other unintended occurrence results?

Hurricanes that make landfall are devastating to life and property. Even the threat of landfall can be expensive—to save lives the US government may evacuate miles and miles of coastline at a cost of perhaps $1 million per mile. The American Meteorological Society (AMS) policy statement "Hurricane Research and Forecasting" (AMS, 2000b) summarizes the hazards of tropical cyclones over land: loss of life and nearly $5 billion (in 1998 dollars) annually in damage due to the storm surge, high winds and flooding. The economic cost continues to rise due to growing population and wealth in coastal regions. If we really knew where a hurricane will strike we could much more efficiently protect lives and property. Or even better, could we lessen the effects of a hurricane by changing its track or strength?

Hurricane forecasts have been improving but there is still much room for improvement. Hurricanes are very difficult to forecast even with modern satellite data and computers. One reason is that the evolution of a hurricane is very sensitive to details of its structure and environment. Small deviations in the ocean temperature, the location of the jet stream, or differences from the typical circular symmetry about the hurricane eye may have very large impacts on the future track and strength of the hurricane.

Of course forecasting hurricanes depends on observing the hurricane—but hurricanes present a very hostile environment for making observations. This is clearly a problem for ship or aircraft measurements. Satellite sensors are also at a disadvantage, since many cannot see through clouds and most cannot see through rain. New passive and active microwave instruments carried on NOAA and NASA research aircraft demonstrate partial solutions to this problem. (NOAA is the US National Oceanic and Atmospheric Administration and NASA is the US National Aeronautics and Space Administration.)

And it is not just our ability to specify a particular hurricane that is affected. Our knowledge of physical processes in hurricane conditions is limited by the amount and quality of *in situ* measurements that are the ground truth for calibrating satellite sensors and validating models. Even now there are debates about how to represent the effects of sea spray and foam on the exchanges of heat, moisture, and momentum between the sea surface and the atmosphere at hurricane wind speeds. Such processes are key to the evolution of hurricanes and critical to satellite observations of hurricanes. Satellite sensors do not directly measure winds or temperature as does an anemometer or thermometer. In many cases satellite algorithms for retrieving such environmental variables must be tuned to more direct measurements. Again, under hurricane conditions, the quality and quantity of such direct measurements is severely limited.

Since hurricanes are powered by heating and evaporation at the sea surface, sea surface temperature is critical to hurricane evolution, but the sub-surface temperature structure is also important. There is a two-way interaction between hurricanes and the ocean temperature structure. Hurricanes are thought to be a very important factor in mixing the upper ocean. If a hurricane moves too slowly it will lose strength as it mixes cooler sub-surface water into the warm surface waters. In fact, a hurricane traveling in a loop may lose strength as it crosses its own track. The energy used in the process of evaporation at the ocean surface is later converted to heat in the atmosphere when clouds and rain form.

And finally to close the loop: Errors in observing the hurricane structure and errors in modeling the physical processes in the hurricane are compounded by how quickly errors grow under the influence of chaos in modeled hurricanes. However, this sensitivity to small differences might also one day make it possible to control hurricanes. That is, the extreme sensitivity that is a hallmark of chaos might actually provide a handle to control the weather. Theoretical and model studies have established that the dynamics governing the atmosphere can be extremely sensitive to small changes in initial conditions (e.g., Rabier et al., 1996). This suggests that the earth's atmosphere is chaotic. In a realistic numerical weather prediction (NWP) model, since small differences in initial conditions can grow exponentially, small but correctly chosen perturbations induce large changes in the evolution of the simulated weather. Current operational NWP practices—including data assimilation, generation of ensembles, and targeted observations—illustrate this daily (Hoffman, 2002). A series of perturbations to the atmosphere might therefore be devised to effectively control the evolution of the atmosphere, if the atmosphere is observed and modeled sufficiently well. Hoffman (2002) hypothesized that as we observe and predict the atmosphere with more and more accuracy, we will become able to effect control of the atmosphere with smaller and smaller perturbations. One question addressed in the present study is how to calculate the optimal perturbations. Theory tells us that perturbations must have a special spatial structure to grow significantly. A random perturbation, such as a butterfly flapping its wings in South America, might create a tornado over Oklahoma two weeks later, but such an event has zero probability of occurring.

With funding from the NASA Institute for Advanced Concepts (NIAC), we have been applying this idea to control hurricanes—not real hurricanes but simulations of hurricanes in computer experiments. In these experiments we use a mathematical optimization technique to find the most efficient way of changing the future track or intensity of our simulated hurricanes. In particular, we calculate perturbations to control a simulation of Hurricane Iniki of 1992. We base our approach on a particular data assimilation method known as four-dimensional variational data assimilation (4d-VAR). 4d-VAR determines a small perturbation to the initial estimate of the atmosphere to optimally fit data in a short, say six-hour, assimilation window (Courtier, 1997). Our experiments are idealized both in terms of the simulation of the hurricanes and in terms of the method of control and should be considered prototypes. Our simulations of hurricanes could be improved with more sophisticated physical parameterizations and higher model resolution.

Further, we introduce instantaneous perturbations to the model atmosphere while practical perturbations would necessarily be introduced over a finite time interval. For example, we cannot change the temperature throughout a hurricane all at once. Instead one might heat or cool the atmosphere, thereby changing the temperature over a period of time. In the future we hope to conduct experiments in which we hold the initial state fixed but calculate the precise pattern and strength of heating needed to control the hurricane. The energy required would be huge, but solar power stations in space in large enough numbers could do the job. Sometime this century, space solar power might provide a small but significant fraction of our energy needs. One way to transmit the energy down to earth is to beam it down as microwaves. Depending on the microwave frequency used, the atmosphere is transparent or absorbing, and so a secondary use of space solar power might be to heat the atmosphere to control the weather. Microwave frequencies of interest, however, do not penetrate rain. Therefore, in one experiment we allowed only changes to temperature outside of the center of the hurricane. The final results are similar, but not surprisingly the initial temperature perturbations outside of the central area are larger in amplitude. In the future, combined with more realistic simulations, the vector of control parameters that is optimized might be a description of the temporal and spatial patterns of feasible forcing. For example, these parameters might describe additional heating supplied to the atmosphere by a space solar power downlink in the 183 GHz water vapor spectral region.

In this chapter we hope to convince the reader that there is a scientific basis to believe that controlling the weather, even the most powerful storms, may be possible in the future. In spite of various simplifications, our experiments demonstrate the control of simulated hurricanes. To explain our technique we will first describe 4d-VAR, the process used at some of the main weather forecast centers to estimate current atmospheric conditions for the purpose of initiating a computer model weather forecast. We modified this technique to calculate the most efficient perturbation to control our simulated hurricanes. We will see that the structure of these most efficient perturbations is very complex, and it will require enormous investments in infrastructure to be able to calculate and generate them quickly enough.

While the amount of energy required is very large, we do find that the area-weighted magnitude of the perturbations decreases as the resolution of the simulation system increases. Sensitivity experiments show, in general, that increasing the degrees of freedom decreases the overall amplitude of the perturbations and that wind perturbations are more effective than other fields. At the end we will hazard a guess for the future of weather modification.

2. BACKGROUND

Most scientists agree that the atmosphere is chaotic. Certainly numerical weather prediction models are extremely sensitive to the starting conditions for the forecast. This sensitivity implies it is difficult to make accurate weather forecasts. This extreme sensitivity also allows the possibility that very small (in a relative sense) planned perturbations to the atmosphere will provide a measure of control. There are two caveats. First, because the atmosphere is so difficult to predict, it follows that it is difficult to predict the result of any perturbation we apply. Highly accurate observing and forecasting capabilities are required. Second, the smallest scales have the shortest predictability times. Swarms of trained butterflies just will not do. For weather control to succeed we must be able to continuously create perturbations, that, while small compared to the atmosphere, are large in human terms.

Our first experiment with Hurricane Iniki was an immediate success. In September 1992, Hurricane Iniki passed directly over the island of Kauai, killing several people and causing enormous property damage and blowing down thousands of trees (Lawrence and Rappaport, 1994). In our computer simulation, we decided to try to change the track of Iniki starting 30 hours before landfall. To do this we created a model "target" 24 hours before landfall that positioned Iniki approximately 100 km to the west of the observed track. We then calculated the smallest possible change to the hurricane 30 hours before landfall that would nearly match the target at 24 hours before landfall (Hoffman et al., 2002). In this first experiment perturbations to all fields were permitted. Because the changes to the initial conditions were designed to be small, the difference between the original and perturbed versions of Iniki was not discernable when looking at the full fields. However, the new version veers off to the west for six hours, passes close to the target, then tracks due north in the subsequent model run. In this case Kauai escaped the most damaging winds of Iniki.

We are not the first group to dream of altering hurricanes. Hurricane modification in the early 1960s was the subject of a series of courageous (or foolhardy depending on your view point) but inconclusive cloud seeding experiments called Project Stormfury. As described by Simpson and Malkus (1964) the idea tested in Project Stormfury was that an increase of precipitation would decrease surface pressure at the radius of maximum wind, thereby increasing the eye radius and decreasing the hurricane intensity. Our ability to forecast hurricanes has improved dramatically since the 1960s but as we have already noted there is still much room for improvement. From today's vantage point Stormfury seems overly ambitious and potentially risky. In the early 1960s with more can-do and less worry about unintended consequences, Stormfury probably seemed like the prudent thing to do. See Kwa (2001) for an interesting historical summary of how attitudes have evolved towards weather modification.

Diverting hurricanes is certainly not the first type of weather modification that scientists have pursued. Seeding clouds to enhance precipitation is the most common form of weather modification. These same seeding techniques were used in Project Stormfury, but in these cases the precipitation is for farmers' fields, not to weaken hurricanes. Our ability to model precipitation processes, in general, and in hurricanes in particular, is limited. While much is known (Pruppacher and Klett, 1997), there is considerable uncertainty in the physical processes that take place in clouds. These processes include how droplets and ice crystals form, how cloud particles interact and evolve by collision and breakup processes, how winds and electric fields in a cloud evolve, how these winds and electric fields effect the growth and interaction of cloud particles, how individual clouds interact, and so on. Clouds and rain in turn depend on presence and composition of dust in the atmosphere. In any real cloud there is always a mixture of different particle sizes and types. Individual dust particles can be composed of almost anything (Holmes, 2002). Here are three common but very different examples of dust: First, tiny grains of sand made of the mineral quartz are lifted by the wind from places like the Sahara Desert. Because of their crystalline structure these dust particles are ideal starting places for ice crystals to grow. Such particles are called ice nuclei. Second, smokestack emissions from the burning of coal create tiny sulfate aerosol particles. The sulfate particles have a very strong affinity for water and as humidity increases, the aerosols collect water, becoming drops of sulfuric acid, and then cloud drops as more water is collected. We say that the sulfate aerosol particles act as cloud condensation nuclei (CCN). Third, salt particles form when drops of water or spray from the ocean remain airborne long enough to evaporate. Salt particles, like sulfate aerosols, are hygroscopic (i.e., readily absorb water) and also serve as CCNs. In a cloud all the different sizes and types of particles grow at different rates in humid air. So differences in atmospheric dust result in differences in cloud particle populations. This leads to different cloud drop populations, which affects the precipitation reaching the ground. For example, in polluted air with many CCNs, fewer drops will grow to the size of rain drops. These factors are also critical for climate models (Ramanathan et al., 2001). For example, polluted clouds have many small drops and reflect more solar energy than unpolluted clouds.

For five decades weather modification practitioners have claimed that adding economically small amounts of a variety of seeding agents would substantially enhance rainfall or reduce damaging hail. Weather modification for these purposes is a profitable business. Evidence that has been collected to date suggests, but does not prove (in a generally accepted statistically valid scientific way) that cloud seeding is effective. One scientifically well established success story is seeding to dissipate cold ($<0\,^{\circ}$C) fog. Experiments demonstrated this in the late 1940s and an operational program has been running for many years at the Salt Lake City airport. A thorough up to date discussion of the problems and prospects for weather modification is given in a recent National Academy of Sciences report (Garstang et al., 2003). This report reviews the history of weather modification, summarizes the uncertainties involved, and suggests that with new knowledge and technology it is time to reinvigorate the science of weather modification.

It is easy to see why it is so difficult to rigorously prove weather modification works. First, large samples, perhaps several years of data, are needed to show the results are significant. Second, rainfall is very difficult to measure accurately. Third, it is difficult to get evenly matched treated and control samples. Clouds are difficult experimental subjects: no two are alike, they move around, merge and dissipate. Clouds that look the same to the eye may be quite different in terms of initial cloud condensation nuclei, drop size distribution, the meteorological conditions of the surroundings, and to what extent the environment of the cloud has been influenced by neighboring clouds. All these factors introduce uncertainty as to whether a particular seeding strategy might work for a particular cloud. Further, operational programs usually have no control samples—every likely cloud is seeded. Fourth, our ability to predict clouds and precipitation is currently limited so we cannot compare a treated case to a prediction of what would have happened without cloud seeding. Large samples and sophisticated statistics have been applied in attempts to overcome this problem but without clear successes to date. New remote sensing and *in situ* techniques, and improved modeling capabilities, should now allow us to more clearly see the differences between clouds and thereby reduce the uncertainty of choosing matched pairs of clouds for experiments; to model the effect of seeding on particular cloud types; and to diagnose the chain of physical processes linking the release of the seeding agent to precipitation falling on the ground.

The scientific evidence for inadvertent weather modification is much stronger than for planned weather modification. It is well accepted that man-made changes—in land use, industrial pollution, aircraft contrails, etc.—affect our weather today. Credible evidence abounds showing the existence of anthropogenic changes in local rainfall patterns. For example, there are man-made sources for each of the three dust examples given earlier. These include desertification due to poor agricultural practices, smokestack emissions from fossil fuel power plants, and drying out of inland seas (like the Aral in Central Asia) which all contribute dust downwind. In the first two cases, rainfall decreases due to the increase in particle density (Rosenfeld et al., 2001; Rosenfeld, 2000), but in the third case, precipitation increases since the large salt particles from the exposed seabed grow very quickly at the expense of other cloud drops (Rudich et al., 2002).

3. METHODOLOGY

The instability of the dynamics of the atmosphere is in fact used by the weather forecast centers in several ways. Our technique is similar to one of these techniques, but our group is the first to explicitly make use of this instability to control hurricanes (albeit simulated hurricanes). To explain what we have done, we first summarize the weather prediction problem.

NWP is an initial value problem. Workable methods to solve this problem were proposed around 1920 by L.F. Richardson (Richardson, 1922), but it was not until the first computers were built that NWP became practical. NWP is based on the "conservation" of mass, energy, momentum, and moisture. Once the state of the atmosphere is defined at one time, the conservation laws are used in an NWP model to predict the state of the atmosphere into the future. The model used in this study, the Penn State/National Center for Atmospheric Research (NCAR) Mesoscale Model 5 (MM5), is described in Section 4. The "conserved" quantities are not exactly conserved and sources and sinks are included in the model. For example, at the sea surface the atmosphere and ocean exchange all four conserved quantities—mass, energy, moisture, and momentum. At low wind speeds this process is mediated by small scale gravity-capillary waves. In a hurricane, droplets of water vapor are continuously torn off the crests of waves, partially evaporated and cooled, and then fall back into the ocean. The atmospheric state is a complete specification of the variables such as pressure, temperature, relative humidity, and wind speed and direction that correspond to the conserved quantities. In most models these variables are defined on a three-dimensional grid. NWP models have a top boundary, usually defined with zero vertical velocity, well above the domain of interest. For each variable for each vertical level one could plot a map. The collection of values of all the variables at all the gridpoints is the model state. To generate a forecast, a NWP model repeatedly advances the model state from one instant through a small time step (of a few seconds to a few minutes depending on the scales of motion resolved by the model). A first model state is needed to initialize this process.

This first model state is not complete and not exact and so forecasts are imperfect. The model state is not complete because, for example, scales smaller than the grid length are not represented. Thus without high resolution the hurricane's structure near the eye wall will be smoothed. Also the only chemistry usually included has to do with H_2O and its phase changes, but many other species are important for radiative heating of the atmosphere and for the production of aerosol particles that are important for cloud formation and thus to precipitation and to the reflection of solar radiation. The model state is not exact because we cannot make perfect observations, nor can we make observations at every location of all the variables. Finally, because the models, like the atmosphere, behave chaotically, errors in the model state grow quickly as the forecast proceeds.

Since even our best estimate of the model state is uncertain, and this uncertainty will grow during the forecast period, all weather forecasts are statements in probability. Today the more advanced forecasts include an estimate of the uncertainty of the forecast. Comparing several forecasts also provides an estimate of this uncertainty.

Estimating the atmospheric state is also a problem in statistics. In order to use all past observations a process called data assimilation is used. At regular intervals, say every six hours, the data assimilation system produces a best estimate of the model state. This is used to initiate a six-hour forecast. All observations collected in the next six hours are then combined with the previous estimate of the model state—the latest six-hour forecast—and the cycle continues. The combination step produces a best estimate by properly accounting for the estimated uncertainties in the observations and the six-hour forecast. The statistical theory for this is clear, but the assumptions and information needed for the proper application of this theory to NWP are only approximate. As a result, practical data assimilation is part art and part science. Ultimately these systems are judged by forecast skill.

The data assimilation method that we have modified for our experiments is called four-dimensional variational data assimilation or 4d-VAR. 4d-VAR is currently used operationally at ECMWF (the European Centre for Medium-range Weather Forecasts) (Rabier et al., 2000). 4d-VAR is different from other data assimilation techniques by using the model equations directly while combining observations with the previous estimate. Specifically, 4d-VAR finds the atmospheric state that satisfies the model equations and that is simultaneously close to both the previous estimate and the observations by minimizing the difference to each source of information, each weighted according to its relative accuracy. This is done by adjusting the model state at the start of the six-hour interval. The model state at the end of the six-hour interval provides the next "previous estimate" for the next six-hour interval. The sensitivity of the NWP model to small changes in the initial model state is key to the success of 4d-VAR. Technically 4d-VAR is very challenging because of the complexity of the NWP model and the relationship between some types of observations and the atmospheric variables. Mathematically 4d-VAR is similar to the problem of fitting a straight line to noisy data using linear regression. 4d-VAR solves this complex nonlinear minimization problem iteratively, making use of the adjoint of a linearized version of the model.

Current 4d-VAR practice finds the smallest global perturbation as measured by the *a priori*, or background, error covariances but it is possible to modify 4d-VAR to find the smallest perturbation measured in some other way. The MM5 implementation of 4d-VAR used in this study is described by Zou et al. (1997). It has been applied to assimilate zenith delay observations from global positioning system (GPS) satellites (De Pondeca and Zou, 2001) and to assimilate cloud-cleared brightness temperatures from geostationary operational environmental satellite (GOES) sounders (Zou et al., 2001). For our hurricane experiments we have modified 4d-VAR so that instead of minimizing the misfit to the observations we minimize the estimated wind damage caused by the hurricane. We still require that the solution be close to the previous estimate at the start of the six-hour interval so that the calculated perturbation will be small.

3.1 Initial Perturbation Cost Function

The experiments reported here are based on a variation of 4d-VAR. In what follows, we consider the unperturbed simulation as reality. To mathematically define the objective function that will be minimized by 4d-VAR, we first define the unperturbed simulation U, from time 0 to 6 h, with corresponding states $U(0)$ and $U(6)$.

We then use 4d-VAR to find an optimal (or controlled) simulation C that simultaneously minimizes the difference from the initial state (i.e., $C(0)$-$U(0)$), and the estimated wind damage later in the forecast period.

In these preliminary experiments, the size of the initial perturbation is measured by a simple quadratic form:

$$J_b = \sum_{x,k} \frac{1}{S_{xk}^2} \left[\sum_{i,j} \{C_{xijk} - U_{xijk}\}^2 \right]. \tag{1}$$

Here x defines the control vector variables (i.e., the temperature T, or the horizontal wind components (u, v), or all variables), and i, j, and k index the grid points in the three spatial dimensions. In Eq. (1), the "coupled" form of the variables is used since the MM5 is based on the so-called "flux" form of the primitive equations governing the evolution of the atmosphere. For example, $p_* u$ is the coupled eastward wind component, where p_* is the reference pressure difference between the surface and top model boundaries. The reference state varies in the vertical only, therefore p_* depends only on the model surface topography. Below, we present the components of J_b, in particular the contribution to J_b of the temperature at different times and model layers, as the square root of the terms in square brackets in Eq. (1) for x equal to temperature, normalized by the number of grid points, and dimensionalized assuming $p_* = 950$ hPa, the value over the ocean.

The scales S_{xk} depends only on variable and layer. The scales are used to equalize the contributions of variables of different quantities and magnitudes. Effectively the scales are the relative costs of introducing perturbations at different levels or in different variables. In the present experiments S_{xk} is calculated as the maximum absolute difference between $U(0)$ and $U(\delta t)$ for each variable at each layer, with δt taken to be 2400 s (40 minutes). (This is similar to the approach taken by Zou et al., 2001.)

Figure 1 shows the vertical profiles of the S_{xk} used in our experiments, again dimensionalized assuming $p_* = 950$ hPa. In general these scales vary smoothly in the vertical.

3.2 Damage Cost Function

In our experiments the total cost function is defined as

$$J = J_b + J_D. \tag{2}$$

Here the subscript D stands for damage. The damage cost function, J_D estimates the likely property damage based on an empirical relationship between surface wind speeds and economic losses. The contribution to the cost function at each grid point is the product of the fractional wind damage (D_{ij}) and the property value (P_{ij}). Thus,

$$J_D = \lambda \sum_{i,j} D_{ij} P_{ij}. \tag{3}$$

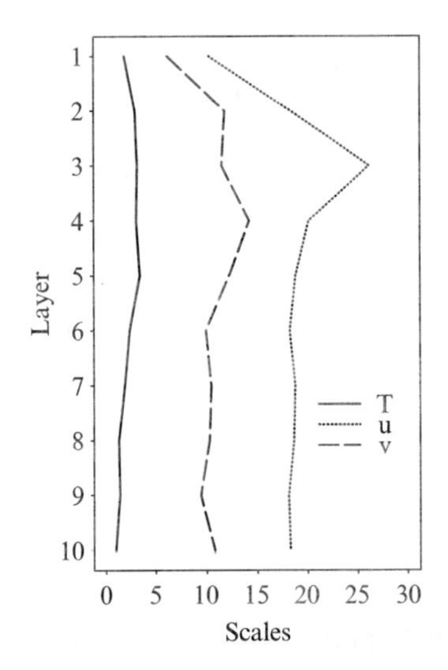

Figure 1. Profiles of scaling factors for temperature (T), eastward wind component (u) and northward wind component (v) plotted using solid, dotted, and dashed lines, respectively. Values are calculated from the scales of the coupled variables and then dimensionalized for this plot assuming $p_* = 950$ hPa. The horizontal scale is °C for temperature and m s^{-1} for the wind components

Here λ is a weighting factor. The property values are unitless. The fractional damage (Unanwa et al., 2000) depends upon two threshold wind speeds; the lower threshold (W_0) is the wind speed at which damage to property first occurs, while the second (W_1) is the wind speed at which complete destruction occurs. Between these two threshold values, we model the increase in damage using a cosine curve

$$D(W) = \frac{1}{2}\left[1 + \cos\left(\pi \frac{W_1 - W}{W_1 - W_0}\right)\right], \tag{4}$$

where W is the simulated horizontal wind speed in the lowest model layer. Note that W and hence D vary with location and time. In the experiments described here, Eq. (3) and Eq. (4) are evaluated only at the end of the 4d-VAR interval (i.e., at t = 6 h). In all cases $W_0 = 25$ m s^{-1} and $W_1 = 90$ m s^{-1}. We experimented with the weighting factor λ; $\lambda = 400\,000$ in the results presented here.

3.3 Control Vector

The control vector is a list of all the quantities that are allowed to be varied by the minimization. An example of an element of the control vector is the temperature at a particular grid point. In 4d-VAR, one could minimize J with respect to $\mathbf{X}$, the entire model state vector (that is, all prognostic variables at all grid points). For the MM5 these are the three-dimensional fields of p_*u, p_*v, p_*T, p_*q, p', and p_*w (coupled eastward and northward wind components, temperature, and specific humidity, perturbation pressure, and coupled vertical velocity, respectively). Note that while all prognostic variables may be allowed to vary, only temperature, horizontal wind, and humidity observations are normally used. In such systems an additional constraint may be included in J to control the excitation of gravity waves. In other data assimilation systems p' and w, and perhaps q, are not allowed to vary. We will present results for $\mathbf{X}$, for temperature only (T), and for temperature only outside the core of the hurricane (T_d). (Here "d" is for doughnut since the region outside the core of the hurricane is topologically equivalent to a doughnut.) As a practical matter, during the minimization, we always calculate the entire gradient of J with respect to $\mathbf{X}$. However we project the gradient onto the control vector by setting elements of the gradient to zero that correspond to variables that are not allowed to vary.

4. EXPERIMENTAL PROCEDURES

We use the Penn State/NCAR Mesoscale Model 5 (MM5) system to simulate Hurricane Iniki of 1992. Central Pacific Hurricane Iniki (1992) caused extensive damage to property and vegetation on parts of the Hawaiian Islands and killed six people (CPHC, 1992; Lawrence and Rappaport, 1994). The storm made landfall on Kauai at 0130 UTC 12 September 1992, with a central pressure of 945 hPa. Maximum sustained winds over land were estimated at 60 m s^{-1} with gusts as high as 80 m s^{-1}.

For Iniki our experiments calculate optimal perturbations at 0600 UTC 11 September 1992. MM5 produces very detailed and accurate simulations of tropical cyclones when high resolution and advanced physical parameterizations are used (e.g., Liu et al., 1999; Tenerelli and Chen, 2001). However the current prototype experiments used coarser, 20 km resolution. Due to the relatively coarse resolution and simple parameterized physics used here, our simulations are only crude representations of observed track and intensity.

4.1 Mesoscale Model

The MM5 used in our experiments is described by Grell et al. (1994) and by Dudhia (1993). In our experiments, the MM5 computational grid covers an approximately 3000×4000 km horizontal domain with ten "sigma" layers in the vertical from the surface to 50 hPa. Hurricane Iniki remains far enough from the domain edges that boundary effects are small during the course of the experiments. The sigma coordinate system is a terrain-following normalized pressure coordinate system (Holton, 1992, section 10.3.1).

The MM5 may be configured in many ways. Except as noted all experiments described here use nonhydrostatic dynamics, a 60 second time step and a 20 km Mercator projection grid. The parameterizations of surface fluxes, radiative transfer, and cumulus convection are currently limited in the MM5 4d-VAR system. For our experiments, the physical parameterizations include the medium-range forecast (MRF) model planetary boundary layer (PBL) and the Anthes Kuo convection schemes.

Large-scale stable (i.e., nonconvective) precipitation occurs whenever a layer reaches saturation. Excess moisture rains out immediately with no re-evaporation as it falls. Long wave radiation uses simple radiative cooling, with cloud effects included. The radiation computation occurs every 30 time steps.

4.2 Hurricane Initialization in the Gridded Datasets

Gridded National Center for Environmental Prediction (NCEP) reanalysis fields (Kalnay et al., 1996) from the NCAR archives, and operational NCEP sea surface temperature (SST) analyses, were used to initialize the MM5 model and provide boundary conditions during the 4d-VAR and forecast periods. The boundary conditions define both the model state along the lateral boundaries and surface parameters such as SST, land use, and others. Other than SST, the surface parameters were derived from the data bases included in the MM5 distribution. Figure 2 shows the SST over the entire computational domain. While the model grid is 158 × 194 for

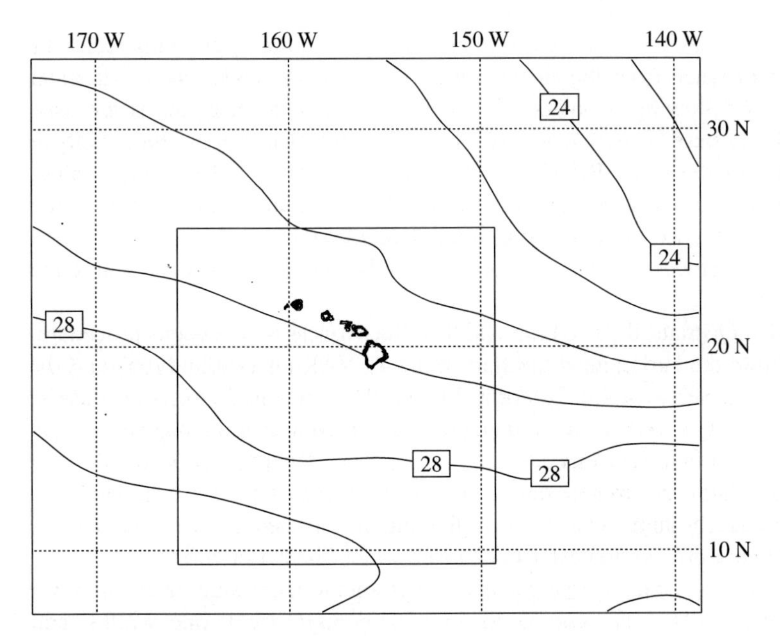

Figure 2. Computation domain showing sea surface temperature (SST, °C) at the start of the 4d-VAR interval. The rectangle plotted indicates the area shown in all other map figures, except for Fig. 4

the experiments reported here, the smaller domain indicated in this figure of 92×95 grid points centered over the storm is the area plotted in all subsequent figures (except for Fig. 4).

The available datasets have only a hint of an actual hurricane's strength and structure. The use of high resolution satellite data to properly initialize a mesoscale model is an area of ongoing research. Consequently we add an analytic representation of a hurricane vortex using the method of Davis and Low-Nam (2001) 6 h before the start of our experiments (i.e., at $t = -6$) and let the model representation of the hurricane equilibrate during the 6 h leading up to the start of the 4d-VAR interval. Even with the bogus procedure the simulated storm is still too weak.

The Davis and Low-Nam (2001) tropical cyclone bogussing system is part of the MM5 pre-processing procedures and was developed by NCAR and the Air ForceWeather Agency (AFWA). Before the bogus storm is added, the representation of the storm in the original data set is removed. The bogus storm is axisymmetric and is based on specifying the storm position and the radius and magnitude of the maximum wind in the lowest model layer.

The maximum wind specified should represent the average wind speed at the radius of maximum wind speed and might be in the range of 75–90% of the best track wind speed. (The best track is the official description of a tropical cyclone based on all available information, collected either in real-time or later.) Given these parameters a Rankine wind vortex (AMS, 2000a) is used to generate the bogus lowest model level wind field. In a Rankine vortex the wind increases linearly with distance from the storm center to the radius of maximum wind and then decreases following a power law in distance. Traditionally the power law exponent is –1, but the exponent used here is $-3/4$. The bogus wind field at upper levels has the same Rankine vortex shape, but the wind speeds decrease according to a specified vertical profile. From the bogus wind field a temperature field is calculated to be in nonlinear gradient balance at all levels. Surface friction is ignored in this process but the surface winds and other variables adjust within the first 1-2 h.

Figure 3 demonstrates the need for, and the effectiveness of, the bogus procedure. Figure 3a shows the hurricane at the start of the 4d-VAR interval in the reanalysis. There is no evidence of a strong storm. Figure 3b shows the bogus storm at the same time. The bogus storm was initiated 6 h earlier. Note that the bogus vortex is well defined, with high wind speeds, and a clear wave number one asymmetry (in azimuth about the storm center) due in part to the fact that the Rankine vortex is added to a southerly large-scale flow. In this plot and in similar plots that follow, the winds plotted are the lowest model layer winds (hereafter referred to as the surface winds), wind speed thresholds that correspond to increasing Saffir-Simpson category (AMS, 2000a) are contoured with increasingly thick line widths, and shading denotes the extent of wind speeds above 25 m s^{-1}—the lower threshold of damaging winds.

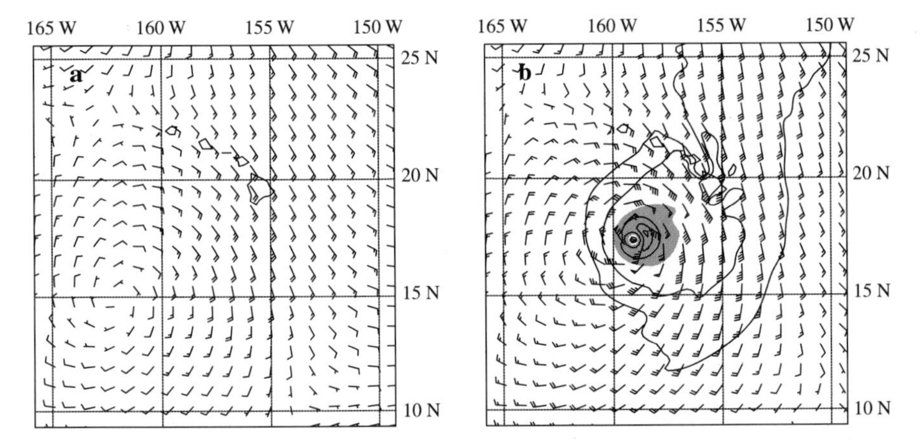

Figure 3. Surface pressure and winds at the start of the 4d-VAR interval for Iniki as depicted by (a) the reanalysis and (b) the bogus procedure. Wind speed thresholds that correspond to increasing Saffir-Simpson category are contoured with increasingly thick line widths. The plotted contours correspond to the lowest wind speeds associated with tropical depressions (12 m s^{-1}), tropical storms (17 m s^{-1}), category 1 hurricanes (33 m s^{-1}) and category 2 hurricanes (43 m s^{-1}). Shading denotes the extent of wind speeds above 25 m s^{-1}—the lower threshold of damaging winds. Wind symbols are in m s^{-1} with a full barb indicating 5 m s^{-1}. [Please see this figure in the color section at p. 311]

4.3 Generation of Property Values

Property values used in the computations are shown in Fig. 4. This basic two-dimensional property value field was generated by smoothing topography. The smoother averages all points within 400 km resulting in a gradient of property values for nearshore water points. This simple definition was found to be suitable for Hurricane Iniki.

5. RESULTS

The experiment names given below refer to both the 4d-VAR analyses and the subsequent MM5 forecasts carried out beginning from the analyses. The forecast using unperturbed initial conditions is denoted U. The 4d-VAR analyses reported here examine the effects of limiting the variables allowed to change in the analysis, and of excluding changes near the center of the storm.

5.1 Summary of Experiments

A summary of the three Iniki experiments described here is provided in Table 1. For the purpose of this discussion we take experiment $C[T]$ as a baseline experiment. Note that in $C[T]$ the control vector is the temperature field only, indicated by T.

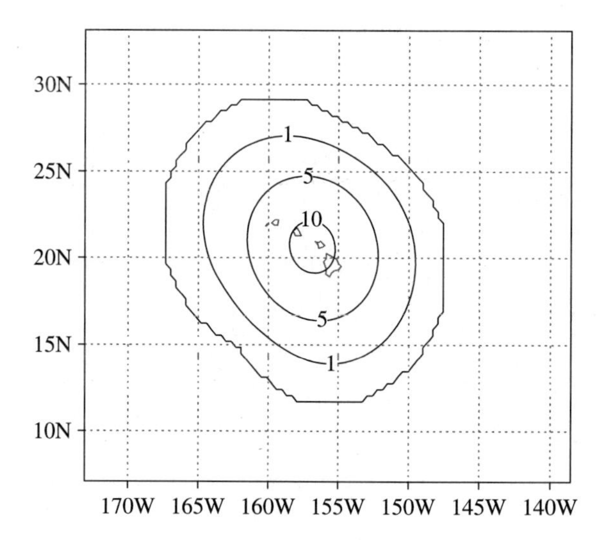

Figure 4. Property values based on smoothing the topography field. Property values are unitless, but might be considered tens of thousands of dollars per square kilometer

Table 1. Hurricane Iniki experiments and cost functions*

	Number of		J			Final J_b components					Gradient		
Exp.	Iter.	Eval.	Initial	Final	%	T	$\mathbf{V}$	w	q	p'	Initial	Final	%
$C[\mathbf{X}]$	10	11	80727	313	0.4	20	259	0	12	12	18255	809	4
$C[T]$	10	12	80727	3742	5	3167	0	0	0	0	8935	2280	26
$C[T_d]$	10	12	80727	8551	11	6522	0	0	0	0	7125	1323	19

* The experiment names in the first column are explained in Section 5.1. Columns 2 and 3 give the number of iterations and function evaluations required by the optimization. The initial and final total cost function J are then listed along with the ratio of the final to initial value expressed as a per cent. Next, the components of the cost function due to the perturbation at the initial time are listed for each model variable. The final three columns list the magnitude of the gradient of the cost function at the start and end of the minimization along with the ratio of the final to initial value expressed as a per cent. As the minimum is approached J should decrease and the gradient should approach zero.

Table 1 shows that 4d-VAR is more successful in preventing wind damage at 6 h when all fields are allowed to vary (i.e., when the control vector is $\mathbf{X}$).

Experiment $C[T_d]$ is identical to $C[T]$ except that grid points in a square region centered on the hurricane center are excluded from the control vector. The maximum iterations allowed were 10 for these experiments. In the table we report the final value of the components of J_b and the initial and final values of J. Note that the initial values for J_b are zero.

5.2 Cost Function Baseline Experiment

Figure 5 shows the cost function versus iteration for experiment $C[T]$. Note the asymptotic leveling off of the total and component cost functions. This indicates a successful minimization. Figure 6 shows the profiles of root mean square (*rms*) difference of perturbation temperature for $C[T]$, $C[\mathbf{X}]$, and $C[T_d]$ experiments. The components of J are converted into *rms* differences as described in the discussion of Eq. (1). The $C[T]$ temperature perturbations are larger in magnitude and concentrated in the upper troposphere, where perturbations are typically 0.4 °C in magnitude. (The discussion of this figure continues in the next section.)

By differencing two computer simulations ($\delta = C - U$) we can clearly visualize the perturbation and how it evolves. Figure 7 shows the structure of the perturbation δT at $\sigma = 0.350, 0.650$, and 0.950 for experiment $C[T]$. Over the ocean these layers correspond to 382.5, 667.5, and 952.5 hPa. At $\sigma = 0.950$, the lowest model level, there is cooling close to the eye (or center) of the hurricane and heating to the west. At mid and upper levels there is a complex pattern of stronger heating and cooling. These patterns are not correlated between $\sigma = 0.650$ and $\sigma = 0.350$. Patterns in the intervening layers show that these features twist and amplify with altitude. As is the case in most experiments, there is evidence of a banded structure away from the hurricane center increasing with altitude.

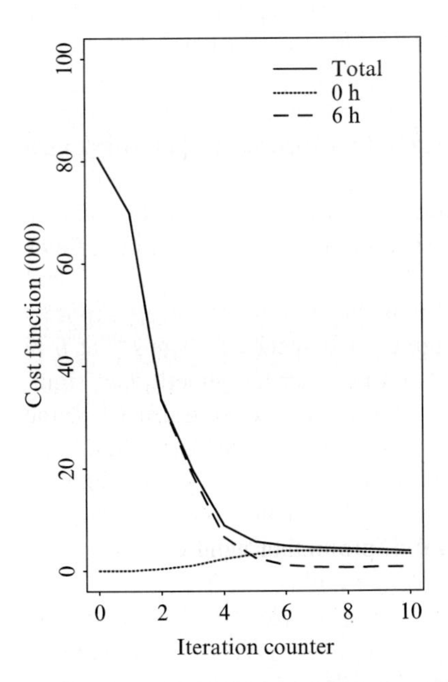

Figure 5. Cost function versus iteration for experiment $C[T]$. The total cost function (in thousands) and the individual parts of the cost function at $t=0$ and 6 h are shown as solid, dotted, and dashed lines, respectively

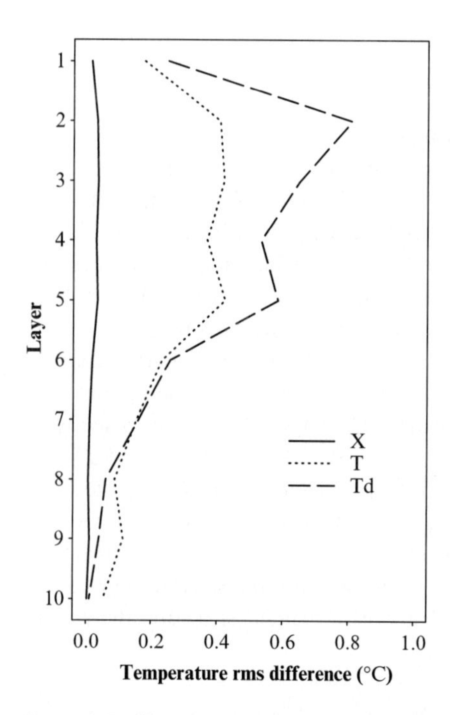

Figure 6. Profiles of *rms* difference of temperature with respect to the unperturbed simulation U for experiments $C[\mathbf{X}]$ (solid line), $C[T]$ (dotted line), and $C[T_d]$ (dashed line). The components of J_b are converted into rms differences as described in the text

Figure 8 shows the evolution of the perturbation for experiment $C[T]$. Horizontal slices of δT and δw at selected levels are shown at 2 h intervals. Although only temperature has been changed at the initial time all variables are quickly affected. The structure of the perturbation at 6 h shows extraordinary changes in both fields close to the hurricane center at this time.

The effect of these changes on the full wind fields (Fig. 9) is to effectively suppress the winds to near or below the critical damaging wind speed of 25 m s^{-1} at 6 h, and at 6 h only. Figure 9 shows the surface wind field for the unperturbed simulation U and for experiment $C[T]$. The figure shows that in experiment $C[T]$, the hurricane has accelerated along its track relative to the unperturbed case. At 4 h the counterclockwise rotation of the wave number one asymmetry is evident. Although the areal extent of the damaging winds (>25 m s^{-1}) has changed little, significant reductions in peak wind speeds are evident at 6 and 8 h and especially at 6 h.

5.3 Sensitivity Experiments

We conducted two additional Iniki experiments that allowed temperature to vary—$C[\mathbf{X}]$ and $C[T_d]$. Profiles of *rms* temperature increments for these temperature sensitivity experiments (see Fig. 6) show that the temperature increments are smallest

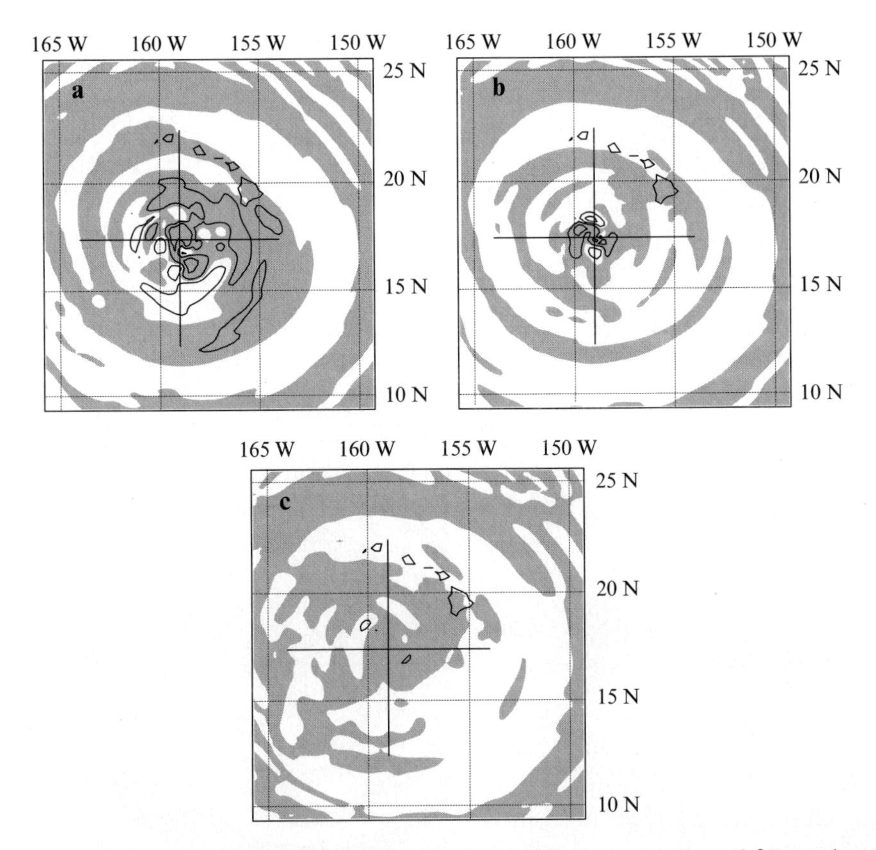

Figure 7. Structure of the perturbation for experiment $C[T]$. Horizontal slices of δT are shown at (a) $\sigma = 0.350$, (b) $\sigma = 0.650$, and (c) $\sigma = 0.950$. Negative perturbations are shaded, and contours indicate magnitudes of temperature differences exceeding 1 (thin lines) and 3 °C (thick lines). The crosshairs here and in subsequent figures indicate the positions of the center of the storm in experiment U at the appropriate times. [Please see this figure in the color section at p. 312]

when the control vector is **X** since fields other than temperature are also adjusted. Comparing $C[T_d]$ to $C[T]$, we see that preventing increments near the storm center results in considerably larger increments away from the storm. In experiment $C[T]$ there are larger increments at lower levels near the storm center. Since $C[T_d]$ increments in this part of the domain are necessarily zero, the $C[T_d]$ *rms* increments are smaller than those for $C[T]$ at lower levels, but $C[T_d]$ more than makes up for this at upper levels.

Figure 10 and Fig. 7a show the patterns of perturbation temperature δT at $\sigma = 0.350$ for these experiments. Plots at other levels are qualitatively similar. First, note that the $C[T]$ temperature increments are 2-4 times larger than those of $C[\mathbf{X}]$. In $C[T]$, a large cold temperature increment is present directly over the center of Iniki. This temperature increment is in direct opposition to the "warm

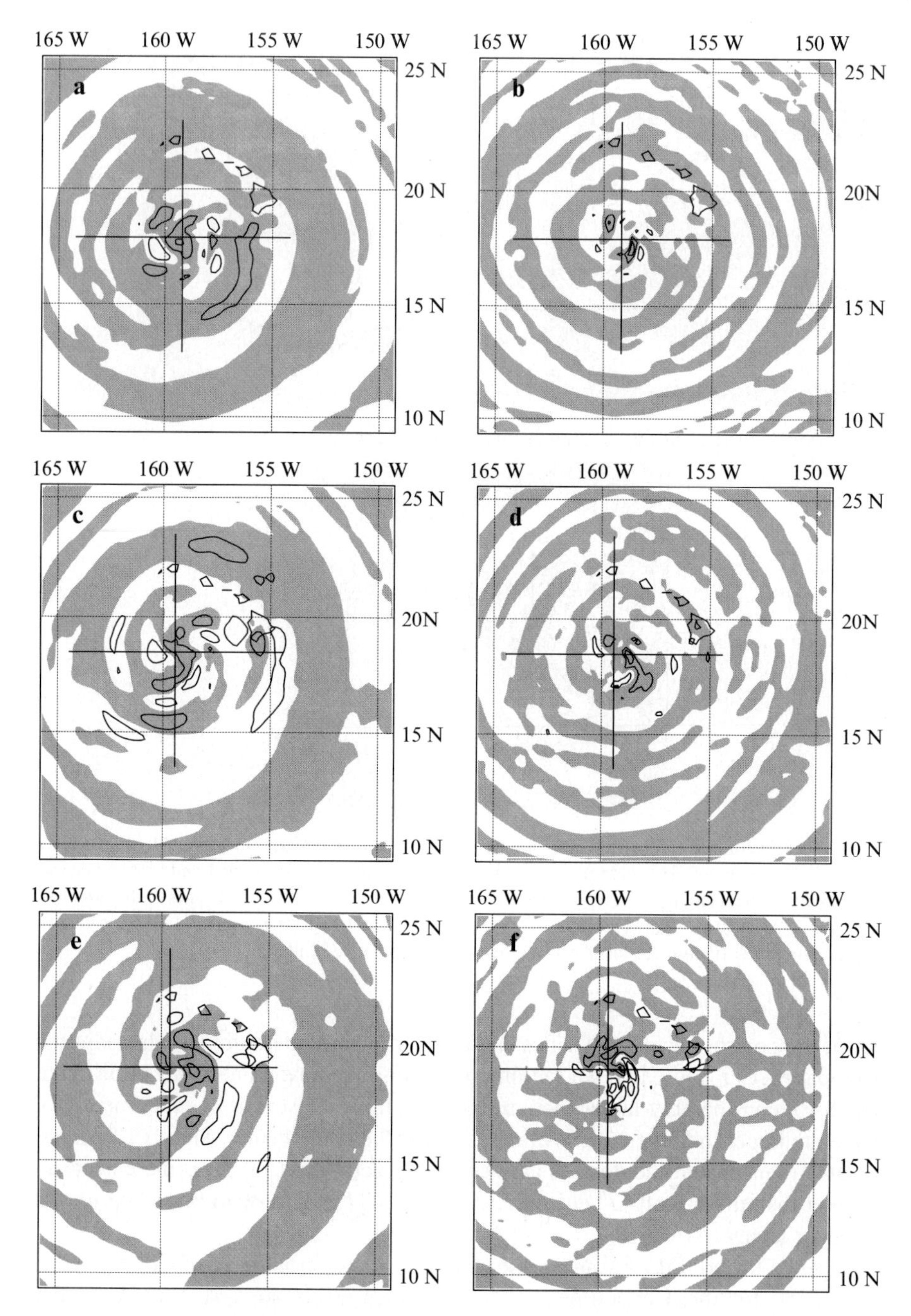

Figure 8. Evolution of the perturbation for experiment $C[T]$. Horizontal slices of δT and δw are shown at $\sigma = 0.350$ and at $\sigma = 0.650$ respectively (from left to right), at 2, 4, and 6 h (from top to bottom). Negative perturbations are shaded. Thin and thick contours indicate magnitudes of differences greater than 1 and 3 °C for temperature and 30 and $60 \cdot 10^{-2}$ m s^{-1} for vertical velocity

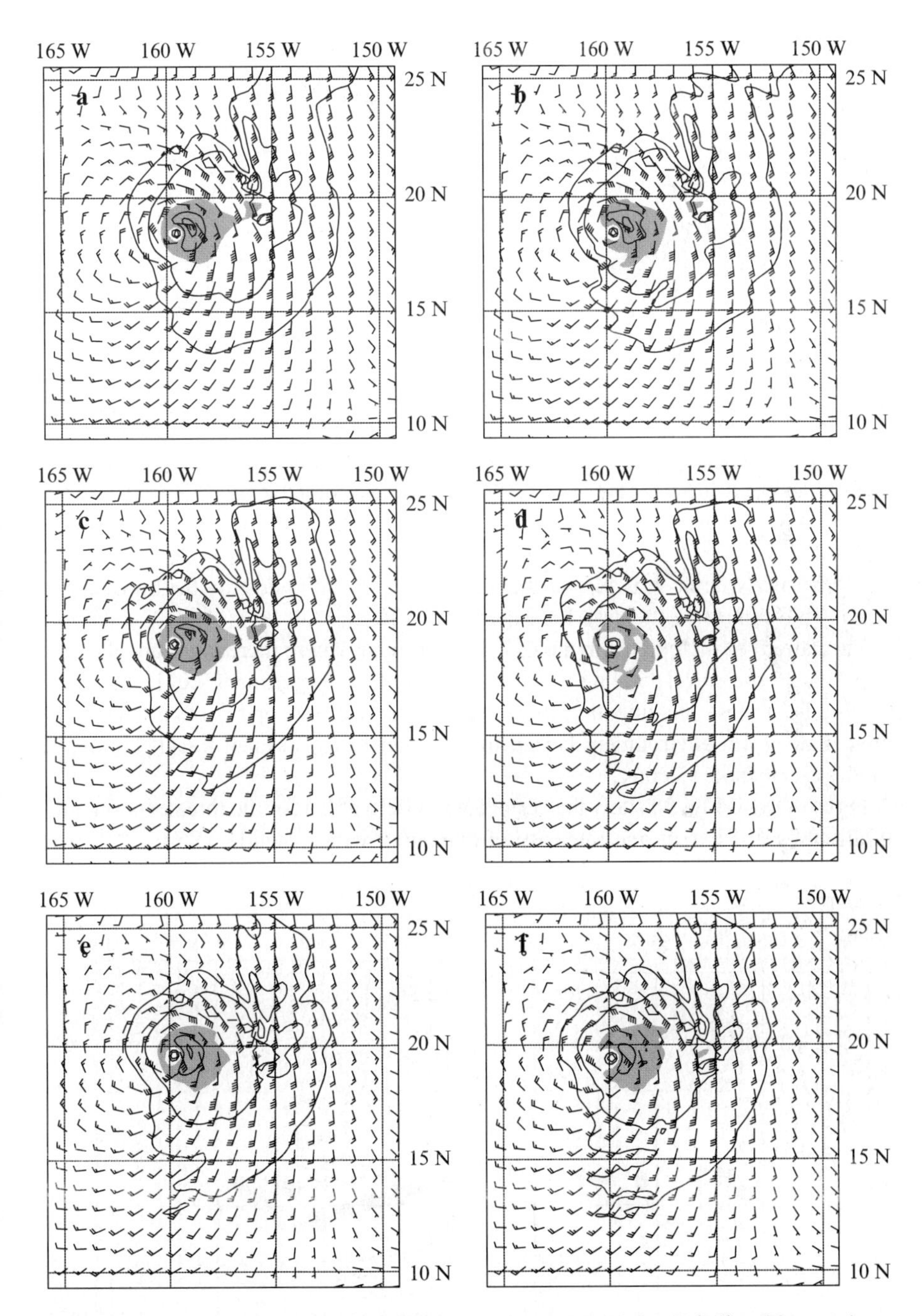

Figure 9. Evolution of the surface wind field for the unperturbed simulation U (left) and for experiment $C[T]$ (right) at 4, 6, and 8 h (from top to bottom). Plotted as in Fig. 3. [Please see this figure in the color section at p. 313]

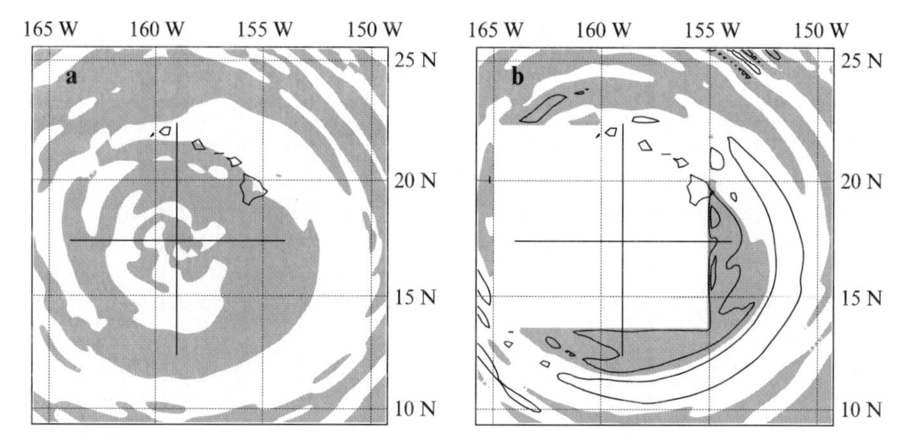

Figure 10. Perturbation temperature δT at $\sigma = 0.350$ for experiments (a) $C[\mathbf{X}]$ and (b) $C[T_d]$. Compare to Fig. 7a. Plotted as in Fig. 7. [Please see this figure in the color section at p. 314]

core" thermodynamic structure of the hurricane and acts to destroy the hurricane in place. This cannot be the case for experiment $C[T_d]$.

The pattern of increments in the $C[T_d]$ experiment, where increments are allowed, is similar to the corresponding increments in the $C[T]$ experiment, but in $C[T_d]$ the amplitude of these increments is much larger. The warming and cooling outside the immediate storm environment is four times larger in $C[T_d]$ than in $C[T]$ south and southwest of Iniki's center.

Figure 11 and Fig. 9d show the surface wind field at 6 h for the three experiments. In all cases, the extent and intensity of damaging winds (winds >25 m s^{-1}) were

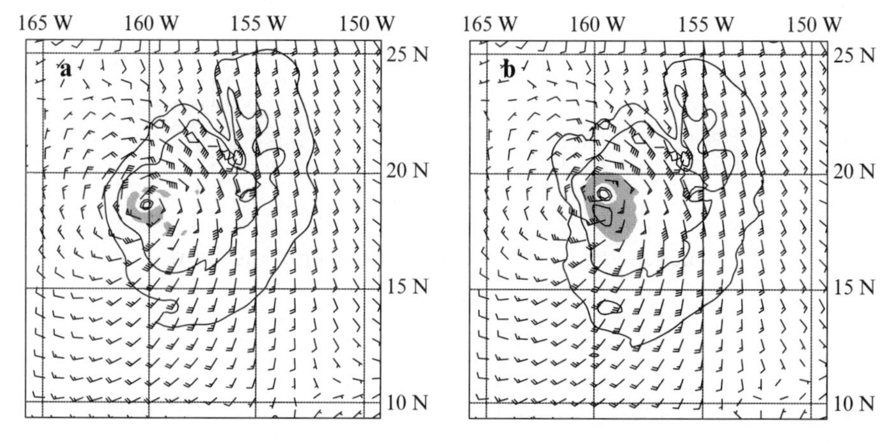

Figure 11. Surface wind field at 6 h for experiments (a) $C[\mathbf{X}]$ and (b) $C[T_d]$. Compare to Fig. 9c,d. Plotted as in Fig. 3. [Please see this figure in the color section at p. 314]

drastically reduced in the hours near the evaluation time of the wind damage cost function, 1200 UTC 11 September (compare to Fig. 9c).

It should be noted that the extent and intensity of the winds increased rapidly in the hours following this time in these experiments. All three experiments were extremely successful at limiting the number of grid points with damaging winds at 6 h. In experiment $C[\mathbf{X}]$ the storm remains nearly stationary for 6-9 h. This is followed by a northwesterly motion of the vortex, eventually resulting in the storm passing the latitude of Hawaii much later than in the unperturbed case and much farther west. In some experiments, the sea-level pressure field (which is related to the overall temperature through the depth of the atmosphere) appeared to be temporarily mispositioned with respect to the wind field close to the time of the wind damage cost function evaluation. This effect was most noticeable in experiment $C[T_d]$.

The $C[T_d]$ experiment is less successful in reducing wind damage compared to the $C[T]$ experiment. This is expected since, when allowed, 4d-VAR focuses the increments in the proximity of the storm; in $C[T_d]$, of course, increments in this location are not permitted.

Figure 12 shows the evolution of the perturbation for experiment $C[T_d]$. The perturbation fields for δT and δw at selected levels are shown at 2 h intervals. Qualitatively there are many similarities between this figure and Fig. 8 beyond 2 h, and even at 2 h outside the central "doughnut" area. Although similar in structure the $C[T_d]$ increments are noticeably more intense. See especially δT and δw at 4 h. The temperature increments show the inward radial movement of the ring of positive temperature increments. Coincident with the collapse of the ring into a centrally located bubble of warm increments is a sudden and rapid decrease in sea-level pressure because sea-level pressure is proportional to the weight of the atmosphere above, which decreases as temperature increases. The increments temporarily disrupt the wind field at the appropriate time (Fig. 11) but there is a rapid increase in wind speed near the surface after 6 h. Thus, it appears that in experiment $C[T_d]$, and generally in the successful C experiments, the kinetic wind energy is temporarily converted into thermal (potential) energy. At the same time that the perturbations focus on the hurricane center, other wave-like perturbations propagate radially outward at greater distances from the storm center.

The surface wind increments (not shown), initially identically zero, increase in magnitude and propagate radially inward. Near the storm center at 6 h, the wind increments are very large just east of the center to reduce the hurricane wind speed (the area of most intense wind at 6 h in experiment U shown in Fig. 9c). But there is an area south of the center, farthest from land, where the wind speed actually increases (the thicker contour south of the storm center in Fig. 11b).

5.4 Findings from Preliminary Experiments

We carried out preliminary 4d-VAR experiments analogous to $C[\mathbf{X}]$, $C[T]$, $C[T_d]$ at a resolution of 40 km and with simpler physical parameterizations (fixed soil temperature, no land surface fluxes, and no interactions between clouds and radiation). Perturbations from these experiments had qualitatively similar characteristics

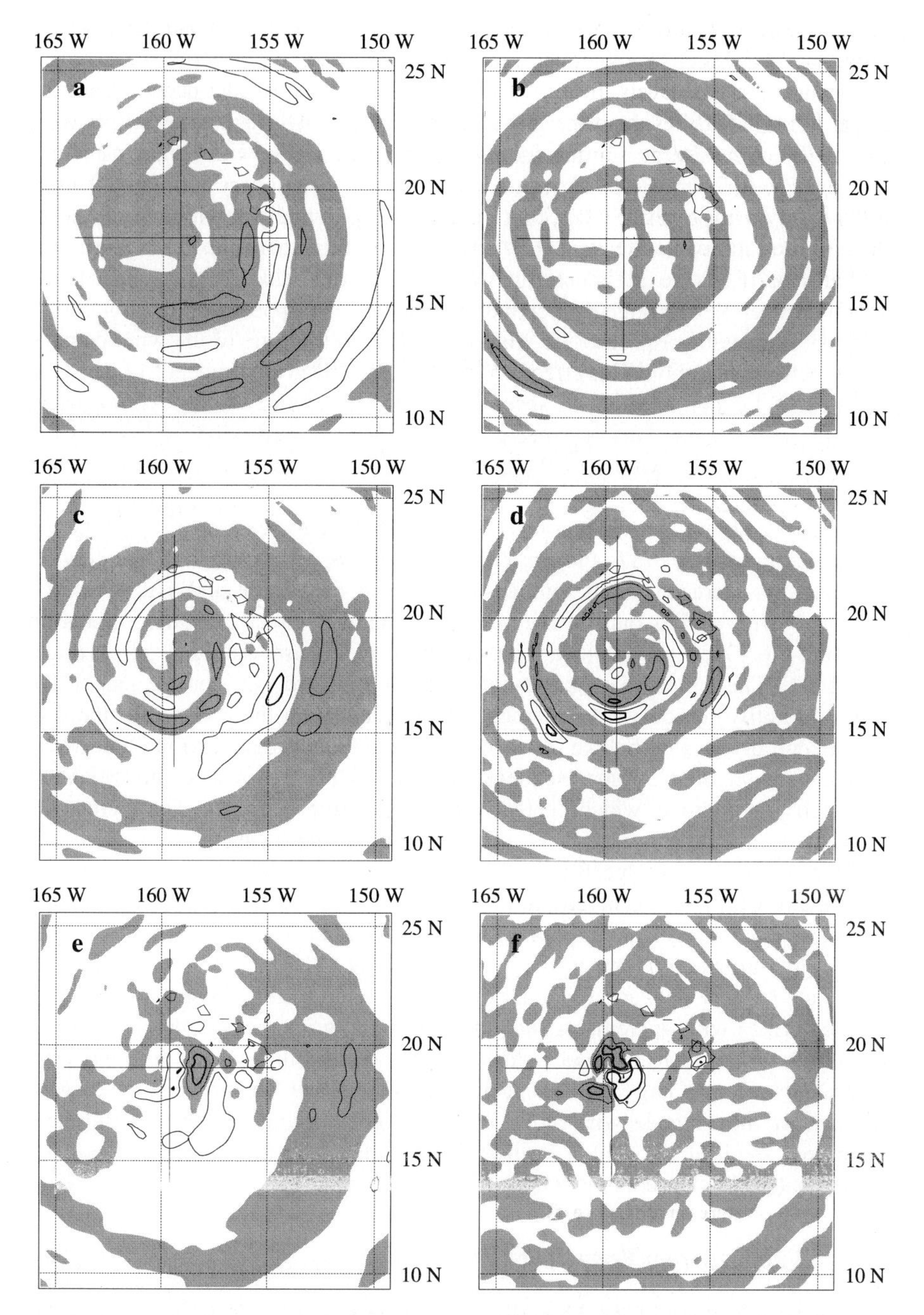

Figure 12. Evolution of the perturbation for experiment $C[T_d]$. Horizontal slices of δT and δw are shown at $\sigma = 0.350$ and at $\sigma = 0.650$ respectively (from left to right), at 2, 4, and 6h (from top to bottom). Plotted as in Fig. 8

to the experiments already described with concentric rings of positive and negative perturbations peaking in the upper troposphere and large increments near the center of the storm at low levels. In the 40 km experiments the optimal perturbations are larger in scale with less spatial complexity and noticeably larger (roughly speaking two to three times larger) in magnitude. Compare Fig. 13 to Fig. 6. The range for temperature is more than three times larger in Fig. 13. Note that in Fig. 13 results are plotted for experiment $C[T_d]$ (the same curve appears in Fig. 6) and for another 20 km experiment similar to $C[T_d]$ but using the same simplified physics as the 40 km experiments.

5.5 Findings from Hurricane Andrew Experiments

In a set of experiments reported elsewhere (Henderson et al., 2005), we calculated temperature increments to limit the surface wind damage caused by Hurricane Andrew over the six-hour period beginning 00 UTC 24 Aug 1992. In this experiment we used the modified form of 4d-VAR to simultaneously minimize the initial perturbation and the damaging winds over the last two hours of the six-hour interval.

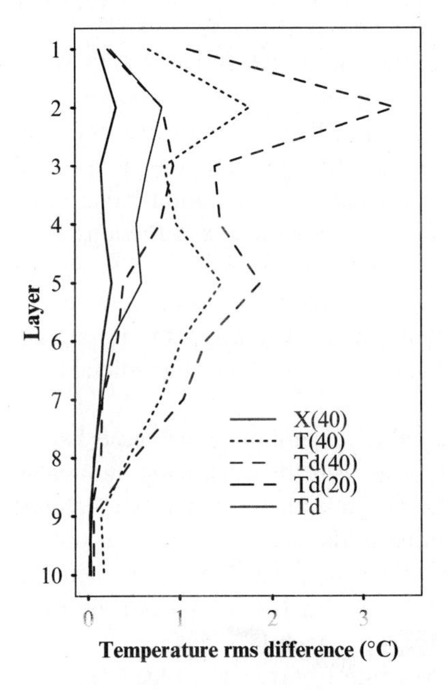

Figure 13. Profiles of *rms* difference in temperature for various C preliminary experiments for Iniki. The components of J_b are converted into *rms* differences as described in the text. The *rms* differences for an experiment using a full model state control vector is drawn with a full line, that with temperature only with a dotted line. Both of these are at 40 km resolution. Results for doughnut experiments at both 20 and 40 km are shown as short and long dashed lines. For reference experiment $C[T_d]$ is plotted as a thin line

The initial temperature increments are strongest near the surface close to the eye and at higher levels from 6 to 12 km above the ocean within a radius of 300 km from the eye. At the end of six hours in the unperturbed case, damaging winds (winds greater than 25 m s^{-1}) cover the Miami metro area, but they do not in the controlled case. Very similar results are obtained when we use the same perturbation but change the model to use higher resolution and more sophisticated approximations of the sub-grid-scale physical processes. This is evidence that the method is robust. As in the experiments reported here damaging winds reappear beyond six hours.

6. SPACE SOLAR POWER TO GENERATE PERTURBATIONS

Our working hypothesis is that chaos allows the control of a variety of weather phenomena provided we can observe and predict such phenomena with high accuracy. In the future we expect scientific advances in understanding and modeling the atmosphere, and technological advances in observing the atmosphere and in communications and computer systems, will result in much more accurate weather predictions. Two additional technological factors will then be needed for weather control—a means to calculate perturbations to obtain the desired result and a means to generate the calculated perturbations. The experiments described in the previous sections are prototypes of how to calculate perturbations.

There are many possible means to generate perturbations (Hoffman, 2002). These range in scale from seeding an individual cloud from a single-engine aircraft to very large space based reflectors. In the introduction, we touched on the use of space solar power to create perturbations. Even butterflies create tiny perturbations to the wind currents as they flit from flower to flower. For weather modification larger scale and larger amplitude perturbations are required because such perturbations have longer time scales. The perturbations must be predictable. Generally smaller spatial scales have shorter time scales, and predictability beyond two or three lifetimes for a feature in a chaotic system is severely limited. Put another way, any uncertainty associated with tiny perturbations implies their effects will be unpredictable in short order.

Evidence that cloud seeding experiments enhance rainfall or decrease hail is accumulating but not on a rigorous scientific basis, in part because we are unable to accurately predict the precipitation processes in actual (unmodified) clouds (section 2). In addition to cloud seeding, changes to the heating of the atmosphere and changes to the surface properties which affect the exchanges of momentum, heat, and moisture between the atmosphere and the land or ocean are the most plausible means of generating perturbations.

A number of ideas have been advanced to control hurricanes. One idea is to directly limit the hurricane's source of energy by coating the ocean surface with a biodegradable oil that limits evaporation (Simpson and Simpson, 1966). However as wind speed increases, such a film will break up. Hurricanes might also be controlled at longer range, by introducing perturbations days in advance and thousands of miles away that create changes in the large-scale wind patterns at jet stream level.

These wind patterns can have large effects on a hurricane's intensity and track. (We could not look into this strategy with our current experimental setup using a limited area model.) It is possible that appropriate perturbations might be generated by relatively minor changes in our normal activities. For example aircraft contrails affect heating in the atmosphere and we already control the height at which aircraft fly. Or if we control the timing and location of crop irrigation we would affect the moistening of the atmosphere.

Based on our current experiments however, the most plausible approach to control hurricanes is to use a space-based system to produce precise heating of the atmosphere. With regard to space-based reflectors, the atmosphere is mainly transparent to solar energy except for clouds which mainly reflect solar energy. Thus the principal effect of reflected solar energy is to change the surface temperature. For example, at the beach on a sunny day the temperature of the exposed sand can be quite a bit higher than sand sheltered by an umbrella. However, such a system has little use over the ocean. Over the ocean, solar energy is primarily absorbed by the ocean or reflected by clouds, but it will be difficult to quickly effect changes to ocean surface temperature using solar reflectors due to the effective high heat capacity of the ocean. (It is not just the ocean surface that gets heated because unlike beach sand the sunlight penetrates the water and the upper layer of the ocean mixes.)

A viable alternative is space solar power (SSP). SSP has been proposed as a nonpolluting inexhaustible source of energy (Glaser, 1968; Glaser et al., 1998). SSP would collect solar energy, and beam it down to earth (Fig. 14). A downlink in microwave frequencies would be selected in a part of the spectrum where the atmosphere is transparent, to minimize losses due to heating of the atmosphere. For weather control, other frequencies would be chosen.

Space solar power could provide an energy source that would be modulated in time and directed at different locations. In the vertical, the energy deposition and hence heating is controlled by the transmission frequency and by the distribution of absorbing species, mainly water vapor and oxygen.

Figure 15 illustrates this. In Fig. 15 we plot the time rate of change of temperature as a function of altitude for different frequencies in the microwave spectrum for an SSP satellite that provides a top of the atmosphere power flux density of 1500 Wm^{-2}.

A single nominal-design SSP station might provide 6 GW of power which would cover an area 2×2 km at 1500 Wm^{-2}. The heating rates are calculated for radiation vertically incident at the top of a standard tropical atmosphere. MonoRTM, a very accurate line-by-line radiative transfer model described by Clough et al. (2005), is first used to calculate transmissivities ($\tau(z)$) every 0.5 km of altitude (z). These transmissivities are the ratio of the transmitted to incident radiation, and therefore decrease monotonically from the top of the atmosphere to the surface. Energy not transmitted through a layer is deposited in that layer.

Thus the time rate of change of temperature may be calculated as

$$\frac{dT}{dt} = -\frac{F_0}{C_p \rho(z)} \frac{d\tau(z)}{dz}. \tag{5}$$

Figure 14. In the future solar power satellites may be also used to precisely heat parts of the atmosphere by transmitting at a frequency that is absorbed by water vapor. Image copyright by Pat Rawlings SAIC. Used with permission. [Please see this figure in the color section at p. 314]

Here T is temperature, t is time, F_0 is the flux at the top of the atmosphere, C_p is the specific heat of air, and ρ is the density of air. There are four major absorption bands in the microwave spectrum: the 22 GHz water vapor band, the 60 and 118 GHz oxygen bands, and the 183 GHz water vapor band. Figure 15a shows that while oxygen and water both absorb strongly, the absorption by oxygen occurs mainly at levels in the atmosphere high above the bulk of the troposphere. The 183 GHz water vapor band presents the greatest opportunity to apply heating to the troposphere (approximately 0–12 km). Since water vapor in the atmosphere is very variable, the heating profiles will be a function of the meteorology, but this variation could be included in calculating the optimal orientation and power of the downlink. The 183 GHz band allows the vertical distribution of the heating to be controlled by tuning the transmission frequency in this part of the spectrum. This can be seen more clearly in Fig. 15b which shows dT/dt profiles for selected frequencies near 183 GHz. As the frequency approaches 183 GHz the atmosphere becomes more opaque and more energy is absorbed at higher levels. (The peak values of dT/dt also increase because at higher levels the air density ρ is lower.)

While SSP is a plausible system to control the atmosphere by heating, the prototype experiments reported here do not directly correspond to SSP. Instead, for this

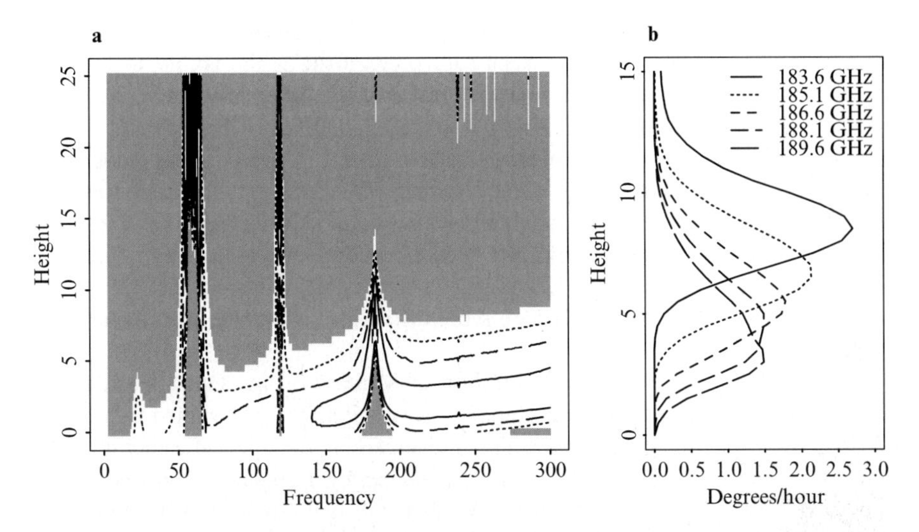

Figure 15. The time rate of change of temperature (degrees Celsius/hour) as a function of frequency (GHz) and height (km). Calculations are for the standard tropical atmosphere assuming vertically incident radiation from an SSP satellite with power flux density of 1500 W m^{-2}. The whole microwave spectrum and heights to 25 km are shown in (a). Values less than $0.125°\ h^{-1}$ are grey. Dotted, dashed, and solid contours are drawn for values of 0.25, 0.5, and $1°\ h^{-1}$. Values greater than $2°\ h^{-1}$ are black. Selected profiles near the 183 GHz water vapor resonance are plotted in (b). Here as height increases, density decreases, so that peak values of *dT/dt* are larger for more opaque frequencies. [Please see this figure in the color section at p. 315]

study we considered instantaneous changes to the atmospheric state. In experiment $C[T]$ we determined changes to the temperature structure of the atmosphere only—closely related to but surely not the same as heating of the atmosphere. In experiment $C[T_d]$ we excluded changes to the temperature of the central part of the hurricane—an option we wished to examine since rain drops strongly absorb 183 GHz radiation. Our methods could be extended so that the control variables describe heating as a function of time and position instead of perturbations of the atmospheric state. A further extension would use frequency and intensity of radiation at the top of the atmosphere as control variables. Ultimately a model of the SSP station would be added and the control variables for the optimization would be the actual control parameters determining the power, frequency, and orientation of the downlink.

7. THE FUTURE

The preliminary study described here shows that 4d-VAR can be used to calculate "optimal" perturbations to control the intensity of a simulated tropical cyclone. Clearly it will be a long time before it is possible to control a tropical cyclone in reality. While the results reported here are preliminary in many aspects, they point the way towards further work. A necessary prerequisite is the ability to forecast tropical cyclones accurately. Beyond this, advances in several technical areas are needed. These are discussed below.

In addition, a number of problems must be solved in the political, economic, and legal realms. For inhabitants of New Orleans, eliminating a hurricane threat to that city may take precedence over all else, yet farmers in the middle of the US might suffer without the resulting rain. This example shows that many competing factors must be considered in defining the cost function to be optimized. These "social engineering" issues may prove more difficult to overcome than the science and hardware engineering issues.

The U.N. Convention on the Prohibition of Military or any Other Hostile Use of Environmental Modification Techniques was negotiated and signed in the late 1970s. (Visit www.unog.ch/frames/disarm/distreat/environ.pdf.) Weather modification for beneficial purposes (e.g., for managing hydrological resources) is not banned, however. Most changes to the weather will have some positive and some negative effects. It is not too soon to think about the political and social issues involved. Since the atmosphere is a coupled system, one nation practicing large-scale weather modification may change the weather of its neighbors. The same holds on a smaller scale for provinces and metropolitan regions.

In the future, weather modification will give mankind substantial power to control our environment. But weather modification should NOT be considered the primary method of managing water resources and mitigating atmospheric hazards. As with all environmental resources, increasing population exerts pressure on water resources and land use. Composting toilets and precision irrigation can result in very large savings of water. Improved building codes and prohibition of building in particularly vulnerable areas can greatly reduce loss to property. Improved forecast and warning systems will protect lives.

Returning now to the required technological advances, items 1, 2, and 3 below relating to the calculation of the perturbations could be usefully examined now with computer simulations that would be a natural follow on to this study. Advances in items 4 and 5—improved models and observations of the atmosphere—will occur naturally as we improve NWP. Item 6, the creation of perturbations will require engineering new systems. The last two items—improved observations and the creation of perturbations—will benefit from new space-based assets.

1. **Calculation of realistic perturbations.** Solving for the optimal perturbation using a more realistic model is difficult due to the number of degrees of freedom required to represent the atmosphere adequately and the nonlinear and sometimes discontinuous nature of the physics governing the atmosphere. With higher resolution and more degrees of freedom, effective perturbations are expected to require smaller magnitudes but more detailed structure. Incremental 4d-VAR (Lorenc, 1997; Rabier et al., 2000) will allow the most sophisticated physics to be used for the trajectory calculation, but simpler physics for the 4d-VAR calculation. The incremental approach eliminates the need to use full resolution

in the linear models, and the use of limited physics eliminates the need to code the adjoint of the most complex packages. These changes can increase the speed of the gradient calculation.

2. **Calculation of feasible perturbations.** The 4d-VAR methodology could be extended for this purpose. A control vector could be developed first in terms of heating perturbations continuous in time over a three- or six-hour period, later in terms of the radiation perturbations at the top of the atmosphere, and finally in terms of the orientation, power, and frequency of the SSP downlink antenna.
3. **Overcoming chaos.** The control must be effected at significant time lags to minimize the size of the perturbations, yet the system is inherently unpredictable at long lead times. In general, theoretical predictability studies (Lorenz, 1969) suggest that doubling the resolution of the observations will only increase predictability by an amount similar in magnitude to the timescale of the motions of the smallest resolved phenomena. For example, since the timescale for the evolution of a thunderstorm is smaller than one hour, observing details of individual thunderstorms will improve predictability by hours at best. Therefore controlling small-scale phenomena may be difficult.

   One method is to continuously monitor and control the system by adding perturbations regularly. An alternative is to control the environment of the phenomena of interest. This approach has some validity for the case of hurricanes. Internal hurricane dynamics have a timescale of a day or less. This limits how far back in time we can go to calculate optimal perturbations. But hurricane tracks are greatly affected (one could say "steered") by the large-scale upper level winds and hurricanes cannot maintain intensity and structure in the presence of environmental vertical wind shear. So, an alternative is to control the large-scale wind field several days or even a week in advance to affect the hurricane's path or intensity, or to prevent a hurricane from forming.
4. **Improved numerical weather prediction.** Projecting future computer and space technology trends is difficult. However, the technical roadmap for improving NWP and data assimilation is well established, and the timing of future progress has been estimated (e.g.. ECMWF, 1999).
5. **Improved atmospheric observations.** Satellites provide a large volume of information, but not always in the right place, or of the right variable, or sufficiently accurate. New instruments on the NASA Terra, Aqua, and Aura satellites hold the promise to fill some of these gaps. Future space-based lidar sensors should be valuable by providing more direct and very accurate measurements of atmospheric properties including winds.
6. **Creation of perturbations.** Optimal perturbations, while small in amplitude, may be large in scale and require substantial amounts of energy. The costs of controlling a hurricane in our simulation experiments in terms of energy required are enormous. In preliminary experiments we did find that halving the grid size more than halved the energy required. If this trend continues down to sub-kilometer-scales (scales that we would like to use for more accurate

forecasting in any case), then control of large-scale weather in the future becomes much more feasible.

Mechanisms do not yet exist to create large-scale perturbations. Global weather control is by nature opportunistic. Useful technologies will typically have multiple other primary uses. For example, solar reflectors could be used to increase power from solar electric farms, increase the growing season for agriculture, and provide lighting for arctic cities, and space solar power has electric power generation as its primary mission. Eventually weather perturbations may become a commodity with a Global Weather Control Authority purchasing rights to run wind farms in reverse or to control the downlink frequency of space solar power, or negotiating in real time for modest aircraft flight path changes.

With regard to demonstrating effective control of weather, we first note that in spite of our desire for perfection, observations and predictions are always somewhat uncertain. Modern data assimilation uses estimation theory to treat NWP, whether on the global scale or some smaller scale, in a probabilistic sense. We can keep track of uncertainty with Kalman filters or ensemble methods so that we can tell if the predicted impact of some treatment is small or large compared to the predicted uncertainty. Then, if we also simulate the effect of the perturbation, we can perform significance testing before the weather control activity begins!

Our method is not just for hurricanes. It should be applicable to any atmospheric phenomenon that can be observed and modeled sufficiently accurately and that depends sensitively on details of the initial structure and environment. In fact, the method proposed here should first be applied to rainfall enhancement or other smaller-scale phenomena. Most weather modification techniques try to exploit the sensitivity of precipitation to the ambient dust by seeding clouds from aircraft or from the ground, with some special type of dust. Our approach might help to optimize the timing, location, type, and amount of such seeding. A test bed in a relatively small region could be densely instrumented. With current observation systems, it may be possible to take this approach with cloud-scale models in the next several years. Real time applications may be far off since the time scales associated with cloud scales are so small compared to current computation resources. However, for the purpose of weather modification experimentation, one could make a probabilistic forecast, say, for untreated cases and validate these probabilistic forecasts with observations. Then having the capability to make validated probabilistic cloud-scale forecasts, it becomes possible to state the significance of the difference between an observed treated result and the corresponding simulated untreated result. Furthermore, if one models the effect of the treatment, then one could also compare a simulated treated case and the untreated observation.

This probabilistic approach could be applied now for simple stratiform rain situations. An optimistic assessment is that it will likely be 5 and perhaps 10 years before the current state of the art for observing and simulating cumulus clouds has advanced enough that the uncertainty of a probabilistic forecast is sufficiently small so that useful conclusions may be drawn. In the time range of 10–20 years, our ability to forecast hurricanes may have advanced so much that we will have sufficient confidence to begin control experiments using aircraft contrails or aircraft-dispersed surface oils or other methods. If successful, such experiments may provide additional impetus to speed the development of space solar power. Active control of the large-scale weather patterns to reduce the severity of droughts, to decrease the number of severe tornadoes, and to reduce damage due to hurricanes, may then become a reality 40–50 years from now.

ACKNOWLEDGEMENTS

The NASA Institute for Advanced Concepts (NIAC) supported this research. NCEP analyses were obtained from NCAR. We made extensive use of community supported publicly available models and data sets.

REFERENCES

AMS (2000a) Glossary of Meteorology, 2nd edn. American Meteorological Society, 850 pp

AMS (2000b) Hurricane research and forecasting. Bull Am Meteorol Soc 81:1341–1346

Clough SA, Shepard MW, Mlawer EJ, Delamere JS, Iacono MJ, Boukabara S, Brown PD (2005) Atmospheric radiative transfer: a summary of the AER codes. J Quant Spectrosc Radiat Transfer 91:233–244

Courtier P (1997) Dual formulation of four-dimensional variational assimilation. Quart J Roy Meteor Soc 123:2449–2461

CPHC (1992) Tropical cyclones report for the central Pacific. Technical Memorandum NWS-PR-38, Central Pacific Hurricane Center, NOAA, Washington, DC. http://www.prh.nuaa.gov.cphc/summaries/1992.php

Davis C, Low-Nam S (2001) The NCAR-AFWA Tropical Cyclone Bogussing Scheme. Technical Memorandum, Air Force Weather Agency (AFWA), Omaha, NE. http://www.mmm.ucar.edu/mm5/mm5v3/tc-report.pdf

De Pondeca MFV, Zou X (2001) A case study of the variational assimilation of GPS zenith delay observations into a mesoscale model. J Appl Meteor 40:1559–1576

Dudhia J (1993) A nonhydrostatic version of the Penn State-NCAR mesoscale model: Validation tests and simulation of an Atlantic cyclone and cold front. Mon Wea Rev 121:1493–1513

ECMWF (1999) A strategy for ECMWF, 1999–2008. Miscellanenous publication, ECMWF, Reading, United Kindom, 16 pp

Garstang M et al. (2003) Critical issues in weather modification research. National Research Council, National Academy of Sciences, Washington, DC. http://www.nap.edu/books/0309090539.html/

Glaser PE (1968) Power from the sun: Its future. Science 162:957–961

Glaser PE, Davidson FP, and Csigi KI (eds) (1998) Solar power satellites: A space energy system for earth, Wiley-Praxis Series in Space Science and Technology. John Wiley & Sons, Hoboken, New Jersey, U.S.A., 654 pp

Grell GA, Dudhia J, and Stauffer DR (1994) A description of the fifth-generation Penn State/NCAR mesoscale model (MM5). Technical Note 398+1A, NCAR, 122 pp

Henderson JM, Hoffman RN, Leidner SM, Nehrkorn T, Grassotti C (2005) A 4D-VAR study on the potential of weather control and exigent weather forecasting. Quart J Roy Meteor Soc 131:3037–3052

Hoffman RN (2002) Controlling the global weather. Bull Am Meteorol Soc 83:241–248

Hoffman RN, Henderson JM, Leidner SM (2002) Using 4d-VAR to move a simulated hurricane in a mesoscale model. 19th Conference on Weather Analysis and Forecasting/15th Conference on Numerical Weather Prediction. American Meteorological Society, Boston, MA, San Antonio, Texas, J137–J140, paper JP4.4

Holmes H (2002) The secret life of dust. Wiley, 240 pp

Holton JR (1992) An introduction to dynamic meteorology, 3rd edn. Academic, New York, 511 pp

Kalnay E, Kanamitsu M, Kistler R, Collins W, Deaven D, Gandin L, Iredell M, Saha S, White G, Woollen J, Zhu Y, Leetmaa A, Reynolds B, Chelliah M, Ebisuzaki W, Higgins W, Janowiak J, Mo KC, Ropelewski C, Wang J, Jenne R, Joseph D (1996) The NCEP/NCAR 40-year reanalysis project. Bull Am Meteorol Soc 77:437–471

Kwa C (2001) The rise and fall of weather modification: Changes in American attitudes toward technology, nature, and society. In: Miller C, Edwards P (eds) Changing the atmosphere—Expert knowledge and environmental governance. MIT Press, Cambridge, Massachusetts, U.S.A, pp 135–165

Lawrence MB, Rappaport EN (1994) Eastern North Pacific hurricane season of 1992. Mon Wea Rev 122:549–558

Liu Y, Zhang D-L, Yau MK (1999) A multiscale numerical study of Hurricane Andrew (1992). Part II: Kinematics and inner-core structures. Mon Wea Rev 127:2597–2616

Lorenc AC (1997) Development of an operational variational assimilation scheme. J Meteor Soc Jpn 75:339–346

Lorenz EN (1969) The predictability of a flow which possesses many scales of motion. Tellus 21:289–307

Pruppacher HR, and Klett JD (1997) Microphysics of clouds and precipitation. Springer, Berlin, 976 pp

Rabier F, Järvinen H, Klinker E, Mahfouf J-F, Simmons A (2000) The ECMWF operational implementation of four-dimensional variational assimilation. I: Experimental results with simplified physics. Quart J Roy Meteor Soc 126:1143–1170

Rabier F, Klinker E, Courtier P, Hollingsworth A (1996) Sensitivity of forecast errors to initial conditions. Quart J Roy Meteor Soc 122:121–150

Ramanathan V, Crutzen PJ, Kiehl JT, Rosenfeld D (2001) Aerosols, climate, and the hydrological cycle. Science 293:2119–2124

Richardson LF (1922) Weather prediction by numerical process. Cambridge University Press, London, UK, reprinted by Dover, 1965, 236 pp

Rosenfeld D (2000) Suppression of rain and snow by urban and industrial pollution. Science 287:1793–1796

Rosenfeld D, Rudich Y, Lahav R (2001) Desert dust suppressing precipitation: a possible desertification feedback loop. Proc Natl Acad Sci USA 98:5975–5980

Rudich YO, Khersonsky O, Rosenfeld D (2002) Treating clouds with a grain of salt. Geophys Res Lett 29 (22) 2060, doi: 10.1029/2002GL016055

Simpson RH, Malkus JS (1964) Experiments in hurricane modification. Scientific Am 211:27–37

Simpson RH, Simpson J (1966) Why experiment on tropical hurricanes?. Trans New York Acad Sci 28:1045–1062

Tenerelli JE, Chen SS (2001) High-resolution simulations of Hurricane Floyd using MM5 with vortex-following mesh refinement. 14th Conference on Numerical Weather Prediction, American Meteorological Society. Boston, MA, Fort Lauderdale, Florida, J52–J54

Unanwa CO, McDonald JR, Mehta KC, Smith DA (2000) The development of wind damage bands for building. J Wind Eng Ind Aerodyn 84: 119–149

Zou X, Vandenberghe F, Pondeca M, Kuo Y-H (1997) Introduction to adjoint techniques and the MM5 adjoint modeling system. Technical Note 435-STR, NCAR, Boulder, CO

Zou X, Xiao Q, Lipton AE, Modica GD (2001) A numerical study of the effect of GOES sounder cloud-cleared brightness temperatures on the prediction of Hurricane Felix. J Appl Meteor 40:34–55

CHAPTER 8

SPACE TOWERS

ALEXANDER A. BOLONKIN
C&R Co., 1310 Avenue R, #6-F, Brooklyn, NY 11229, USA
http://Bolonkin.narod.ru, E-mail: aBolonkin@gmail.com

Abstract: The author proposes two new revolutionary macro-engineering projects: inflatable pneumatic high altitude towers (height up to 100 km) and kinetic cable space towers (height up 160,000 km). The second method allows building of space elevator without rocket flights to space. Related to the first macro-project, the author provides theory and computations for building inflatable space towers. These macro-projects are not expensive and do not require rockets. They require thin strong films composed of artificial fibers and fabricated by current industry. They can be built using present technology. Towers can be used (for tourism, communication, etc.) during the construction process and provide self-financing for further construction. The tower design does not require outdoor work at high altitudes; all construction can be done at the Earth's surface. The transport system for a tower consists of a small engine (used only for friction compensation) located at the Earth's surface. The tower is separated into sections and has special protection mechanisms in case of damage. Related to the second macro-project, the author discusses a revolutionary new method to access outer space. A cable stands up vertically and pulls up its payload into space with a maximum force determined by its strength. From the ground the cable is allowed to rise up to the required altitude. After this, one can climb to any altitude using this cable or deliver a payload at altitude. The author shows how this is possible without infringing the law of gravity. The Section 2 contains the theory and computations for four macro-projects (towers that are 4, 75, 225 and 160,000 km in height, respectively). The first three macro-projects use the conventional artificial fiber produced by current industry, while the fourth project requires nanotubes currently made in scientific laboratories. The chapter also shows in a fifth macro-project how this idea can be used to launch a load at high altitude

Keywords: space tower, pneumatic tower, cable tower

V. Badescu et al. (eds.), Macro-Engineering: A Challenge for the Future, 121–149.

1. PNEUMATIC SPACE TOWERS

1.1 Introduction

The idea of building a tower high above the Earth into the heavens is very old. The writings of Moses, about 1450 BC, in Genesis, Chapter 11, refer to an early civilization that in about 2100 BC tried to build a tower to heaven out of brick and tar. This construction was called the Tower of Babel, and was reported to be located in Babylon in ancient Mesopotamia. Later in chapter 28, about 1900 BC, Jacob had a dream about a staircase or ladder built to heaven. This construction was called Jacob's Ladder (Smitherman, 2000). More contemporary writings on the subject date back to K.E. Tsiolkovski (see his manuscript "*Speculation about Earth and Sky and on Vesta*," published in 1895; Tsiolkovski, 1959). This idea inspired Sir Arthur Clarke to write his novel, *The Fountains of Paradise* (Clarke, 1978) about a space tower (elevator) located on a fictionalized Sri Lanka, which brought the concept to the attention of the entire world. Landis and Catarelli (1999) re-examined Tsiolkovski tower. The Russian scientist G. Pokrovskii (1964) suggested a rigid conic tower of height 160 km having a base diameter 100 km and filled by hydrogen.

Today, the world's tallest construction is a television transmitting tower near Fargo, North Dakota, USA. It stands 629 m high and was built in 1963 for KTHI-TV. The CNN Tower in Toronto, Ontario, Canada is the world's tallest building. It is 553 m in height, was finished in 1975, and has the world's highest observation deck at 447 m. The tower structure is concrete up to the observation deck level. Above is a steel structure supporting radio, television, and communication antennas. The total weight of the tower is 3,000,000 tons. The Ostankin Tower in Moscow is 540 m in height and has an observation desk at 370 m. The world's tallest office building is the Petronas Towers in Kuala Lumpur, Malasia. The twin towers are 452 m in height.

Current materials make it possible even today to construct towers many kilometers in height. However, conventional towers are very expensive, costing tens of billions of dollars. When considering how high a tower can be built, it is important to remember that it can be built to any height if the base is large enough.

The pneumatic towers proposed in this chapter are cheaper by factors of hundreds. They can be built on the Earth's surface and their height can be increased as necessary. Their base is not large. The main innovations in this macro-project are the application of helium, hydrogen, or warm air for filling inflatable structures at high altitude and the solution of a stability problem for tall (thin) inflatable columns, and utilization of new artificial materials (Bolonkin, 2002d, 2003i).

The inflatable high towers (3–100 km) have numerous applications for government and commercial purposes:

- Entertainment and observation platform.
- Entertainment and observation deck for tourists. Tourists could see over a huge area, including the darkness of space and the curvature of the Earth's horizon.

- Drop tower: tourists could experience several minutes of free-fall time. The drop tower could provide a facility for experiments.
- A permanent observatory on a tall tower would be competitive with airborne and orbital platforms for Earth and space observations.
- Communication boost: A tower tens of kilometers in height near metropolitan areas could provide much higher signal strength than orbital satellites.
- Solar power receivers: Receivers located on tall towers for future space solar power systems would permit use of higher frequency, wireless, power transmission systems (e.g. lasers).
- Low Earth orbit (LEO) communication satellite replacement: Approximately six to ten 100-km-tall towers could provide the coverage of a LEO satellite constellation with higher power, permanence, and easy upgrade capabilities.

Further methods proposed by the author for access to space are given in the references (Bolonkin, 2002–2005).

1.2 Concept Description

1.2.1 Design of pneumatic tower

The simplest tourist tower includes (Fig. 1): inflatable column, top observation deck, elevator, stabilizing cables, and control pressure and stability. The tower is separated into sections by horizontal and vertical partitions (Fig. 2) and contains entry and exit air (gas) hoses and control devices.

1.2.2 Used gas

The compressed air which fills in the pneumatic tower is the heaviest part of the tower. Air density decreases at high altitude and it cannot support a top tower load. The author suggests filling the towers with a light gas, for example, helium, hydrogen, or warm air. Suggested tower design provides filling up by gas in any section in any time (see item 13, 14 in fig. 2).

The computations for changing pressure of air, helium and hydrogen are presented in Fig. 3 of Bolonkin (2003i) (see also Eq. 1 in this chapter). If all the gases have the same pressure (0.11 MPa, 1.1 atm) at Earth's surface, their columns have very different pressures at 100 km altitude.

Air has the pressure 0 atm, hydrogen – 0.4 atm (0.04 MPa), and helium – 0.15 atm (0.015 MPa). A pressure of 0.4 atm means that every square meter of a tower top can support 4 tons of useful loads. Helium can support only 1.5 tons (Bolonkin, 2003i, Fig. 3).

Unfortunately, hydrogen is dangerous as it can burn. The catastrophes involving dirigibles are sufficient illustration of this. Hydrogen can be used only above an altitude of 13–15 km, where the atmospheric pressure decreases by 10 times and the probability of hydrogen burning is small.

The average temperature of the atmosphere in the interval from 0 to 100 km is about 240 °K. If a tower is made of dark color material, the temperature inside the tower will

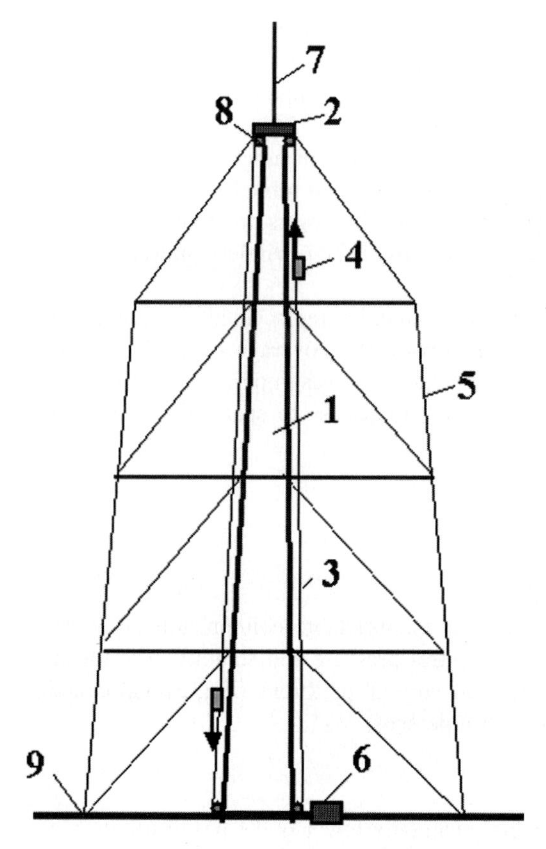

Figure 1. Pneumatic tower of height 3 km. 1 – Inflatable column of radius 5 m; 2 – observation desk; 3 – load cable elevators; 4 – passenger cabins; 5 – stabilizing cables; 6 – engine; 7 – radio and TV antenna; 8 – rollers of cable transport system; 9 – stability control

be higher than the temperature of the atmosphere at a given altitude in day time, so that the tower support capability will be greater (see Eq. 1).

The observation radius versus altitude is presented in (Bolonkin, 2003i, Figs. 4–5), (see also Eq. 23 in this chapter).

1.2.3 Construction material

The author relies only on old information about textile fiber for inflatable structures (Harries, 1973). This refers to DuPont textile Fiber B and Fiber PRD-49 for tire cord. They are six times as strong as steel (maximum tensile stress is 312 kg/mm^2) with a specific density of only 1.5 g/cc, and ultimate elongation is 4% (B) and 1.8% (PRD-49).

The tower parameters vary, depending on the strength of the textile material (film), specifically the ratio of the safe tensile stress σ to specific density γ. Current industry widely produces artificial fibers that have tensile stress

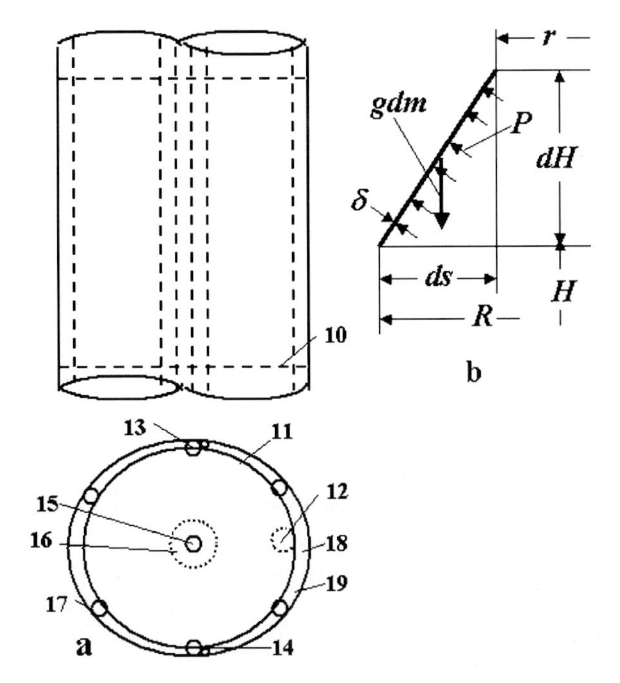

Figure 2. (a) Section of pneumatic tower. Side view is in top of figure, cross section area is in lower part of figure. 10 – horizontal film partitions; 11 – light internal film (internal cover, membrane); 12 – free fly air balls which close a hole when the cover is damaged; 13 – entrance line (hose) of compression air (gas) and pressure control; 14 – exit line (hose) of air and control; 15 – control laser beam; 16 – sensors of laser beam location; 17 – control cables and devices; 18 – section volume, 19 – tower cover (film, casing, membrane). (b) Scheme for computation an optimal tower cone and a tower cover thickness

σ = 500–620 kg/mm^2 and density $\gamma = 1800\,\mathrm{kg/m^3}$. The ratio of these quantities is $k = \sigma/\gamma$ or $K = \sigma/\gamma/10^7 = 0.28-0.34$ $[(\mathrm{m^2/s^2})/10^7]$ (Using the ratio K is more convenient, because it is seven order of magnitudes lower than k). There are whiskers (in industry) and nanotubes (in scientific laboratories) with K = 1–2 (whisker) and $K = 5-11$ (nanotubes). Theory predicts fiber, whisker and nanotubes could have K values ten times greater (Galasso, 1989; Dresselhous, 2000; Carbon Fibers, 1995).

The tower parameters have been computed for K = 0.05 – 0.3 (Bolonkin, 2002, 2003), with a recommended value of $K = 0.1$.

1.2.4 Safety of tower

One may think that inflatable construction is dangerous, on the basis that a small hole (damage) could deflate the tower. However, this assumption is incorrect. The tower may be built with multiple vertical and horizontal sections, double walls (covers, membranes), and special devices (e.g. air balls) which will temporarily seal a hole. If a tower section sustains major damage, the tower height is only decreased by one section. This modularity is similar to combat vehicles – bullets may damage its tires, but the vehicle continues to operate. The special hoses and devices control and support the tower internal pressure. If surface cover is damage, the internal film temporary closes the hole.

1.2.5 Stability of pneumatic tower

Stability is provided by stabilizing cables (tensile elements). The verticality of the tower can be checked by laser beam and sensors monitoring beam location (Fig. 2). If a section deviates from vertical control cables, control devices, and pressure changes restore the tower position. The stabilizing cables also support the tower in a windy weather.

1.2.6 Tower design

The tower building will not have conventional construction problems such as lifting building material to high altitude. All sections are identifiably. New sections are put in at the bottom of the tower, the new section is inflated, and the entire tower is lifted. It is estimated the building may be constructed in 2–5 months. A small tower (up to 3 km) can be safety located in a city. The tower may be illuminated by color lights and it can be the city symbol.

1.2.7 Estimation of tower cost

The pneumatic tower does not require high-cost building materials. The tower will be a hundred times cheaper than conventional solid towers 400–600 m tall.

1.3 Theory of Pneumatic Towers

Equations developed and used by author for estimations and computation are provided below (in the metric system).

1.3.1 The pressure of gas

The given molecular weight, μ, the temperature, T, of an atmospheric gas mixture, the gravitational acceleration, g, and the atmospheric pressure, $P(H)$, versus altitude, H, may be calculated by using:

$$(1) \qquad P = P_0 \exp\left(-\frac{\mu g H}{RT}\right) \qquad \text{or} \qquad P_r = \frac{P}{P_0} = \exp\left(-aH\right),$$

where P_o is the pressure at the planet's surface (for the Earth $P_o \approx 10^5$ N/m^2), $R = 8314$ J/(kmol·K) is ideal gas constant and P_r is a relative (dimensionless) pressure. For air: $\mu = 28.96$ kg/kmol, for hydrogen: $\mu = 2$ kg/kmol, for helium $\mu = 4$ kg/kmol. Also, $g = 9.81$m/s^2 and $a = \mu g/(RT)$.

1.3.2 Optimal cover thickness and tower radius (optimal variable tower cone)

Let us consider a small horizontal cross-section of a tower element (fig. 2b). Using the known formulas for mass and stress, we write

$$Pds = gdm, \tag{2}$$

where

$$\begin{aligned} &dm = 2\pi\ r\gamma\delta\ dH, \qquad ds = \pi(R^2 - r^2), \\ &R = r + dr, \qquad ds \approx 2\pi r\ dr. \end{aligned} \tag{2a}$$

Here, m – the mass of the cover (i.e. the membrane which holds the internal pressure), γ – cover specific mass [kg/m^3], σ – cover tensile stress [N/m^2], s – tower cross-section area which supports a tower cover [m^2], $R = R(H)$ – external radius of tower [m] at height H (r is tower radius at height $H + dH$ – see Eq. 2a), P is the extra internal gas pressure over outside atmosphere pressure, while ds, dm, dH, dr are differentials of s, m, H, R, respectively. The optimal cone shape of the tower is such that the internal pressure supports the tower cover at any altitude H.

Substituting the above formulas in Eq. (1), one gets

$$Pdr = g\gamma\delta\ dH. \tag{3}$$

From equations for stress we find the cover thickness, δ [m],

$$2\pi RPdH = 2\delta\sigma\ dH \quad \text{or} \quad \delta = \frac{\pi RP}{\sigma}. \tag{4}$$

If we substitute Eq. (4) in Eq. (3) and integrate, we find

$$R = R_0 \exp\left(-\frac{\pi g H}{k}\right) \quad \textit{or} \quad R_r = \frac{R}{R_0} = \exp\left(-\frac{\pi g H}{k}\right), \tag{5}$$

where R_r is relative (dimensionless) radius and R_o is base tower radius.

1.3.3 Computation of tower lift force F

The tower lift force can be computed by using the relationships:

$$F = PS, \qquad S = S_r S_0, \tag{6a}$$

where

$$S_r = \frac{\pi\,(R_r R_0)^2}{S_0}, \qquad S = S_0 R_r^2. \tag{6b}$$

Substituting Eq. (6b) in Eq. (6a), we got finally

$$F = PS_0 R_r^2, \tag{7}$$

where $S_o = \pi R_o^2$ is a cross-section tower area at $H = 0$, $S_r = S/S_o$ is the relative (dimensionless) cross-section of the tower area, $S = S(H)$ is a variable cross-section tower area at H. If we substitute Eqs. (1) and (5) in Eq. (7) we find

$$(8) \qquad F = P_0 S_0 \exp\left[-(a+\frac{2\pi g}{k})H\right] \quad \text{or} \quad F_r = \frac{F}{P_0 S_0} = \exp\left[-(a+\frac{2\pi g}{k})H\right],$$

where F_r is the relative (dimensionless) force.

1.3.4 *Tower base area*

Now we intend to estimate the tower base area for a given top load W [kg]. The required base area S_o (and the associated radius R_o) for given top load W may be found from Eq. (8) under the condition $F = gW$:

$$(9) \qquad P_0 S_0 = \frac{gW}{F_r(H_{max})} \quad \text{and} \quad R_0 = \left(\frac{S_0}{\pi}\right)^{1/2}.$$

1.3.5 *Mass of tower cover*

From Eqs. (2) and (2a) we get

$$(10) \qquad dm = 2\pi R \gamma \delta dH.$$

If we substitute Eqs. (1), (4) and (5) in Eq. (10) we find

$$(11) \qquad dm = \left(\frac{2\pi}{k}\right) P_0 S_0 \exp\left[-(a+\frac{2\pi g}{k})H\right] dH.$$

Integrate this relation from H_1 to H_2, we get the mass M [kg].

$$(12) \qquad M = \frac{2\pi P_1 S_1}{k}\left(a+\frac{2\pi g}{k}\right)\left[F_r(H_1) - F_r(H_2)\right],$$

and the relative (dimensionless) mass (for $H = 0$) is

$$(13) \qquad M_r = \frac{M}{P_0 S_0} = \frac{2\pi}{k}\left(a+\frac{2\pi g}{k}\right)(1-F_r).$$

1.3.6 *The thickness of cover*

The cover thickness may be found from Eqs. (4), (5) and (1)

$$(14) \qquad \delta = \frac{\pi}{\gamma k} P_0 R_0 \exp\left[-\left(a+\frac{2\pi g}{k}\right)H\right].$$

The relative (dimensionless) cover thickness is

$$(15) \qquad \delta_r = \frac{\delta}{P_0 R_0} = \frac{\pi}{\gamma k}\exp\left[-\left(a+\frac{2\pi g}{k}\right)H\right].$$

1.3.7 Computation of bending moment

A wind tries to bend the tower. The maximum safety bending moment which the tower can keep is found using the relations [see Eqs. (8) and (5)]

$$M_b = FR = R_0 P_0 S_0 R_r F_r \tag{16}$$

and the relative (dimensionless) bending moment is

$$M_{b,r} = \frac{M_b}{R_0 P_0 S_0} = R_r F_r . \tag{17}$$

1.3.8 Estimation of gas mass M_g into tower

Let us write the gas mass in a small volume and integrate this expression for altitude:

$$dm_g = \rho dV, \quad dV = \pi R^2 dH, \quad \text{or} \quad \rho = \frac{\mu P}{RT} = \rho_0 P_r, \tag{18}$$

where V is gas volume [m^3], ρ is gas density [kg/m^3], ρ_1 is gas density at altitude H_1. If we substitute P_r from Eq. (1), integrate, and substitute F_r from Eq. (8), we have

$$M_g = \frac{\pi \rho_1 R_1^2}{a + \dfrac{2\pi g}{k}} \left[F_r(H_1) - F_r(H_2) \right], \tag{19}$$

where lower index (subscript) "1" means values for lower end of tower and subscript "2" means values for top end of tower. The relative (dimensionless) gas mass is

$$M_{g,r} = \frac{M_g}{\rho_1 R_1^2} = \frac{\pi}{a + \dfrac{2\pi g}{k}} \left[F_r(H_1) - F_r(H_2) \right] \tag{20}$$

1.3.9 Computation of base radius

We get the base tower radius from Eq. (8) under the condition $F = gW$:

$$R_1 = \left(\frac{gW}{\pi P_1 R_r} \right)^{1/2}, \tag{21}$$

where W is the top load.

1.3.10 Computation of tower mass M

The mass of the tower, including the inside gas, is given by

$$M = \pi R_1^2 P_1 . \tag{22}$$

1.3.11 Distance L of Earth view from the tower

The Earth view distance is important for communication, cell-telephones, radiolocation and tourists. This distance may be computed for a high tower by using the relationship:

$$(23) \qquad L \approx \left(2R_e H + H^2\right)^{1/2},$$

where $R_e = 6,378$ km is the Earth's radius. Results of computations are presented in (Bolonkin, 2003i, Figs. 4 and 5).

1.4 Macro Project 1. An Air Tower of 3 km Height

Design values: Base radius = 5 m, $K = 0.1$.

This inexpensive project provides experience in design and construction of a tall pneumatic tower, and of its stability. The project also provides funds from tourism, radio and television. The pneumatic tower has a height of 3 km. Tourists will not need a special suit or breathing device at this altitude. They can enjoy an Earth panorama of a radius of up 200 km. The bravest of them could experience 20 seconds of free-fall time followed by 2g overload.

1.4.1 Results of computations

Assume the additional air pressure is 0.2 atm ($1\text{atm} = 1.013 \cdot 10^5\ \text{N/m}^2$), air temperature is 288 K (15°C, 60°F), base radius of tower is 5 m, $K = 0.1$. If the tower cone is optimal, the tower top radius must be 4.55 m. The maximum useful tower top lift is 92 tons. The cover thickness is 0.174 mm at the base and 0.114 mm at the top. The outer cover mass is only 23 tons (Bolonkin, 2003i, Fig. 9). If we add light internal partitions, the total cover weight will be about 32–36 tons (compared to 3 million tons for the 553 m CNN tower in Toronto). Maximum safely bending moment ranges from 780 ton· meter (at the base) to 420 ton·meter at the tower top. Other variants and more detail computations may be found in Bolonkin, (2003i, Figs. 6–10).

The expansion load is approximately 15 tons for a wind speed 30 m/s. It decreases the maximum useful top load to $77 = 92\text{-}15$ tons.

1.4.2 Economic efficiency

Assume the cost of the tower is \$5 million, its life time is only 10 years, annual maintenance \$1 million, the number of tourists at the tower top is 200 (15 tons), time/tourist at the tower top is 0.5 hour, and the tower is open 12 hours per day. Then 4800 tourists will visit the tower per day, or 1.7 million per year. The unit cost of one tourist is $(0.5 + 1)/1.7 = 1$ \$/person. If a ticket costs \$9, the profit is $1.7 \times 8 = \$13.6$ million per year. If for a drop from the tower (in a special cabin, for a free-fall (weightlessness) time of 20 seconds, followed by a overload of 2g) costs each individual \$5 and 20% of tourists take it, the additional profit will be \$1.7 million.

1.5 Macro Project 2. Helium Tower of Height 30 km

Design values: Base radius = 5 m, $K = 0.1$.

Let us take the additional pressure over atmospheric pressure as 0.1 atm ($1 \cdot 10^4$ Pa). For $K = 0.1$, the radius is 2 m at an altitude of 30 km. For $K = 0.1$ the useful lift force it is about 75 ton-force at an altitude of 30 km, thus it is a factor of two times greater than the 3 km air tower. It is not surprising, because the helium is lighter than the air and it provides a lift force. The cover thickness changes from 0.08 mm (at the base) to 0.42 mm at an altitude of 9 km and decreases to 0.2 mm at 30 km. The outer cover mass is about 370 tons. Required helium mass is 190 tons. The tourist capability of this tower is twice than that of the 3 km tower, but all tourists must stay in an-tight cabins. The reader will find other variants and more detail computations in Bolonkin, (2003i, Figs. 11–16).

1.6 Macro Project 3. Air-hydrogen Tower of Height 100 km

Design values: Base radius of air part = 25 m, the hydrogen part has base radius = 5 m.

This tower consists of two parts. The lower part (0–15 km) is filled with air. The top part (15–100 km) is filled with hydrogen. It makes this tower safer, because the low atmospheric pressure at high altitude decreases the probability of fire. Both parts may be used for tourists.

1.6.1 Air part, 0–15 km

The base radius is 25 m, the additional pressure is 0.1 atm, average temperature is 240 °K, and the stress coefficient $K = 0.1$. This tower can be used for tourism and as an astronomy observatory. For $K = 0.1$, the lower (0–15 km) part of the project requires 570 tons of outer cover and provides 90 tons-force of useful top lift force. The reader may find other variants and more detail computations in Bolonkin (2003i, Figs. 17–21).

1.6.2 Hydrogen part, 15–100 km

This part has the base radius of 5 m, the additional gas pressure is 0.1 atm. A stronger cover is needed, with $K = 0.2$. The useful top tower load can be about 5 tons, maximum, for $K = 0.2$. The cover mass is 112 tons, the hydrogen lift force is 37 tons-force. The top tower will press on the lower part with a force of only $112 - 37 + 5 = 80$ tons-force. The lower part can support 90 tons. Other variants may be found in Bolonkin (2003i, Figs. 22–27). The proposed projects use an optimal change of the radius, but designers must find the optimal combination of the air and gas parts.

2. CABLE SPACE TOWERS

2.1 Introduction

This section suggests a very simple and inexpensive method and installation for lifting and launching payload into space. This method is different from the centrifugal method (2002, 2002a, 2002b, 2002c; 2003b, 2003d, 2003e; 2005d) in which a cable, circular or semi-circular in shape, and a centrifugal force, are used to keep the space station at high altitude. In the present method there is a straight line vertical cable connecting the space station to the Earth's surface. The space station is held in place by reflected cable and cable kinetic (shot) energy. The present method spends less energy in air drag.

This is a new method and transport system for delivering payloads and people into space. This method uses a cable and any conventional engine (mechanical, electrical, gas turbines) located on the ground. After completing an exhaustive literature and patent search, we decided that no similar method was proposed previously. However, earlier proposals to use kinetic energy to support a tower have been made by Yunitskii (1982), Lofstrom (1983, 1985), Hyde (1988), and Forward (1995). Differences between these earlier concepts and the designs analyzed here are discussed in section 2.5. Also, note that current access to outer space is described in Bolonkin (2002a–2005a), Smitherman (2000) and Space Technology (1996–1997).

2.2 Technical Solution

2.2.1 *Brief description*

The installation includes (see notations in Figs. 3 to 6): a strong closed-loop cable, rollers, a conventional engine, a space station (top platform), a load elevator and support stabilizing cables (expansions).

The installation works in the following way. The engine rotates the bottom roller and permanently moves the closed-loop cable at high speed. The cable reaches a top roller at high altitude, turns back and moves to the bottom roller. When the cable turns back it creates a reflected (centrifugal) force. This force can easily be calculated by using a centrifugal theory or a reflection (momentum) theory. The force keeps the space station suspended at the top roller; and the cable (or special cabin) allows the delivery of a load to the space station. The station has a parachute that saves people if the cable or engine fails.

The theory shows that currently produced artificial fibers allow the cable to reach altitudes up to 100 km (see Projects 1 and 2 below). If higher altitude is required a multi-stage tower must be used (Fig. 4, see also Project 3). If a very high altitude is needed (geosynchronous orbit or more), a very strong cable made from nanotubes must be used (see Project 4).

The tower may be used for a horizon launch of the space apparatus (Fig. 5). The vertical cable towers support horizontal closed-loop cables rotated by the vertical cables. The space apparatus is lifted by the vertical cable, connected to horizontal cable and accelerated to the required velocity.

The closed-loop cable may have variable length. This allows the system to start from zero altitude, and gives its workers/users the ability to increase the station altitude to a required value, and to spool the cable for repair. The innovation device for this action is shown in Fig. 6. The spool can reel up and unreel in the left and right branches of the cable at different speeds and can alter the length of the cable.

The safety speed of the cable spool is the same with the safety speed of the cable because the spool operates as a free roller. The conventional rollers made of the composite cable material have the same safety speed with the cable. The proposed spool is an innovation, because it consists only of cable (no core) and it allows reeling up and unreeling simultaneously with different speed. This is necessary to change the tower altitude.

The lower (ground) conventional drive roller is shown in Fig. 3 (see Chapter 10). The small drive rollers 25 Fig. 3 (Chapter 10) press the cable to main (large) drive roller, provide a high friction force between the cable and the drive rollers and pull (rotate) the cable loop.

The drive roller 3 in Fig. 6 (variable cable spool) is a row of double drive rollers which press the cable between them, create the high friction force and rotate the cable loop.

2.2.2 *Advantages*

The proposed towers and launch system are more advantageous as compared to the currently proposed towers and rocket systems:

1. They allow a very high altitude (up to geosynchronous orbit and more) to be reached, which is structurally impossible for solid material towers.
2. They are cheaper by some thousands of times than the current low high towers. No expensive rockets are required.
3. The cable towers may be used for tourism, power, TV and radio signal relay over a very wide area, as a radio locator, as well as space launchers.
4. The proposed towers and space launchers decrease the delivery cost at high altitude up to \$2–\$10 per kg.
5. The proposed space tower launcher may be built in a few months, whereas the modern rocket launch system requires years for development, design, and building.
6. The proposed cable towers and space launcher do not require high technology and may be produced by any non-industrial country from currently available artificial fibers.
7. Rocket fuel is increasingly expensive. The proposed cable towers and space launcher can use the cheapest sources of energy such as wind, water or nuclear power, or the cheapest fuels such as gas, coal, peat, etc., because the engine is located on the Earth's surface. The flywheels may be used as an accumulator of energy.
8. There is no necessity to have highly qualified personnel such as rocket specialists with high salaries.
9. We can launch thousands of tons of useful payloads annually.

The advantages of the method are the same as for the centrifugal and cable launcher (Bolonkin, 1965–Bolonkin, 2005). The suggested method is approximately half the cost of the semi-circle launcher because it uses only one double vertical cable. It also has approximately half the delivery cost (up to \$2–10 per kg), because it has half the air drag and fuel consumption.

2.2.3 *Cable properties*

The reader may find details about the cable in Chapter 10 and about the cable characteristics in Galasso (1989), Dresselhous (2000), Carbon Fibers (1995). In the projects 1–3 below we are using only cheap artificial fibers widely produced by current industry.

2.3 Theory of the Cable Tower and Launcher

2.3.1 *Lift force of the cable tower*

Take a small part of the rotational circle and write the mechanical equilibrium relation (see Fig. 7 in Chapter 10):

$$\frac{2SR\alpha\gamma V^2}{R} = 2S\sigma \sin\alpha\,, \tag{24}$$

where V is rotational cable speed, R is circle radius, α is the angle of the circular part, S is cross-section of cable areas, σ is cable stress, γ is cable density. When $\alpha \to 0$ the relationship between the maximum rotational speed V and the tensile stress of the closed loop (curve) cable is

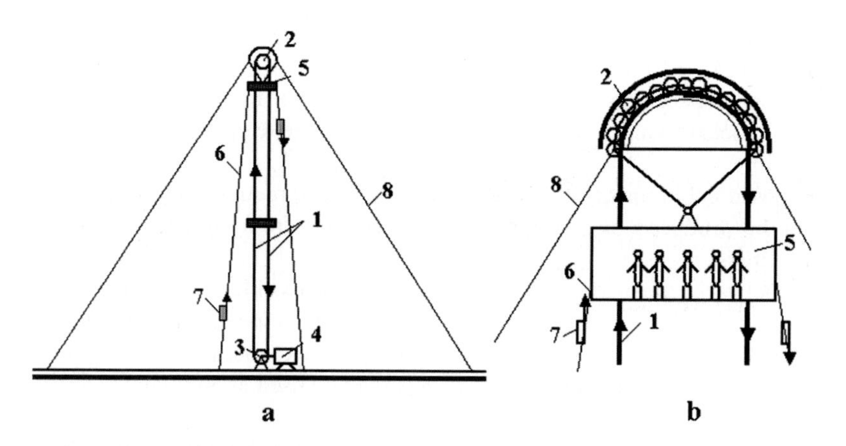

Figure 3. (a). Offered cable tower: 1 – mobile closed loop cable; 2 – top roller of the tower; 3 – bottom roller of the tower; 4 – engine; 5 – space station; 6 – elevator; 7 – load cabin; 8 – tensile element (stabilizing rope). (b). Design of top roller

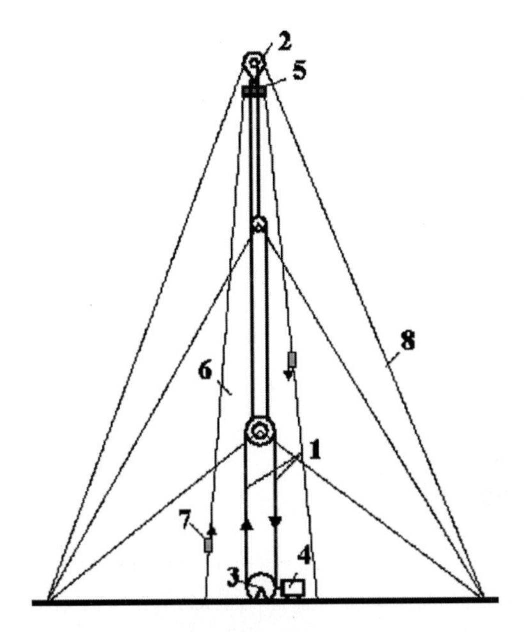

Figure 4. Multi-stage cable tower. Notations are same as Fig. 3

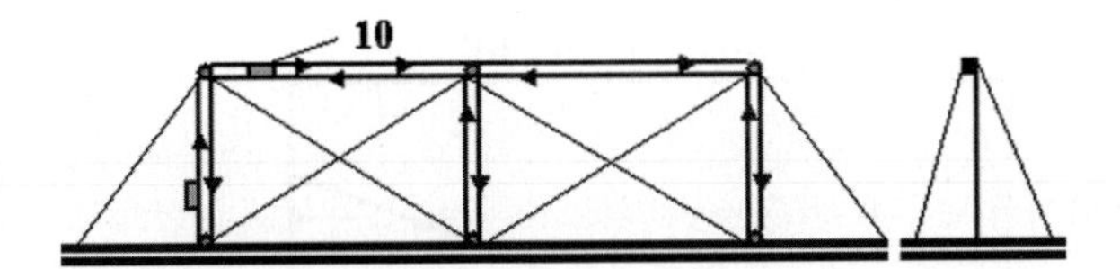

Figure 5. a. Kinetic space installation with horizontally accelerated parts. b. 10 – missile

$$V = \sqrt{\frac{\sigma}{\gamma}} = \sqrt{k}\,, \quad \text{the lift force is} \quad F = 2\sigma S, \tag{25}$$

where F is the lift force, $k = \sigma/\gamma$ is the relative cable stress. The computations of the speed for intervals K = 0 to 1, and 1 to 10, respectively, ($K = k/10^7$) are presented in Figs. 7 and 8. We shall find the lift force of the proposed installation from ordinary mechanics. Writing the momentum of the reflected mass per unit time (one second) we find:

$$F = mV - (-mV) = 2mV, \quad m = \gamma SV, \quad \text{and finally} \quad F = 2\gamma SV^2 \tag{26}$$

where m is the cable reflected mass per unit time (one second). If we substitute Eq. (25) in Eq. (26), the same expression for the lift force $F = 2\sigma S$ will be found.

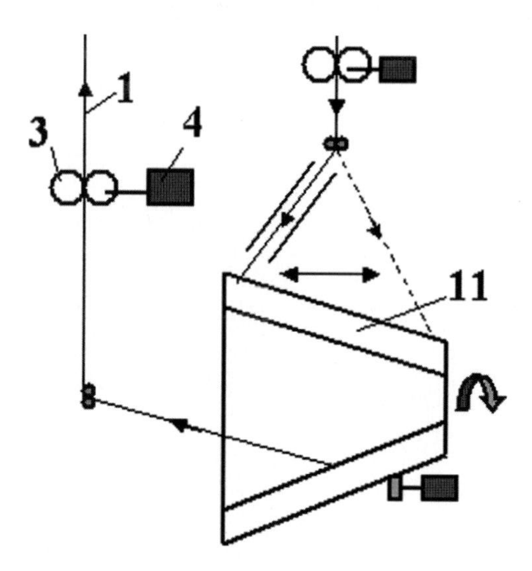

Figure 6. Variable cable spool at ground. 1–cable, 3–rollers, 4–engines, 11–cable spool

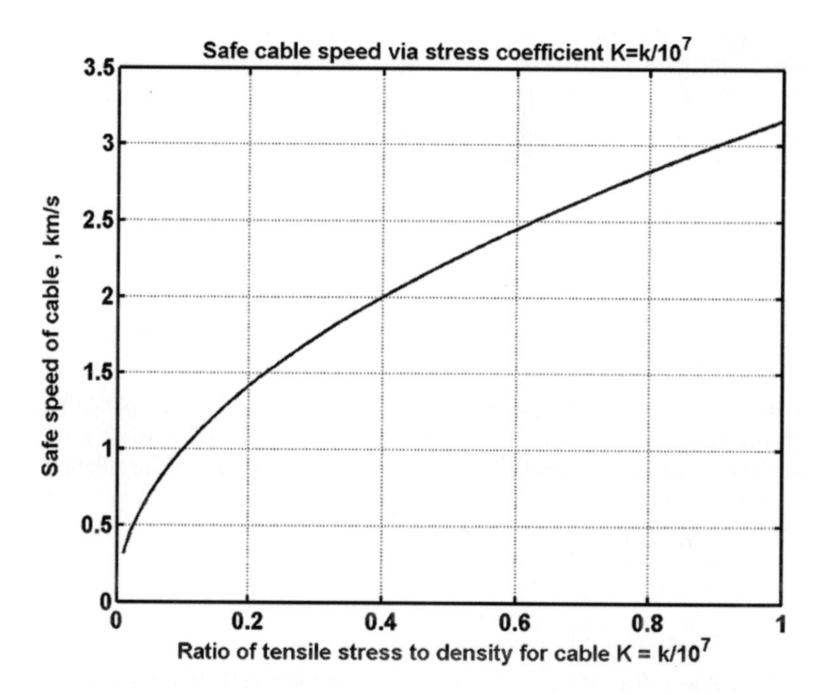

Figure 7. Safe cable speed versus safe stress coefficient $K = 0–1$. K is in $(\mathrm{m}^2/\mathrm{s}^2)/10^7$

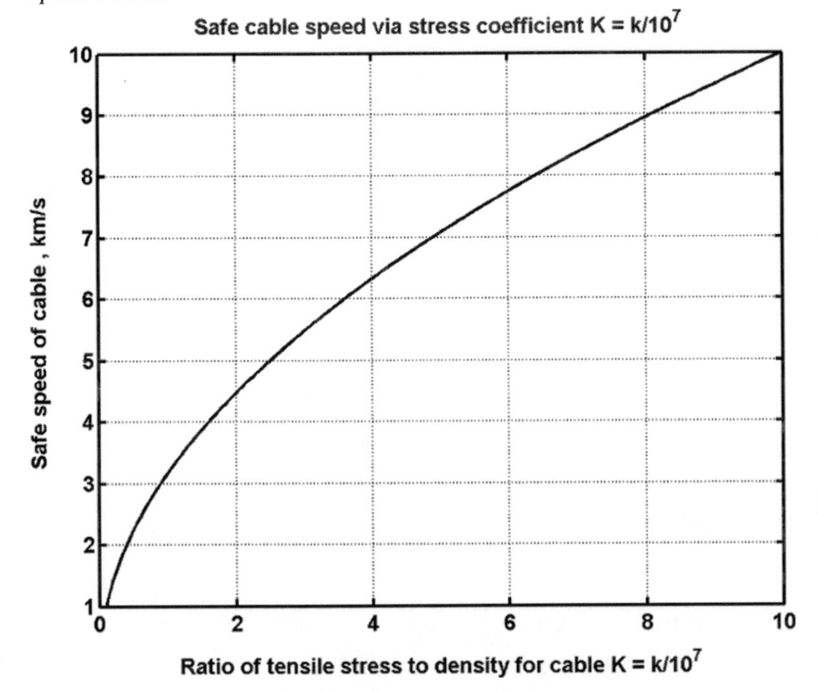

Figure 8. Safe cable speed versus relative stress coefficient $K = 1$–10, $(\mathrm{m}^2/\mathrm{s}^2)/10^7$

2.3.2 *Lift force in constant or variable gravity fields*

In a constant gravity field without air drag, the lift force of the proposed device equals the centrifugal force F minus the cable weight W:

$$F_g = F - W = F - 2\gamma gSH = 2\gamma S(V^2 - gH) =$$
$$2S(\sigma - \gamma gH) = 2S\gamma(k - gH), \tag{27}$$

where H is the altitude of the cable tower.

The maximum tower height and the minimum cable speed in a constant gravity field are given by, respectively, (see Eq. 27):

$$H_{\max} = \frac{\sigma}{g\gamma} = \frac{k}{g}, \quad V_{\min} = \sqrt{gH}. \tag{28}$$

Results of computations are presented in Figs. 9 and 10. In this case the installation does not produce a useful lift force and will support only itself.

Now we find a kinetic lift force, F_g, in a variable gravity field and for the rotating Earth. The weight of cable in differential form is

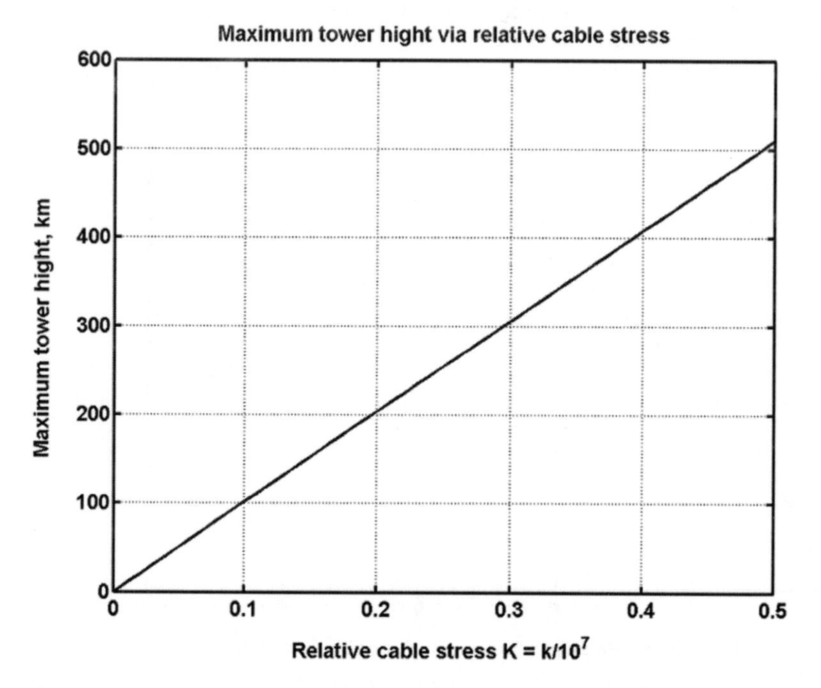

Figure 9. Maximum tower height versus relative cable stress. K is in $(\mathrm{m}^2/\mathrm{s}^2)/10^7$

$$\text{(29a,b)}\quad dP = \left(g - \frac{V^2}{R}\right) dm, \quad g = g_0\left(\frac{R_0}{R}\right)^2,$$

where

$$\text{(30)}\quad \frac{V^2}{R} = \omega^2 R\,, \quad dm = \gamma S dR\,, \quad R = R_0 + H$$

The following integral is computed:

$$\text{(31)}\quad P = \int_{R_0}^{R}\left[g_0\left(\frac{R_0}{R}\right)^2 - \omega^2 R\right]\gamma S dR = g_0\left(R_0 - \frac{R_0^2}{R}\right) - \frac{\omega^2}{2}\left(R^2 - R_0^2\right),$$

and the final equation for the kinetic lift force is

$$\text{(32)}\quad F_g = 2\sigma S - 2P = 2\gamma S k - 2P = 2\gamma S\left[k - g_0\left(R_0 - \frac{R_0^2}{R}\right) + \frac{\omega^2}{2}\left(R^2 - R_0^2\right)\right],$$

where

$$\text{(33)}\quad k = \frac{\sigma}{\gamma} = V^2,\ R = R_0 + H.$$

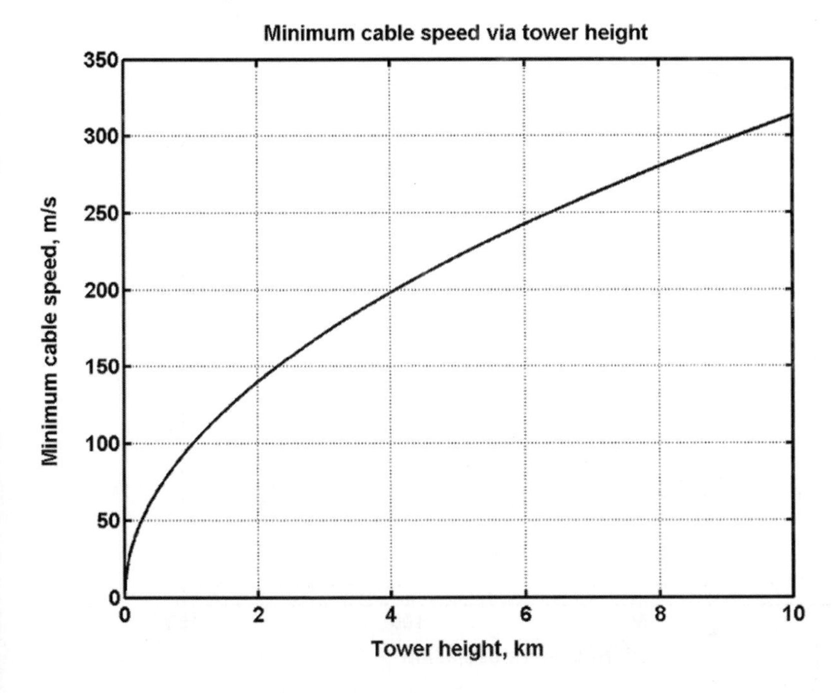

Figure 10. Minimum caple speed versus tower height

Here P is an auxiliary force resulting from the cable weight for a variable gravity field and the rotational Earth while ω is the Earth angular speed. Minimum cable stress and minimum cable speed of a variable rotating planet equals

$$k_{\min} = g_0\left(R_0 - \frac{R_0^2}{R}\right) - \frac{\omega^2}{2}\left(R^2 - R_0^2\right), \tag{34}$$

$$V_{\min}^2 = g_0\left(R_0 - \frac{R_0^2}{R}\right) - \frac{\omega^2}{2}\left(R^2 - R_0^2\right). \tag{35}$$

Results obtained by using these equations for Earth are presented in Figs. 11 and 12. If $K > 5$ the height of the cable tower may be beyond the Earth's geosynchronous orbit. For Mars, this condition is fulfilled for $K > 1$, while in case of the Moon the condition is $K > 0.3$. From Fig. 11 one can see that the proposed tower, of height 145,000 km, can be maintained without a cable rotation, and if the tower height is more 145,000 km, the tower has a useful lift force that allows a payload to be lifted using an immobile cable.

2.3.3 *Estimation of cable friction in the air*

Estimation of cable friction in air is difficult, mainly because there are no experimental data for air friction for an infinitely thin cable (especially at hypersonic

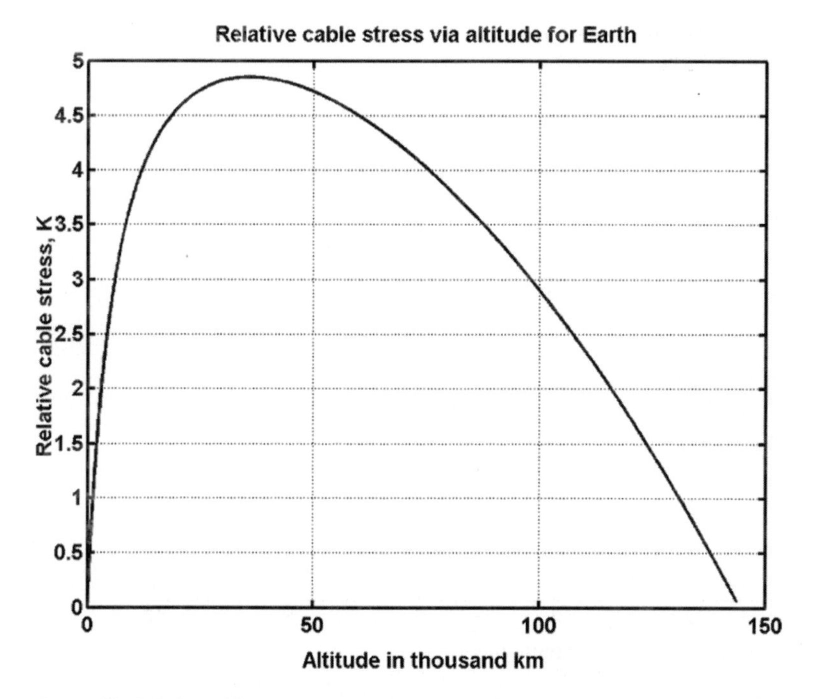

Figure 11. Relative cable stress versus altitude for rotational Earth with variable gravity

speeds). A computational method for plates at hypersonic speed described in the book by Anderson (1989, p. 287), was used. The computation is made for two cases: laminar and turbulent boundary layers.

The results of this comparison are very different. Turbulent friction is greater than laminar friction by hundreds of times. About 80% of the friction drag occurs in the troposphere (height from 0 to 12 km). If the cable end is located on a mountain at 4 km altitude, the maximum air friction will be decreased by 30%.

It is postulated that half of the cable surface will have the laminar boundary layer because a small wind or trajectory angle will blow away the turbulent layer and restore the laminar flow. The blowing away of the turbulent boundary layer is usual in aviation and it is used to restore laminar flow and decrease air friction. The laminar flow decreases the friction in hypersonic flow by about 280 times. If half of the cable surface has a laminar layer, it means that we must decrease the air drag calculated for full turbulent layer by minimum two times.

Below, the relationships proposed in Anderson (1989) to compute the local air friction for a two-sided plate are summarized:

$$\text{(36)} \qquad \frac{T^*}{T} = 1 + 0.032M^2 + 0.58\left(\frac{T_w}{T} - 1\right),$$

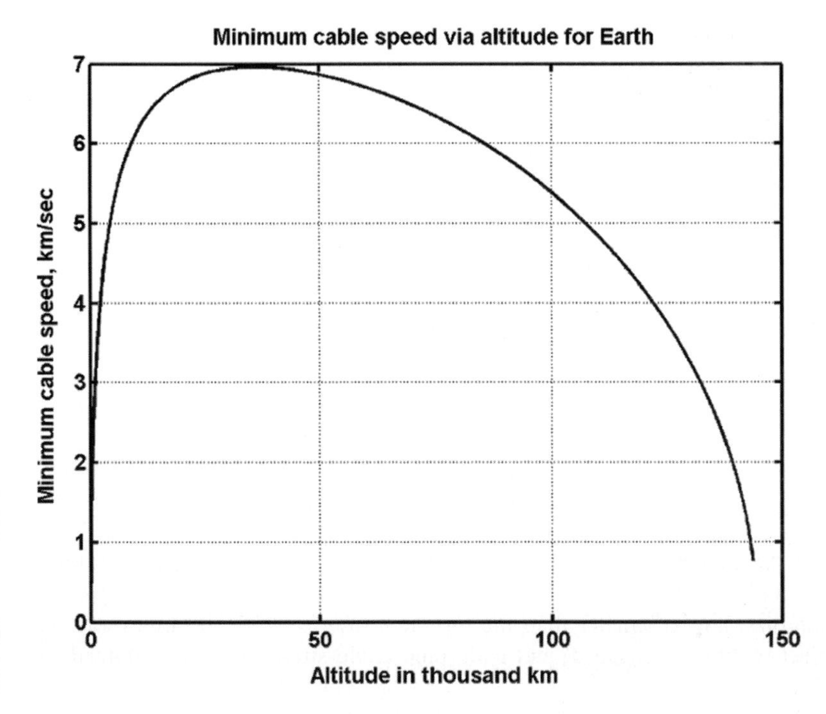

Figure 12. Minimum cable speed versus altitude for rotational Earth with variable gravity

(37) $$M = \frac{V}{a}, \quad \mu^* = 1.458 \times 10^6 \frac{T^{*1.5}}{T^* + 110.4},$$

(38) $$\rho^* = \frac{\rho T}{T^*}, \quad R_e^* = \frac{\rho^* V x}{T^*},$$

(39) $$C_{f,L} = \frac{0.664}{(R_e^*)^{0.5}}, \quad C_{f,T} = \frac{0.0592}{(R_e^*)^{0.2}},$$

(40) $$D_L = 0.5 C_{f.l} \rho^* V^2 S, \quad D_T = 0.5 C_{f.t} \rho^* V^2 S.$$

Where T^*, Re^*, ρ^*, μ^* are the reference (evaluated) temperature, Reynolds number, air density, and air viscosity, respectively. M is the Mach number, a is the speed of sound, V is speed, x is the length of the plate (distance from the beginning of the cable), T is flow temperature, T_w is body temperature, $C_{f.l}$ is a local skin friction coefficient for laminar flow, $C_{f,t}$ is a local skin friction coefficient for turbulent flow. Also, S is the area of skin of both plate sides, so this means for the cable we must take $0.5S$.

It can be shown that the general air drag D for the cable is obtained from linear superposition between the turbulent drag D_T and the laminar drag D_L (i.e. $D = 0.5D_T + 0.5D_L$).

From Eq. (40) we can derive the following relations for turbulent and laminar drags of the *vertical cable*:

$$D_T = \frac{0.0592\pi d}{4}\rho_0^{0.8}\left(\frac{T}{T*}\right)^{0.8}\mu^{0.2}V^{1.8}\int_{H_0}^{H} h^{-0.2}e^{0.8bh}dh =$$

$$0.0547d\left(\frac{T}{T*}\right)^{0.8}\mu^{0.2}V^{1.8}\int_{H_0}^{H} h^{-0.2}e^{0.8bh}dh, \quad (41)$$

$$D_L = \frac{0.664\pi d}{4}\rho_0^{0.5}\left(\frac{T}{T*}\right)^{0.5}\mu^{0.5}V^{1.5}\int_{H_0}^{H} h^{-0.5}e^{0.5bh}dh =$$

$$0.5766d\left(\frac{T}{T*}\right)^{0.5}\mu^{0.5}V^{1.5}\int_{H_0}^{H} h^{-0.5}e^{0.5bh}dh, \quad (42)$$

where d is the diameter of the cable and $\rho_o = 1.225\ \mathrm{kg/m^3}$ is air density at altitude $H = 0$.

The laminar drag is smaller than the turbulent drag by 200–300 times and we can neglect it. Engine power P and additional cable stress may be computed by conventional relationships:

$$P = 2DV, \quad \sigma = \pm\frac{D}{S} = \pm\frac{4D}{\pi d}. \quad (43)$$

The factor 2 entering Eq. (43) is a result of the two branches of the cable: one moves up and the other moves down. The drag does not decrease the lift force because in different branches the drag is in opposite directions. Computations were performed and some results are presented in Figs. 13 and 14.

2.3.4 *Security of visitors*

If the cable is damaged, people can be rescued using a parachute with variable area. The reader will find below relations describing the possibility of saving persons on the tower.

The parachute area is changed so that bodily overload does not go beyond a given value ($N < 5$g).

The differential equations describing the process of moving for passenger cabin are:

$$\frac{dH}{dV} = -V, \quad \frac{dV}{dt} = g - \frac{D}{m}, \quad (44)$$

$$\text{where } \frac{D}{m} = C_D\frac{\rho a V}{2p}, \quad p = \frac{0.5C_D\rho a V}{gN}, \quad (45)$$

$$p \geq 0, \quad \frac{D}{mg} \leq N, \quad (46)$$

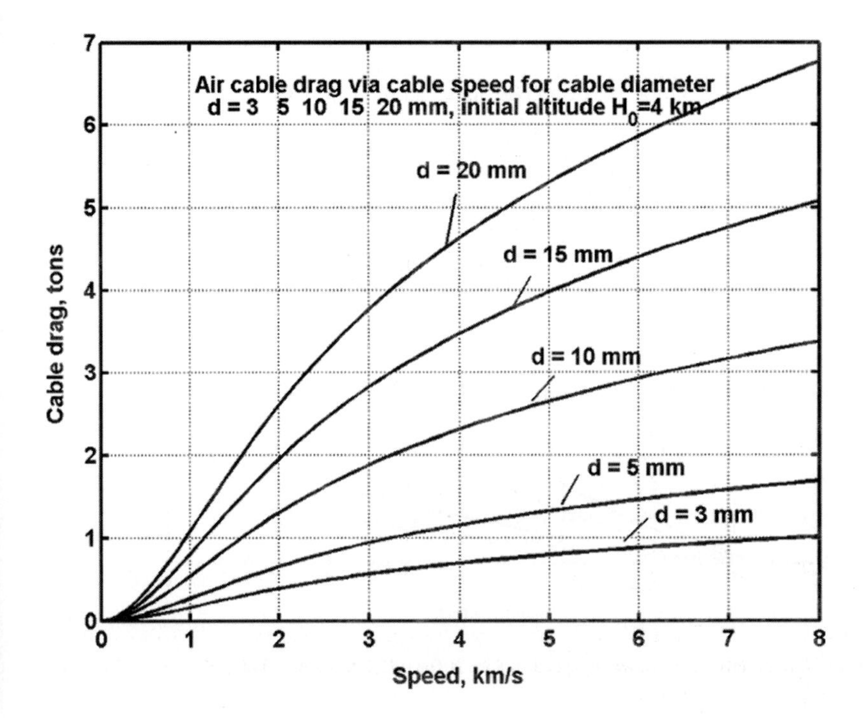

Figure 13. Air cable drag versus cable speed 2–8 km/s for different cable diameter

Here t is time, m is mass, N is the overload and p is the parachutes specific load. The equations for air density are (with H entering in meter):

(47) for $H = 0 - 10$ km, $\rho = 1.225e^{-H/9218}$,

(48) for $H > 10$ km, $\rho = 0.414e^{-(H-10000)/6719}$.

Tourists can be rescued from altitudes up to 250–300 km. The professional astronauts can endure an overload up 8g and may be rescued from greater altitudes.

2.4 Macro-projects

2.4.1 Project 1. Cable tower of height 4 km

To build a 4 km height tower we may use a conventional artificial fiber widely produced by industry with the following cable properties: admissible (safe) stress is $\sigma = 180$ kg/mm^2 (maximum $\sigma = 600$ kg/mm^2, safety coefficient $n = 600/180 = 3.33$), density is $\gamma = 1800$ kg/m^3, cable diameter $d = 10$ mm.

The special stress is $k = \sigma/\gamma = 10^6$ m^2/s^2 ($K = k/10^7 = 0.1$), safe cable speed is $V = k^{0.5} = 1000$ m/s, the cable cross-section area is $S = \pi d^2/4 = 78.5$ mm^2, useful

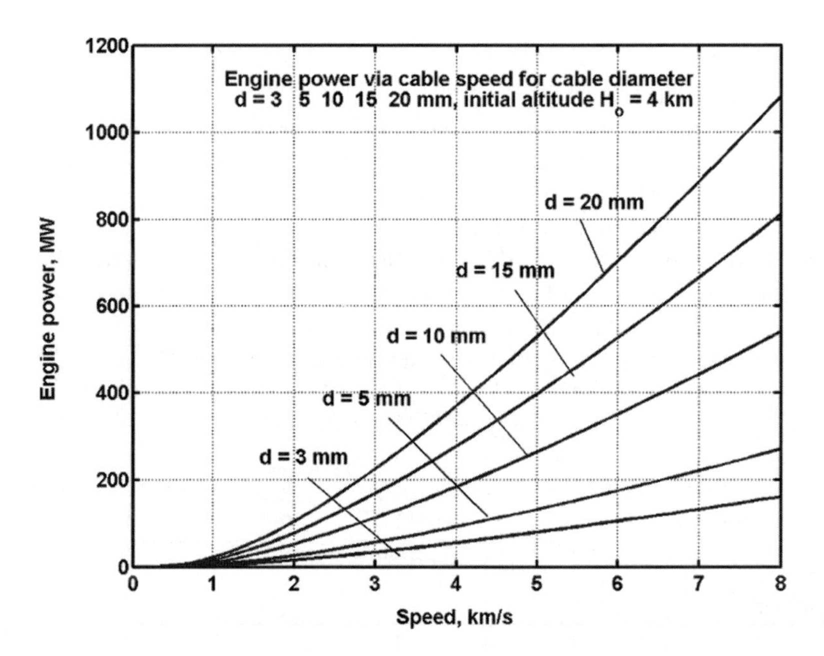

Figure 14. Engine power versus cable speed 2–8 km/s for different cable diameter

lift force is $F = 2S\gamma(k-go) = 27.13$tons-force. Requested engine power is $P =$ 16MW (Eq. 10), cable mass is $M = 2S\gamma H = 2\cdot 78.5\cdot 10^{-6\cdot}1800\cdot 4000 = 1130$ kg.

Assume that the tower is used for tourism with a payload of 20 tons. This means $20000/75 = 267$ tourists may visit the station simultaneously. We take 200 tourists every 30 minutes, i.e. $200\cdot 48 = 9600$ people/day. Let's say 9000 tourists/day which corresponds to $90000\cdot 350 = 3.15$ million/year.

Assume the cost of installation is \$15 million, the life time is 10 years, and the maintenance cost is \$1 million per year. The cost to ensure services for a single tourist is $2.5/3.15 = \$0.8$ per person.

The required fuel $G = Pt/\varepsilon\eta = 16\cdot 106\cdot 350\cdot 24\cdot 60\cdot 60/(42\cdot 106\cdot 0.3) = 38.4\cdot 10^6$ kg. If the fuel cost is \$0.25 per kg, the annual fuel cost is \$9.6 millions, or 9.6/3.15 = \$3.05 per person. Here t is annual time [s], ε is fuel heat capability [J/kg], and η is the engine efficiency.

The total production cost is $0.8+3.05 = \$3.85$ per tourist. If a trip costs \$9, the annual profit is $(9-3.85)\cdot 3.15 = 16.22$ millions of US dollars.

2.4.2 *Macro-project 2. Cable tower of height 75 km*

To build a 75 km height tower one takes the admissible cable stress $K = 0.1$, the cross-section area $S = 90$ mm^2 ($d = 10.7$ mm), the cable density $\gamma = 1800$ kg/m^3. Then, the lift force is $F = 2S\gamma(k-gH) = 7$tons-force. The required engine power is $P = 11$ MW (Eq. 43, Fig. 14), cable mass is $M = 2S\gamma H = 2\cdot 90\cdot 10^{-6}.\cdot 75000 = 24.3$ tons and the cable speed is 1000 m/s.

2.4.3 Macro-project 3. Multi-Stages Cable tower of height 225 km

Current industry produces only a cheap artificial fiber with maximum stress $\sigma = 500$– 620kg/mm^2 and density $\gamma = 1800$ kg/m^3. We take an admissible stress $\sigma = 180$ kg/mm^2 (safety coefficient is $n = 600/180 = 3.33$), $\gamma = 1800$ kg/m^3. Then $k = \sigma/\gamma = 1000000$ N/m^2 and $K = k/10^7 = 0.1$. Using this cable one may design a one-stage cable tower with a maximum height 100 km (payload = 0). Assume we design a tower of height 225 km by using the present-day material. We may choose a 3- stage tower with each stage of height $H = 75$km and useful load capability $M_{3,p} = 3$tons at the tower top.

In this case the 3rd (top) stage (150–225 km) must have a cross-section area $S_3 = M_{3,p}/[2\gamma k - gH] = 33.3$ mm^2 ($d = 6.5$ mm) and the cable mass of the 3rd stage is $M_{3,c} = 2S_3\gamma H = 9$tons. Total mass of third stage is $M_3 = 9 + 3 = 12$ tons.

The 2nd stage (75–150 km) must have a cross-section area $S_2 = M_3/[2\gamma(k - gH)] =$ 133 mm^2 ($d = 13$ mm) and the cable mass of 2nd stage is $M_{2,c} = 2S_2\gamma H = 36$tons. Total mass of third + second stages is $M_2 = 12 + 36 = 48$ tons.

The 1st stage (0–75 km) must have cross-section area $S_1 = M_2/[2\gamma(k–gH)] =$ 533 mm^2 ($d = 26$ mm) and the cable mass of the 1st stage is $M_{1,c} = 2S_2\gamma H = 144$tons. Total mass of third + second+ first stages is $M_0 = 48 + 144 = 192$ tons.

2.4.4 Macro-project 4. Cable tower with height 160,000 km

Assume that a nanotube cable is used, with $K = 6$ (for this height K must be more than 5, see Fig. 11). This means the admissible stress is $\sigma = 6,000$ kg/mm^2 and the cable density is $\gamma = 1000$ kg/m^3. Presently, the scientific laboratories produce nanotubes with $\sigma = 20,000$ kg/mm^2 and density $\gamma = 0.8$–1.8 kg/m^3. Theory predicts $\sigma = 100,000$ kg/mm^2. Industrial nanotubes production will become fully operable during the next 5–10 years.

Take a cross-section cable area of mm^2. The required speed is $V = (k)^{0.5} =$ $(6 \cdot 10^7)^{0.5} = 7.75$ km/s, the mass of cable is $M = S\gamma H = 320$ tons. When full altitude is reached the engine can be turned off and the centrifugal force of the Earth's rotational motion will support the cable. Moreover, the installation has a lift force of about 1000 kg-force, so a useful load can be connected to the cable, the engine can be turned on or slow speed and the load can be delivered into space.

2.4.5 Macro-project 5. Cable tower as Space Launcher

The installation of Fig. 5 may be used as a space launcher. The space apparatus is lifted to high altitude by the left cable tower, connected to the horizon line and accelerated. The required acceleration distance depends on the admissible acceleration. For a projectile it may be 10–50 km (overload is $N = 64$–320g), for astronauts it may be 400 km ($N = 8$g), for tourists it may be 1100 km ($N = 3$g).

2.5 Discussion and Historical Priorities

The project propose here offers a new, simpler, cheaper method for space launches. The project does not need expensive rockets as current methods do, or rockets to launch a counterbalance into space and thousands of tons of nanotubes cable as the NASA space elevator does. It only needs conventional fiber cable and a conventional engine located on a planetary surface.

Pneumatic (inflatable) space towers were proposed and theorized for the first time by Bolonkin (2002d, 2003i). Cable towers were proposed and theorized for the first time by 2002 (2002a, 2002b, 2002c, 2002e, 2003c, 2003d, 2003e, 2003f, 2003j, 2004a, 2005a, 2005b, 2005d). Other applications of these ideas may be found in papers published by the author during 1965–2005. Some of these works are accessible in the Internet (http://Bolonkin.narod.ru/p65.htm).

Other scientists contributed to this field and their findings are briefly presented in the following.

R. L. Forward (1995) offered the idea of a space installation. The project is called "the Space Fountain", for it holds objects up in space in the same way that a water fountain supports a ball bobbing at the top of its vertical jet. This would be done with a stream of pellets that would be shot from a space platform hovering motionless up at 2000 kilometers altitude to another platform partway around the Earth. When the projectiles reach the top of the tower, they are turned around by a large bending magnet. The projectiles reach the bottom of the tower with almost the same velocity that they had when they were launched. The stream of high speed projectiles is then bent through 90 degrees by a bending magnet so that it is traveling horizontally to the surface in an underground tunnel (4 km diameter). The projectile stream is then turned in a large circle by more bending magnets and energy is added by electromagnetic drivers to bring the projectiles back up to the original launch velocity. The installation needs in a gigantic vacuum rigid tower more 100 km for leave out atmospheric drag (see Fig. 5 in Forward, 1995). Some problems still remain to be solved. For example, the projectile has a speed 4 km/s at top station. Computations show that for turning down this projectile, the top station needs a magnet track some kilometers, weighting hundreds of tons-force and requiring a huge energy to operate. Designers have to find solutions how to support this very large weight and an appropriate energy source.

K. H. Lofstrom (1983, 1985, 2002) suggested a space launcher which uses free segmented iron ribbon. Lofstrom's launch idea is different from the cable tower project proposed here. The most important differences are:

1. The present project employs *a fiber cable (rope) which uses tensile stress*. The Lofsrom's idea employs a set of the free segmented iron plates (bullets of size $0.05 \cdot 0 \cdot 0076 \cdot 2\ \text{m}^3$) which hardly could keep tensile stress. Lofsrom's fabric ribbon also hardly could keep the tensile stress. The assumption of keeping the tensile stress is very important for Lofsrom's design because different parts of the bullets (chain ribbon) have different speed during a launch time. Both Lofstrom (and Forward) used the well-known idea of kinetic energy of *bullet* (projectile) for pressing. The present author uses the idea of kinetic energy of

cable for pressing (to our knowledge, there is no other device already built on the base of this idea).
2. Both Lofstrom's bullets and a support electromagnetic system are located into the double vacuum tubes. They need in electromagnets, a special electronic stability control along of all tracks, vacuum pumps, linear electromagnetic engines. These systems are very complex, have a high cost, and require a large amount of electric energy in operation. The present project requests *just* the strong open cable and this cable is used to move any sort of (mechanical or electrostatic) engine.
3. Lofsrom's construction is a *horizontal launcher* having 2000 km length and located at 80 km altitude. The suggested structure is variable *vertical tower* which may change altitude from zero to many kilometers.

An idea similar to Lofsrom's was early suggested by the Russian scientist A. Yunitskii (1982). Yunitskii's space system has also the two moved ribbons located into vacuum tube and moved by electromagnetic engines. The system can lift itself (and useful load) to space, to launch payload, and return to ground. Yunitskii's power space launcher is located as a gigantic ring around the Earth.

The present author offered the simplest space keeper and launcher as a rotated cable around the Earth in Bolonkin (2003d). Other research and comments on these topics may be found in Pokrovskii (1964), Poliakov (1977), Lofstrom (1983, 1985), Mayboroda (1988), Hyde (1988), Landis (1998, 2005) and Landis and Cafarelli (1995, 1999).

3. CONCLUSIONS

The presented theory and design method show that an inexpensive tall tower can be constructed and can be useful for industry, government and science. Cable (kinetic) towers are prospective as space towers and may be used in other fields of activity, too (Bolonkin, 2004a). The computation procedure developed here should be seen as a starting point for further research and development of the associated problems.

ACKNOWLEDGMENTS

The content of this chapter was presented as papers by A. Bolonkin, COSPAR-02, C1.1-0035-02 and IAC-02-IAA, 1.3.05 at Would Space Congress-2002, 10, 19 October, Huston, TX, USA.

REFERENCES

Anderson JD (1989) Hypersonic and high temperature gas dynamics. McGraw-Hill Book Co, New York

Bolonkin AA (1965) Theory of flight vehicles with control radial force. Collection researches of flight dynamics. Mashinostroenie Publisher, Moscow, 1965, pp 79–118 (in Russian)

Bolonkin AA (2002a) Transport system for delivery tourists at altitude 140 km. IAC-02-IAA.1.3.03. 53rd International Astronautical Congress, The World Space Congress – 2002, Houston, Texas, USA. http://Bolonkin.narod.ru/p65.htm. Cited 10–19 Oct 2002

Bolonkin AA (2002b) Non-rocket Earth-Moon transport system, COSPAR-02 B0.3-F3.3–0032-02, 02-A-02226. 34th Scientific Assembly of the Committee on Space Research (COSPAR), The World Space Congress – 2002, Houston, Texas, USA, 10–19 October 2002

Bolonkin AA (2002c) Non-rocket Earth-Mars transport system, COSPAR-02 B0.4-C3.4–0036–02. 34th Scientific Assembly of the Committee on Space Research (COSPAR), The World Space Congress – 2002, Houston, Texas, USA, 10–19 October 2002

Bolonkin AA (2002d) Optimal inflatable space towers with 3–100 km height. Paper presented COSPAR-02-C1.1–0035–02. 34th Scientific Assembly of the Committee on Space Research, IAC–02–IAA.1.3.03. Presented at World Space Congress 2002, Houston, TX, USA, 10–19 October

Bolonkin AA (2002e) Inexpensive cable launcher of high capability, IAC-02-V.P.07. 53rd International Astronautical Congress, The World Space Congress–2002, Houston, Texas, USA, 10–19 October 2002

Bolonkin AA (2003a) Hypersonic space launcher. J Actual Problems Aviation Aerospace Systems, 8(1): pp. 45–58, Kazan, Daytona Beach

Bolonkin AA (2003b) Asteroids as propulsion systems of space ships. J Br Interplanet Soc 56(3–4):98–107

Bolonkin AA (2003c) Space cable launchers. Paper 8057 at International Air & Space Symposium – the Next 100 Years, Dayton, Ohio, USA, 14–17 July 2003

Bolonkin AA (2003d) Centrifugal keeper for space stations and satellites. J Br Interplanet Soc 56:314–322. http://Bolonkin.narod.ru/p65.htm

Bolonkin AA (2003e) Non-rocket Earth-Moon transport system. Adv Space Res 31(11):2485–2490

Bolonkin AA (2003f) Earth accelerator for space ships and missiles. J Br Interplanet Soc 56(11/12):394–404

Bolonkin AA (2003g) Air cable transport and bridges, TN 7567. International Air & Space Symposium – the Next 100 Years, Dayton, Ohio, USA, 14–17 July 2003

Bolonkin AA (2003h) Air cable transport system. J Aircraft 40(2):265–269

Bolonkin AA (2003i) Optimal inflatable space towers with 3–100 km height. J Br Interplanet Soc 56(3/4):87–97

Bolonkin, AA (2003j) Non-rocket transportation system for space travel. J Br Interplanet Soc 56:231–249

Bolonkin AA (2004a) Kinetic space towers and launchers. J Br Interplanet Soc 57(1/2): pp. 33–39

Bolonkin AA (2004b) Multi-reflex propulsion system for space and air vehicles. J Br Interplanet Soc 57(11/12):379–390

Bolonkin AA (2005a) High speed catapult aviation, AIAA-2005–6221. Atmospheric Flight Mechanic Conference – 2005, USA, 15–18 August, 2005

Bolonkin AA (2005b) Kinetic Anti–Gravitator, AIAA-2005–4504. 41th Propulsion Conference, Tucson, Arizona, USA, 10–12 July 2005

Bolonkin AA (2005c) Electrostatic Solar Wind Propulsion System, AIAA-2005–3857. 41th Propulsion Conference, Tucson, Arizona, USA, 10–13 July 2005

Bolonkin AA (2005d) Sling Rotary Space Launcher, AIAA-2005–4035. 41th Propulsion Conference, Tucson, Arizona, USA, 10–13 July 2005

Carbon and High Performance Fibers (1995) Directory, 6th edn. Chapman & Hall, London

Clarke AC (1978) Fountains of paradise. Harcourt Brace Jovanovich, New York

Dresselhous MS (2000) Carbon nanotubes. Springer, New York

Forward RL (1995) Indistinguishable from magic. Baen books, New York

Galasso FS (1989) Advanced fibers and composite. Gordon and Branch Science Publisher, London

Harris JT (1973) Advanced material and assembly methods for inflatable structures. AIAA, Paper no. 73–448

Hyde RA (1988) Earthbreak: A review of Earth-to-space transportation. In: Mark H, Wood L (eds) Energy in physics, war and peace. A festschrift celebrating Edward Teller's, 80th birthday. Kluwer Academic Publishers, Dordrecht, pp 283–307

Koell DE (2000) Handbook of cost engineering. TCS, Germany

Landis GA (1998) Compression structures for earth launch, Paper AIAA-98–3737. 24th AIAA/ASME/SAE/ASEE Joint Propulsion Conference, Cleveland OH, July 13–15

Landis GA (2005) Correction. J Br Interplanet Soc 58:58

Landis A, Cafarelli C (1995) The Tsiolkovski tower, Paper IAF-95-V.4.07. 46th International Astronautics Federation Congress, Oslo, Norway, 2–6 October 1995

Landis A, Cafarelli C (1999) The Tsiolkovski tower re-examined. J Br Interplanet Soc 32:176–180

Lofstrom KH (1983) The launch loop. Analog 103:67–80

Lofstrom KH (1985) The launch loop: A low cost Earth-to-high orbit launch system, Paper AIAA-85–1368. AIAA 21st Joint Propulsion Conference

Lofsrom KH (2002) http://www.launchloop.com/launchloop.pdf

Mayboroda A (1988) Zero gravity on earth. Yuniy Tehnik (Young Technician), No. 10 (in Russian)

Pokrovskii GI (1964) Space tower, *TM* (*Technology for Youth*), No. 10 (in Russian)

Poliakov G (1977) Space Necklace of Earth, TM (Technology for Youth), no. 4 (in Russian)

Smitherman DV Jr (2000) Space elevators. NASA/CP-2000–210429

Space Technology & Applications (1996–97) International forum, parts 1–3, Albuquerque, MN

Tsiolkovski KE (1959) Grezi o zemle i nebe, Moskva, Izd-vo AN SSSR. Speculations about Earth and Sky on Vesta, Academy of Sciences, USSR, Moscow, p 35

Yunitskii A(1982) General planetary transport system, TM (Technology for Youth), No. 6 (in Russian). (see Russian works in http://www.ipu.ru/stran/bod/ing/sovet2.htm, http://www.ipu.ru/stran/bod/ing/sovet_ris.htm)

CHAPTER 9

EXTREME CLIMATE CONTROL MEMBRANE STRUCTURES

Nth Degree Macro-engineering

RICHARD BROOK CATHCART[1] AND MILAN M. ĆIRKOVIĆ[2]

[1] *Geographos, 1300 West Olive Avenue, Suite M, Burbank, California 91506-2225, USA*
[2] *Astronomical Observatory, Volgina 7, 11160 Belgrade, Serbia*

Abstract: We consider some of the implications of the radical macro-engineering efforts in medium-to-long-term future of humanity. In addition to a particular macro-project of Earth, the "Air Bag" Shell based, in part, on the inspiration drawn from Yves Klein's austere visionary "Architecture of the Air", we discuss some ramifications of such a wraparound effect endeavor for humanity's prospects and its cultural outlook, including studio and outdoor art forms. Essentially, we propose an inflated building macro-project to protect Earth from some threats (small asteroids, solar flares, molecular clouds in space) posed by its unaltered trajectory through interplanetary and interstellar space. This "shell-forming" can also be applied to other planets and smaller solid bodies, and can be understood as a generalization of the conventional terraforming. Applied to our planet, it would be a traveling "hibernaculum" for humans in which weather, local time is unimportant and humankind encumbers the Earth with a light touch

Keywords: macro-engineering; future studies; terraforming; physical eschatology; wraparound inflatable constructions

1. INTRODUCTION

It is natural to rationally speculate upon the extension of our current influence on the terrestrial environment. In general, living beings change their environments and are changed by their environments in turn. This truism has become especially pertinent within the framework of the nascent discipline of *astrobiology*. Even before the onset of the explosive development of this field we are witnesses of the fact that even simple forms of life can influence its physical and chemical environment on the planetary scale has been widely known (chemical composition

V. Badescu et al. (eds.), Macro-Engineering: A Challenge for the Future, 151–174.

of planetary atmospheres, etc). Since it deals with the very widest conceivable environments, the question of relevant timescales for changes induced by life and intelligence is very relevant to the astrobiological endeavor. Human existence so far is too brief for the conclusions in this respect to be drawn from experience. In order to consider the impact of humans (and, by analogy, other intelligent communities), on the surrounding universe, we need to investigate what physics and technology may tell us about ways of world-making at grand scale: i.e., about macro-engineering.

In the words of the historian and philosopher of science Steven J. Dick (2003), we need an infusion of "Stapledonian thinking"—the type of cogitation initiated by William Olaf Stapledon (1886–1950)—if we wish to frankly confront the difficult issue of what is generic or typical fate of intelligent technological communities in the Universe. The present discussion of when macro-engineers describe something as taken to the "Nth Degree" they mean it is very great of extreme. The expression issues from mathematics where the letter "N" denotes an indefinite number (in modern English; *numerus* in Latin). Humankind's commencing Age of Molecular Nanotechnology (Phoenix and Drexler, 2004) offers intriguing technological prospects—quite beyond any colossal macro-engineering Earth-sited thing (e.g., Grigsby, 2003) our species has heretofore built—for scientifically predicted planetary change and scientifically imagined global Nature stability. Human spatial migration on a scale larger than the Solar System using inhabited interstellar spacecraft or Spaceship Earth could create a future survival situation in which humans perceive planets as merely suitable bivouac sites. "Bivouac" is French, derived from German "Beiwache" (from "bei", by or near plus "wachen", to watch)—meaning a watching or guarding—and, ordinarily, means a nighttime encampment by people, usually under little or no shelter. The ultimate artificial settlement/bivouac that can be provided by far-future Macro-engineering's unchained efforts, capable of containing the maximum human population of a radically changed Earth-biosphere, is a Fremlin Shell (Badescu and Cathcart, 2006a).

Speculative Nth degree macro-engineering may afford humanity the near-term future luxury of Green-promoted outdoor living on a geographically large-scale. From 1957, when the Space Age started, until his death, Yves Klein (1928–62) dreamed his austere Architecture of the Air (for a comprehensive review, see Noever and Perrin, 2004). During lectures at the Sorbonne in Paris, France, on 3 and 5 June 1959, Klein averred:

> "Air, gas, fire, sound, odors, magnetic forces, electricity, electronics are materials. They must have two main functions, namely: to protect against the rain, the wind and atmospheric conditions in general and to create thermal air conditioning."

He advocated the use of technologies to—literally—convert the Earth-biosphere from a natural state to an unnatural state, especially making Earth's atmosphere equivalent to the aluminum panel skin of the "Atomium" showcased at the 1958 World's Fair in Brussels, Belgium. In other words, Klein was a popular proponent of artistic geo-engineering. Building on a tentative macro-project plan—previously only applied to solid objects other than Earth by its inventors—we seek a wise plan to roof our homeland world with an 8.4 km-high inflated shell, causing an abrupt and anthropocentric global climate regime to ensue subsequently due to a wraparound effect.

2. EXTANT AND NEAR-FUTURE TECHNOLOGIES

Whilst attending the Tyndall Centre & Cambridge-MIT Institute Symposium "Macro-engineering Options for Climate Change Management & Mitigation", held 7–9 January 2004 at the Isaac Newton Institute in Cambridge, England, some climate change macro-problem solvers stepped lightly onto "Cloud Nine". Weather experts testify the scientific phrase refers to a high-altitude layer of cumulonimbus clouds emitting thunder; lexicographers solemnly say it is an expression of a normal human's occasional feeling of bliss, a soaring spirit. Overt weather modification has been studied extensively for a very long period (List, 2004) and continues to be a major topic of research in both theoretical and applied science. Ross N. Hoffman asserts that technological progress ought to permit human control of the Earth's weather—not to change its climate regimes but, instead, to influence the precise timing and paths of outdoor weather systems for the protection of human lives and civilization's valuable property (Hoffman, 2002; see also Chapter 7 in this book).

We, however, foresee the enclosure of Earth's entire atmosphere with a "polyvalent roof" (for an early version of the idea, see Davies, 1981)—a material with embedded technology—so that weather and climate becomes the realm of administration for a technically advanced form of the air-conditioning industry (Meyer, 2002). Architects and designers have just begun professional speculations: "Might there be ultimately found a way to selectively condition only the local environment *immediately surrounding an occupant*, instead of whole rooms " (Addington and Schodek, 2004). Usually a clothed person is most comfortable at a "room temperature" of ~22 °C, while an unclothed person probably would prefer a "room temperature" closer to ~37 °C. In many traditional cultures, climate and season uniformity has been regarded as characteristic of – usually long-lost – "Golden Age" (e.g., Godwin 1996). Thus, the Iranian Avesta texts tell of the thousand-year Golden Reign of Yima, the first king, under whose wise rule cold and heat, old age, death and sickness were unknown. This can be regarded as an interesting parallel to the present.

Yves Klein has outlined his utopist's cloisonné world society, the Air Architecture macro-project, that can be established wherein two widely accepted characteristics of human civilization are discarded forever: (1) above-ground buildings and (2) individual and family privacy. A lack of family privacy is already extant in a primitive basic sense in the form of the Internet; extensions of the Internet that involve real-time sensors, actuators will force humans to question exactly what constitutes "infrastructure" (Allenby, 2004). More and more, the advent of "smart materials" (photochromic glass) permits the design of direct and discrete environments for the human individual. David Brin, distinguished SF author and contemporary thinker, has outlined this future of human privacy in *The Transparent Society: Will Technology Force Us to Choose Between Privacy and Freedom?* (Brin, 1998).

After Yuri A. Gagarin (1934–68) had rocketed into outer space aboard the *Vostok 1* space capsule on 12 April 1961—comfortably ensconced and enveloped by an artificial "Mediterranean Climate" regime—the first public display by Yves Klein of Air Architecture instigated a regionalized academic fracas and widespread popular controversy. Klein imaginatively sought to delete one-half of humanity's technology, retaining and hiding all of the supportive machines whilst creating global or near-global environmental conditions (a "Mediterranean Climate" of stasis and homogeneity like Gagarin's space capsule?) permitting the total absence of buildings through demolition. (Without reference to Klein, Malcolm Wells is carrying forward, in a small way, Klein's evoked concept of human civilization supported by "infra-structures" (e.g., Wells, 2002). Neither man ever directly addresses the obvious psychological problems many persons encounter and display as apparent physical discomfort and pronounced restlessness in long-term subterranean facility use.)

Recently, Jesse H. Ausubel queried: "Will the rest of the world live like America?". Ausubel's considered reply: "So, the answer to my question is that we are likely to live in a cleaner world, with golden inequalities" (Ausubel, 2004). During the same time, the distinguished biologist E.O. Wilson warned that the same standards of living would, with the present efficiency of food production, require no less than two additional planet Earths. Obviously, this does not offer much support for optimism. Yves Klein, however, was deeply optimistic. Klein foresaw a globalized political community that was irreversibly emancipatory both in its intent and its macro-planned actions and, as well, a naturist political community dedicated to leisure that required for its perpetual geographical existence an (implausibly) exalted level of personal loyalty overriding all other obligations; all "brainstorming" (Gallupe, et al., 1992) must result in a completely satisfactory consensus. "Groupthink"—coined by Irving Janis in 1972—describes a particular group or society consensus decision-making process in which all the participants are so wedded to the same social and/or scientific (and, specifically geophysical) assumptions and beliefs that they ignore, discount or even ridicule information/data and rational arguments to the contrary.

All this will, in Klein's artistically outlined geographical vision, occur within discontinuous regions of an unnatural world zone (~29% of Earth's surface)—discrete, non-contiguous continent-spanning aerial enclosures formed by human control of fire, water and air flows, wherein more human physical disorientation and mental disaffection becomes effective that would likely further separate people from global Nature (Hecht, 2001). We concur with Klein that Earth can be appreciated as a playing child's gigantic spherical shell sandbox (for the richness of the emerging physical phenomena, see Kakalios, 2005). A debating group of theoretical terraforming experts attending the Astrobiology Science Conference titled the synoptic partial record of their 30 March 2004 meeting "Walking Naked on the Red Planet"; the world public also witnessed, in Athens, Greece, the 2004 Summer

Olympic Games (Mouratidis, 1985). It was Yves Klein's vision that people living on the continents within his controlled climate regions would be "naked". But, "naked" implies embarrassment, shame, and unease; strangely, Klein did not use the word nude that implies just the opposite (see the report of the Task Committee on Outdoor Human Comfort of the Aerodynamics Committee of the American Society of Civil Engineers, 2004). Nudity supposedly enabled ancient Greek athletes to demonstrate publicly the mind-governed control they exerted over their corporeal presence and, additionally, group nudity may have promoted a sense of community identity and societal cohesion. Clearly, Yves Klein sought a future fulfillment of these world civilization-maintaining aspects, such as athletic skills, as a shared community goal.

With his demise in 1962, Yves Klein's Air Architecture plans became a historical footnote in recorded history of ideas. It joined the 1804 spherical guardhouse of Claude-Nicolas Ledoux (1736–1806) and Sir Isaac Newton's cenotaph of 1784, as planned by Etienne-Louis Boullee (1728–99) in the dusty archives of Architecture's forgotten futurism. Currently, some scientists—especially biologists—suspect there may be a limitation on the number of ideas the human mind can ever entertain (Macer, 2002) and, during 2004, the world's first documented human case of a genetic mutation that boosted muscle growth was reported in the scientific and popular news media (Schuelke et al., 2004). In other words, future humans may be physically quite different from people alive today (e.g., Tatem et al., 2004). Only the dispassionate "little gray Aliens" could ever properly select a name-bearing specimen (holotype) describing the human species. Further progress in the development of the biological science of "Neurotechnology" (Naam, 2005) may invalidate soon the postulated barrier to unlimited conceptualization (Lynch, 2004). Naturally, all such deliberate modifications of humans and descendent humans will certainly alter the extent of responsibility and the boundaries of the self in both law and the Earth-biosphere when Nth degree macro-engineering is applied to any selected planet (Langlois, 2004). Tongue-in-cheek science fiction novelist Bruce Sterling predicted the absence by AD 2380 of purely human persons (Sterling, 1999). Property rights in law will be a social convention of the past (Yandle, 1999). The world's first artificial organism with chemistry unlike any living organism present today in global Nature, put together in a laboratory by 2003, augurs a future widespread change of the characteristics of Earth's biota if humans continue to attempt a redefinition and reorganization of all wild life (Mehl et al., 2003).

3. AIR-BAG ARCHITECTURE

The world's first laboratory for "biospherics", *Biosphere 2*, was not Yves Klein's legacy (Allen et al., 2003). However, Klein's aim was to permanently convert professional Architecture into a globalized Environmental Planning profession. Earth's volume of breathable air—up to an altitude of ~5 km where most healthy persons can still breathe without laboring—is ~2,487,899,916 km^3. At any given moment there is only enough oxygen in the air to support the breathing of all living creatures in the Earth for about 2,000 years (Gitelson, Lisovsky, and MacElroy, 2003). Oxygen is an industrial gas (Almquist, 2003). About 9% of the total mass of the Earth-atmosphere, which consists of $\sim 5.148 \cdot 10^{18}$ kg, is displaced by the planet's above-sea level landmass (Trenberth and Smith, 2005). The thermal resistance of the walls of a common well-built European home is ~ 15 times greater than the Earth-atmosphere's. Yves Klein attempted to visualize humankind's future full control of the daily micrometeorology of a boundary layer at least 2 m thick coating Earth's continents (cf., Pattantyus-Abraham, 2004)—in fact, true "space weather" at a geographical scale exceeding Biosphere 2, but only $\sim 0.000238\%$ of the Earth-atmosphere's isothermal scale-height (Kalu, 2002). "Little World" (terrarium)—the transparent, closed container in which selected living plants and small land animals are kept indoors and observed—achieved its modern-day form *circa* 1827 when invented by Nathaniel Bagshaw Ward (1791–1868); terrariums (and such spin-offs as vivariums, aquariums) were developed by the Victorians but really peaked in world public popularity during the last century, from about 1960 until 1980. Klein must have seen the biotic and geologic parallels in thinking while he unfolded his own futuristic viewpoint; Klein may even have read Ward's *On the Growth of Plants in Closely Glazed Cases* (1852). Several modern artists have chosen to observe and portray Earth's air (Polli, 2005) and its seawater (Sturm, 2005) as very big monitored forms of Art materials, sonifications. All this testifies to another interesting strand of thought not investigated sufficiently in the existing literature – the connection between currently envisioned macro-engineering and contemporary art.

Roger LeBaron Hooke, audited by persons attending the Pardee Keynote Symposium-P3, "Geoscientific Aspects of Human and Ecosystem Vulnerability", at the 7–10 November 2004 Geological Society of America Annual Meeting in Denver, Colorado, offered indisputable documentary proof that humans today surpass all natural forces as earth movers. Urbanization seals some of global Nature's land surface, separating humans from the ground (Turner et al., 2004). Ground sealing results from the construction activities resulting in buildings and roads; for example, the USA's estimated impervious ground has almost the same area as the State of Ohio (Elvidge et al., 2004). In the latitudinal band of 30–60 °N, where most cities are located, urban regions decrease surface albedo by 3–5% and decrease surface emissivity by 1–2%. If the internal energy of an air column is increased by any heating process, the air column must expand vertically, thereby increasing its gravitational potential energy; Yves Klein's 20th Century desire to cause an improvement of the performance of the Earth-atmosphere is now known to rest on the untested capability of 21st Century Nth Degree macro-engineering to minimize all its artificially internalized flow resistances, together and simultaneously, in an integrative manner (Bejan, 2004). In fact, Walter De Maria's gallery-lodged Land Art exhibition, "New York Earth Room" (1977) truly symbolizes the larger object—Earth encased by Yves Klein's Air Architecture—first dramatically described during the summer of 1959. (The "New York Earth Room" can be apprehended as a precursor to geologic sequestration trapping mechanism such as mineral carbonation.) The creation and maintenance machinery Klein foresaw as necessary to effectuate his human living concept must consume energy and, in part, produce heat as waste. His solar and fossil fuel-powered air, water and fire pumping equipment, causing an anthropogenic envelopment of conditioned air, was to be carefully installed underground in vertical shafts, meaning that these vital subterranean facilities would require elaborate mechanically operated systems for maintaining the air supply in a healthful condition via ventilation when and if people entered these sub-surface environments. Air Architecture involved Bernoulli levitation (Waltham et al., 2003) for a universal outdoor living style and, as well, white noise in the form of "directed sound" (Schwartz, 2004) to efficiently mask the racket made by operating industrial-grade machinery decorously hidden from common view; "Sound Architecture" was formalized by Bernhard Leitner in 1971 (Leitner, 1971). Yves Klein's naturist people, whilst enjoying their lives of unfettered leisure, would undoubtedly intercept more sunlight than ever before in humanity's recent history (Streicher et al., 2004), risking uncomfortably painful sunburns, eventual skin cancer and skin blistering, especially as Klein unwisely wished people to cavort upon Anthropic Rock, a reflective paving of glass at locales atop or near the numerous machinery shafts. Photochromic glass, which changes from clear to shaded when embedded photo dye is exposed to the Sun's UV rays, did not play a role in Klein's sketched scheme. Photochromic glass, one of the world's most ubiquitous smart materials, would serve to camouflage the machinery and temporary world public housing shafts if stimulated by UV rays. (By 2002, the world's deepest single-drop mining shaft, with a depth of 2991.45 m below the collar, was dug at the South Deep gold mine in South Africa. Natural air pressure 3 km below sea level is ~30% higher. Unnatural air pressure, present since 1978 in subterranean compressed air storage sites, is released through turbines to supply peak electrical power on short notice in Germany and the USA.)

Almost a decade after Yves Klein revealed his version of wraparound Air Architecture, in 1968, Siah Armajani announced and presented "North Dakota Tower". The Armajani macro-project plan centered on a single 28.8 km-high tower at least 3.2 km wide at its top that was intended to cast an extremely long shadow, from East to West, across the length of the State of North Dakota shortly after sunrise; effectively, his "North Dakota Tower" was to become a gigantic USA gnomon for an imagined North Dakota sundial (Goodman, 1987). The gnomon, symbolically, of the solar calendar and sundial, was humankind's first scientific instrument. Very tall towers, extending farther towards outer space than Armajani's proposed superstructure, are mathematically proved possible as viable macro-projects using currently available technologies and materials (Bolonkin, 2002, 2004; see also Chapter 8 in this book). Richard L.S. Taylor, during 1992, first proposed to erect an enveloping world shell 3–5 km above Mars' arid land that would effectively para-Terraform Mars with modestly tall towers anchoring the enrobing, thin, and air-tight shell safely above the planet's ocean-less surface (Taylor, 2001); Taylor's terraformed Mars might be considered to be a maquette for a future Earth. The purpose of his inflexible metal and glass world shell was to contain a gradually increased mass of air, which ultimately he planned to arrange to have a density sufficient for humans to endure.

4. EARTH SHELL AIR-BAG CIVILIZATION (ESAC)

Quite uniquely and ingeniously, Kenneth I. Roy, Robert G. Kennedy III and David E. Fields have devised a macro-project plan to use an *inflated* fabric spherical shell, which when implemented for celestial object envelopment, would change the Natures of sterile planets and moons, making such places habitable for unclothed humans (Roy et al., 2004). Here, we boldly suppose the Roy-Kennedy-Field World Shell technique applied to the Earth. Impact-deployed automatic automobile passenger air bags were first invented (and a version patented) in 1953. One of the best selling points for a World Shell is that a fabric wraparound atmosphere lid (Zhang et al., 2004), loaded with a few meters-thickness of broken rock might prove to be a civilization-saving feature, protecting the encased planet or moon from small asteroid impacts that could otherwise destroy or cripple infrastructure. In other words, every finished World Shell really becomes a planet or moon with a constantly inflated planet-protective anti-asteroid impact "air-bag" safety device. Roy-Kennedy-Fields label it a "sacrificial ablator" causing small asteroids to vaporize on contact. The Earth Shell Air-bag Civilization (ESAC) could form an Earth-life fostering aerial enclosure approximate 8.4 km thick. Since the area of our natural orb is $\sim 3.2587 \cdot 10^{12}\,\text{km}^2$ and the exterior area of the installed Earth Shell World may be $\sim 3.2685 \cdot 10^{12}\,\text{km}^2$, the *inflated* woven or filmic smart fabric Earth covering would be ~0.2179% bigger (by $7.6699 \cdot 10^9\,\text{km}^2$) than the surface upon which it initially rests. The purpose of ESAC would be to compress an existing mass of air, thereby creating a cloisonné civilization permanently resident inside an anthropogenic pressure balloon. In the UK during 1917 Frederick William Lanchester (1868–1946) patented the first design for an air pressure-tensioned membrane building. ESAC would be laid out on the Earth's natural surface before its inflation; air from above it would then be pumped underneath this hermetic cover, causing ESAC's skin to rise, distend and become an inhabitable "bubble"; the ESAC "bubble" might be stabilized laterally *in situ* through the use of 10 km-high *inflatable space tower* (Bolonkin, 2002) from which super-ropes of carbon nanotubes could be strung for stabilization purposes (Marques et al., 2004). If the entire Earth-atmosphere had a density equal to that of ambient air at sea level, it would necessarily extend upwards only to ~8.4 km, or approximately 400 m less than Mount Everest's elevation. According to the Roy-Kennedy-Fields study, $\sim 120\,\text{kg/m}^2$ of appropriately comminuted rocky material would be needed to balance the tensile forces exerted on the loaded fabric barrier from below; in other words, tall tower elevators must loft (and robotic dumping vehicles must distribute) $\sim 37{,}022 \cdot 10^{12}\,\text{kg}$ of solid material atop the *inflated* Roy-Kennedy-Fields and our proposed Earth Shell Air-bag Civilization. (A cubic meter of air at sea level weighs ~1.2 kg—about 1/800 of water's density; sea level air pressure will remain one atmosphere, or $\sim 10{,}329\,\text{kg/m}^2$. Human pulmonary physiology will remain unchanged—that is, the atmosphere's partial pressure of oxygen [~160 mm Hg in dry sea-level air of 760 mm Hg] and sea-level partial pressure of carbon dioxide [0.23 mm Hg] will still result in a partial pressure of oxygen in the lungs of ~103 mm Hg. Also, the carbon dioxide partial pressure in the lungs is still ~40 mm Hg.) Land Art historian David Bourdon (1969) alleges Christo's 1960s wrapped objects and packages symbolized clothed forms in contrast to "nude" objects; just how would Nth Degree macro-engineering successfully complete our proposed Christo-style ESAC Earth wrapping?

With *The Time Machine* (1895), H.G. Wells (1866–1946) possibly penned the first science fiction novel featuring two distinct "worlds" existing in the year 802,701 A.D.: (1) an inhabited, mechanized subterranean barely human world situated in the Earth-crust housing a genetically divergent population of darkness-loving, muscular ape-like industrialized "Morlocks" who control the world's future which he contrasted with (2) an apparently ideal Green world sited on the Earth's land surface populated by "Eloi" who live a work-free Eden-like existence. Machines linked the two groups—machines based on industrializing humanity's discovery of the Earth's atmosphere by the inventive predecessors and co-extant persons of Thomas Newcomen (1665–1729). (ESAC "probes" the difference in density between planetary air and extra-planetary space.) Yves Klein was inspired by Wells' writings and, subsequently, devised Air Architecture—a macro-project plan bruited during 1959 that still has the potential—as hinted by Roy-Kennedy-Fields—to far exceed the known Earth-atmosphere impact of Thomas Midgeley (1889–1944), the inventor of chlorofluorocarbons in 1928, the same year Yves Klein was born. ESAC erases Architecture's contradiction exposed by Yves Klein: its focus on individual buildings as against the Earth-biosphere. Total design of our planet's future biosphere is nowadays critical to human survival.

Completion of the wraparound (Earth) World Shell—as implied by Roy-Kennedy-Fields—could cause permanent darkness. (The imagination's Universal stimuli—sunrise and sunset—will cease to be a daily part of human life suddenly and even Earth's rotation will become less obvious.) Pitch-blackness 24 hrs/day is debilitating, at the very least, upsetting life's circadian rhythms; some life forms need darkness to function normally and these flora and fauna are intensely studied (Zimmer, 2004). As with many life forms, humans are genetically predisposed to need and want sunlight (Erren et al., 2003); the occurrence of day and night is a stimulus (process-event or agent) setting or resetting the human "biological clock": it is a *zeitgeber*. Unquestionably, solar radiation has a prominent ecological role (Cockell and Blaustein, 2002). Consequently, delightful opportunities are then presented to illuminate ESAC's inhabited Earth-crust as well as to tailor its interior weather superior to the Earth-crust, all within the context of a single, world-encompassing climate regime of our choosing, or a world with polar caps and tropical jungles. Because the ocean is such a large reservoir of stored heat, even if all sunlight impinging Earth was suddenly reflected it would be several decades before the planet cooled (Butler and Hoyle, 1979). We imagine Earth's landmass, like Brussels' Atomium exterior metallic panels attached in 1958, to be exactly replicated (larger than the original.) on the outer Universe-facing ESAC exterior (Tan et al., 2004). We also imagine that the Roy-Kennedy-Fields Earth-roofing team will readily admit that our refinement of its Shell World concept, by our suggested substitution of concrete made translucent by the installation of fiber-optic cables set perpendicular to Earth's natural land and ocean surface, thereby catching all impinging sunlight, permits a low-light, normal diurnal schedule of human activity (Smith, 2004). LiTraCon GmbH in Aachen, Germany, is one of the leading purveyors of light-transmitting hybrid concrete and research on the widespread use of sunlight for interior illumination is ongoing at the Oak Ridge National Laboratory in the USA (Minkel, 2004). (Our envisioned roofing team shares a geophysical fantasy that only the ocean, or perhaps just the landmasses, will be lighted via innumerable vertically penetrating fiber optic cables so that from any viewpoint located in outer space exterior to ESAC our planet-size balloon would have the profound appearance to the naked eye of a gigantic, cartographically accurate teaching-classroom Earth globe.) ESAC's designed concrete cap could be emplaced sequentially and would, therefore, have an Earth-crust like formational geological structure and, certainly, could then be modeled using a Sealed Geological Model (Caumon et al., 2004). And, using humankind's space-based industrial assets, such as sunlight-reflecting satellites (Melnikov and Koshelev, 1998), it would be economically feasible to—at least—intensify daylight at will anywhere inside ESAC; it is foreseeable that >180 lumens will be available (equivalent to a torpedo-shaped Bulbright 15 W krypton light bulb)—at noon on a sunny day—using only fiber optic cables, without any artificial supplementation whatsoever. Spot-focused lasers might be useful for twilight augmentations or at night during some regional, possibly global, public emergency (Dickinson and Grey, 1999). Since cranes, geese and swans have been demonstrated to comprehend human-issued instructions (Ellis et al., 2003), other avifauna species may also be humanely trainable, coaxed to adapt successfully to a fully enclosed anthropic planetary environment. Wherever it is eventually determined that translucent smart concrete is too expensive or structurally inappropriate, we tender the opinion that the $\sim$10 cm (or smaller) particle content (particularly metal oxides and silicon dioxide) of the sitting Roy-Kennedy-Fields rock stratum may be fused via sintering into a harder-to-erode "solid" coating by using standing-wave field of radio waves (Komerath and Wanis, 2004). (The average density of Earth-crust rock is $\sim 3.1\,\text{tons/m}^3$ and the application of $\sim$10 GJ vaporizes it. Since one ton of oil is equivalent to 41.868 GJ, it would take approximately $^1/_4$ ton of oil to vaporize one cubic meter of typical Earth-crust rock.) A Sealed Geological Model will, obviously, be representative of our speculated Nth Degree macro-engineering centered idea of future Planetary Geoscience (van Loon, 2004).

Whilst the unburdened smart fabric rests on Earth's natural surface, for an indeterminate constructional pre-inflation period, some humans will have to dwell within the pre-existing vertical machinery shafts (and tunnels) before the inflation of ESAC; some humans may be required to dwell and work beneath bounded temporary tents formed by the future Earth World Shell's non-distended fabric or film skin that extend over several continents (Cathcart and Badescu, 2004); vertical shaft elevators, interestingly, will induce in human users a momentary sense of gravity defied, an experience post-Yves Klein persons might relish, even adore. This residential option might be termed the "Circus Tent Effect": in such instance, air will be the "performer" issuing from the artwork's pedestal (Earth itself), enclosing and sustaining unclothed people. A medically beneficial effect on humans of such "Circus Tent Effect" envelopes will be that such inhabited bulges are positive pressure isolation "rooms", thus protecting dwellers from possible contamination and pathogens which might otherwise enter (Jaenicke, 2005).

The Earth Shell Air-bag Civilization's translucent concrete crown or the Roy-Kennedy-Fields managed cap deposition of non-stratified sorted rocky material, along with the compacted air below will provide a safe radiation shelter for all Earth-life from the destructive effects of solar flares (Fletcher, 2003), nearby supernovas and gamma-ray bursts (Scalo and Wheeler, 2003; Thomas et al., 2005; but see also Collar, 1996). Currently, Earth looses matter at a rate of ~1 to 3 kg/s, the rate and composition of the matter varying with Solar Cycle; this rate will not be the same when outer space impinges fractured rock and/or sturdy concrete instead of low-density Earth-atmosphere gases. In addition to the deposition of specially selected material upon the intact ESAC, excavated spoils may be stored below Antarctica in artificial tunnels melted within the ice cap (Yoshida, 2004) or even formed into garden-like artificial islands. Glass-capped vertical shafts and horizontal tunnels sited anywhere within the Earth-crust both offer even more biological anti-radiation protection for life sequestered within. All sub-aerial spoils heaps on land—much unwanted material might be invisibly located on the seafloor via ocean dumping—may be artfully disguised in the same manner as New York City's Fresh Kills Landfill, which has obliterated 11% of Staten Island (Reeser and Schafer, 2003). For example, spoil derived from 180 vertical shafts of 2 m diameter penetrating the landmass 1 km would form a conical hillock ~100 m high with a base of only ~150 m if put into a pile. We assume Nth Degree macro-engineering geophysics fantasies are possible—though not as radical as, perhaps, Albert L. Stasenko's "New Earth" macro-project (Stasenko, 1999). It is worth noting that South Africa is, in a daily timeframe, still the Earth's only ecosystem-nation subjected to lots of anthropogenic seismic activity (>5 on the Richter Scale), caused by scheduled blasting in very deep gold and diamond mines; the deepest gold mine reaches almost 4 km and almost stripped miners laboring therein must heroically endure temperatures of ~70 °C.

Whilst ESAC lays inert on Earth's natural surface of land and ocean it must have closeable openings or controllably porous regions where air and precipitation (rain or melt water) can get beneath the tarp-like Earth surface cover. Water and air will have to be "trapped", prevented from flowing uncontrolled into undesirable places. Air consists essentially of nitrogen gas (N_2, 78.1% by volume and 80.4% by mass), oxygen (O_2, 20.9%, 18.8%), argon (Ar 0.9%, 0.67%), and carbon dioxide (CO_2, 0.035%, 0.02%). (Predictable Earth out-gassings such as methane (CH_4) eruptions from the seabed or even violent regional volcanic eruptions (Stone, 2004) can also be vented safely to superjacent space. There is ~1% chance of an eruption with a yield $>10^{15}$ kg during the "next 460–7,200 years" (Mason et al., 2004). How will the ESAC—the world's largest planned compliant wraparound structure (Jenkins, 2005)—be safely up-lifted on a fixed schedule?

The Earth Shell Air-bag Civilization will be levitated, in part, by utilizing the magnetic repulsive force present between a set of patterned room-temperature super-conducting cables (Silberglitt et al., 2002) sewn into the pre-assembled fabric ESAC. (An electrical current of ∼14 MA in cryogenically chilled ESAC cables and an oppositely directed electrical current of 280 MA, has already been shown—by Plus Ultra Technologies, Inc. of Shoreham, New York, USA—to produce a repulsive force of ∼4 tons per linear meter up to a maximum altitude of ∼20 kilometers—more than adequate lifting capacity for the ESAC macro-project. Improvements in chilled cabling are expected momentarily (MacManus-Driscoll et al., 2004). Industry forecasters state that the monetary cost of high-temperature superconducting wire will soon be "... at or even below that of conventional copper wire found in power equipment" (Malozemoff et al., 2005). The ideal location for the circum-terrestrial main super-conducting ESAC cable (Malozemoff et al., 2003) would be at the planet's magnetic Equator where the geomagnetic field inclination is zero (Moss, 1999). The Epoch 2000 World Magnetic Chart can be seen in Fig. 1.

Our proposed creation will also be distended by massive injection of air sucked from outside (superjacent space) and pushed underneath the crinkled globe covering

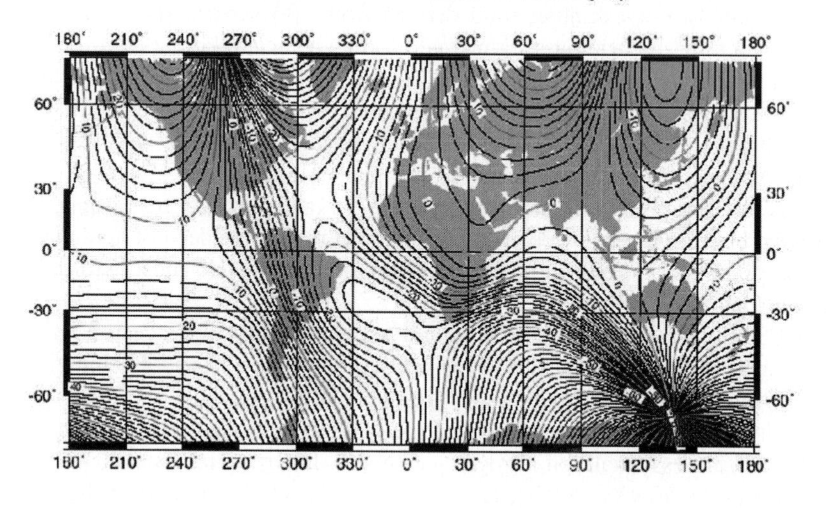

Figure 1. World Magnetic Chart

"tarpaulin". Positive displacement air compressors (reciprocating, rotary screw, rotary sliding vane) work by filling an air chamber with air and then reducing the chamber's volume. Normally, an air compressor works with air at 20 °C compressing the air from one atmosphere absolute ($10^5\,N/m^2$) to ten atmospheres ($10 \cdot 10^5\,N/m^2$): the specific work is 274,200 Nm/kg that when divided by the acceleration of gravity ($9.81\,kg/s^2$) results in a calculated 27,951 m air column head. A continuously functioning dynamically supported vehicle-payload launcher devised by John Knapman can be adapted to perform a single primary function: multiple such devices dedicated to moving air from high altitude (maximally the stratopause at 50 km) to sea-level owing to the constant pump action of vehicles traveling within his rotary sliding vane-like air compression device, with all moved air discharged at his ground-level stations (Knapman, 2005). (About 99% of Earth's air mass is situated in a layer extending from 0 km to ~30 km altitude.) All liberated air—transferred from outside to inside, then becoming a "cryptoclimate" as opposed to a man-made thinner Earth-atmosphere above—will, of course, be filtered before ejection indoors; inevitably, the near-absence of cloud-forming aerosols and water vapor will make clouds less prevalent than before enclosure; however, already free-floating clouds are manufactured rather easily, perhaps someday all indoor clouds (from ground fog to altocumulus) will be rendered as a type of global museum stock item (May, 2003), their exhibition even codified in an operational planning document typified by Section V-15.1 through V-15.14, "The Clouds", in Michael Sorkin's *Local Code: The Constitution of a City at 42°N Latitude* (1993). Any occurrence of lightning will become a rare phenomenon. The total absence of cloud nuclei means that "Global Dimming" caused by clouds, smog and various kinds of aerosols will not diminish the amount of sunlight reaching Earth's roofed surface (Palle et al., 2004). Prior to the Industrial Revolution of the 18th Century, geothermal heat flow was the only net heat source in the Earth (Steffen et al., 2004); 21st Century atmospheric "Global Warming", allegedly caused by atmospheric carbon dioxide gas buildup, is now apprehended as a survival threat to humans and other kinds of extant life present inside the planet's biosphere (Nordell, 2003). Air plucked from outside, compressed and blown indoors as a free-release pump output, means that the gases remaining beyond the ESAC will become increasingly less dense so that all operating compressors will have to perform as vacuum pumps finally. (Air pressures of $5 \cdot 10^{-11}$ torr are not uncommon in school science laboratories.)

The compressors/vacuum pumps must be powered by electricity to do their work. What sources of electricity will be available to practicing macro-engineers? Solar power will always remain a viable option. For example, the "MegaPower" concept studied by the Netherlands Agency for Energy and the Environment is an attractive geographically very large-scale vertical solar power device (Knott, 1996) extending high into the atmosphere, perhaps as much as 7.5 km. But, in addition, there are two more sources that seem to our team to be both attractive and economically feasible: (1) "Power Tube"—USA Patent 6259165 issued 10 July 2001 to Doyle W. Brewington—and (2) vibration-induced electricity generators. Pseudo-quakes caused by human inhabitation of the Earth-crust will provide a beneficial source of temblors. Installation sites may be chosen by very advanced form of reflection seismology made exact by high-performance supercomputing and global positioning systems. Applied to the Earth-crust, "3D detail to a resolution of a few tens of meters over thousands of square kilometers" (Cartwright and Huuse, 2005) is available today because of the advent of three-dimensional seismic technology.

Earth's geothermal resource potential is vast, amounting to $\sim 10^{17}$ W, and shifted to our planet's crust cost-free (Wright, 1998). Liberation of heat from beneath Earth's surface will be necessary since above the height of prevalent humans (personal and species boundary layer) global climate will be a lot cooler than our present-day regime. The highest heat content, with temperatures that range from 650° to 1,300°C, is in magma—the purest form of geothermal "ore". "Power Tube" installations of 1 MW being prototyped are envisioned to each weigh ~30 tons, with a short but variable length, and be inserted into snug drilled shafts with diameters of <70 cm. Initially, the patent owners intend to tap the heat emanating from dry rock but, eventually, they hope to tap the heat extractable from cooling magma "ore"(geomagmatic energy). Still to be resolved is the macro-engineering problem of magma's solidification against the "Power Tube" exterior, causing an insulating layer of Anthropic Rock to form; such Anthropic Rock coating must somehow be induced to slough off, returning to the magma body, otherwise the geomagmatic power generation capacity of the "Power Tube" will decrease drastically within the elapse of a short period.

Vibration-induced electricity generation is a recent technical development. The present study imagines a future Earth-crust embedded with both mechanical and piezoelectric power generators (Anon. 2004). Located in Earth-crust regions with long-recorded histories of natural and anthropogenic micro-seismic activity, such electricity generators could produce a noteworthy, if still rather indefinite, power output. We think the installation costs will be <0.5% of ESAC's final monetary cost of construction. Both power generation equipment types, when emplaced underground, will still respond to Earth-normal tidal and barometric deformations. Still more speculative is the 1-GW nuclear fission reactor, emplaced >100 m underground, that could operate unattended for 30 years after fuel loading and the start of the chain-reaction, with the reactor vessel serving as the final spent-fuel burial coffin, that was designed by Edward Teller (1908–2003) and outlined in 1996 (Teller et al., 1996).

Electrification of human civilization is presumed to be the most important macro-engineered improvement of our species life-style occurring during the 20th Century. Many forecast that, during the 21st Century, radio-controlled objects, devices, vehicles and machinery situated within our planet's uncontained biosphere may become the all-important, global, life-sustaining *cyberelectrosphere* portable homeland for humans in this Solar System. Earth's cyberelectrosphere will become the indoor global totality of technological systems that must be made invulnerable to ESAC's (interior) "space weather"; the infrastructure of the cyberelectrosphere mainly consists of distributed computer, information and communication technology. The Global Ring Network for Advanced Application

Development (GLORIAD) is an existing 10 Gbps optical cable network serving Earth's Northern Hemisphere (USA, The Netherlands, Russia, China) since December 2003. Humans currently endure international geopolitical difficulties with the crowded condition of the radio spectrum, promoting a necessary infatuation with the still-evolving technical possibilities for lasers to convey useful data (Fischer et al., 2004). For example, Air Traffic Control becomes "Cyberelectrosphere Control" following ESAC's completion and, therefore, radio-frequency allocations must be delegated carefully since even commercially grown food and fiber crops may become radio-controlled (Cockell, 2002). Radio-controlled Aerial Nano Devices (AND) may become a useful 21st Century technology to remotely monitor and directly modify virtually all elements comprising our planet's biosphere and its cyberelectrosphere human civilization, which is verging on co-extensiveness. It is probable that global manipulation of life's DNA will also become necessary to avoid selected life's possible future extinction due to sunlight's post-ESAC diminution. Ubiquitous AND technology as well as DNA techniques, initiated and managed planet-wide in real-time, will soon permit people a Nth Degree macro-engineering means to style the ESAC biosphere, shaping and altering it electro-magnetically, to fully accord with humankind's needs and ecologic desires; in effect, Nth Degree macro-engineering holds the prospect of a global radio-stylization of our budding cyberelectrosphere civilization. The unconstrained present-day Earth-biosphere is already a product of massive human intervention—even the planet's so-called "lungs", the Tropic Zone jungles of the Amazon River Basin, are proved artifacts (Willis et al., 2004)—and the future Earth-biosphere promises to be a kind of human holding tank efficiently altered by almost alchemic means directed to creation of a "perfect" global Nature (cf., Newman, 2004). Air's sea level pressure will take on the aspect of being only a relict geophysical feature. One must nearly question if ESAC will qualify Earth to slip from its long-held status, dating from ancient Greece, as a "planet" in astronomy's parlance—surely astrobiologists will invent a category for it.

In a previous study, Cathcart postulated a Mars terraformed in the manner invented by Richard L.S. Taylor but with the significant additional community facility of an encompassing (global) TV screen installation on the ceiling of Taylor's "Mars Worldhouse" (Cathcart, 1998). An ESAC facility of this type is reminiscent of Plato's allegory of the cave, first circulated *circa* 375 BC in *The Republic*. The *IEEE Standard for a Precision Clock Synchronization Protocol for Networked Measurement and Control Systems (IEEE 1588-2002)* defines a protocol that enables the precise synchronization of all clocks in the electronic components of the world's networked, distributed, measurement and control system since 11 September 2002. In other words, ESAC's control by computer has a firm foundation in automation technology. A Plato-inspired cave awaits our inhabitation. Our team irreverently refers to a global measurement and control system for ESAC as "Bubble Brain"—sketched outlines of bubbles placed over cartoon figures' heads symbolize human thought/imagination. The "Bubble Brain" will only be delegated control authority over ESAC's weather and climate regimes. A gigantic 465 m^3 hot air outdoor balloon, dedicated "The World's Largest Brain" owing to its colorful anatomically accurate human brain shape, is ~5 times larger than George Washington's colossal Mount Rushmore "brain". "Bubble Brain" can—literally—rain on anyone's parade. Thin-air computer data displays, such as the patented "Heliodisplay" built by IO_2 Technology, capitalize on the modified properties of air within a region, making it possible to impart information without a hardwired ceiling screen or by supplementing a ceiling screen's pixels. Since the density and cleanliness (Swart et al., 2004) of air within ESAC can be made a steady given, "Heliodisplay"-type public information dissemination media can be counted on to function properly 100% of the time when they are "On". Cash-generating advertisements then become a potential financing basis for ESAC's construction phase, and perhaps afterwards during its inauguration and daily use phases. (USA advertisers alone spent ~US$3 billions during 2003 on outdoor billboard notices.) The world-public always has the option of turning overhead advertisements "Off", and substituting something more beautiful, pleasing and relaxing. Clinging to Yves Klein's projected futuristic Air Architecture civilization, why not pervasive virtual-reality therapy for all troubled and untroubled humans (Lorenzini and Sanmartin, 2004)? Natural and artificial ESAC interior lighting may make a global homogeneous visible field, a *Ganzfeld*, as experiments during the 1930s first conclusively demonstrated (e.g., Metzger, 1930). Then, it is possible to imagine that Klein's unadorned people could believe themselves to be the "elemental spirits"—the air's "sylphs"—concocted by the famous alchemist Teophrastus P.A.B. Paracelsus (*circa* 1493–1541); nowadays, such sylphs (along with gnomes, undines, and salamanders) remain only as technically invalidated romantic figures of ballet, painting, sculpture and literature.

The cost of ESAC's construction, like any proposed tailored Terraforming of Mars using a "Worldhouse" technique, can be off-set by the lack of any further requirement to conserve and, in some costly instances, preserve land and ocean ecosystems. However, in a dawning geological time period, the Anthropocene, when concretions are put together by AND as human-desired accretions and DNA forces change in the planet's essential biota, then it makes sense, logically, to deduct the soundly estimated costs (estimated at ~US$ 1,000,000 per square kilometer) to conserve the chosen flora and fauna of the Earth-biosphere's terrestrial component. Editors of the *Oxford English Dictionary* notified the world-public in July 2004 that they had accepted "ecological footprint" and "nanobot" as proper English words. In the editors trial definition, a "nanobot (n.) [is] a very small self-propelled machine, especially one that has some degree of autonomy and can reproduce". In the recent past, a collection of such atom-assembling nanobots have had the aura of death-dealers, but no longer (Phoenix and Drexler, 2004). Just-in-time product delivery, whether inorganic or organic (Seeman, 2004), rather than being horrifically consumed by out-of-control nanobots and converted into worthless "gray goo" seems to be developing into an ever more likely and desirable humane construction technology. Apropos of Plato's cave, when Nth Degree macro-engineering presents all people with a realized paint-by-numbers wraparound ESAC construction technology, we can hope that the timely published warning by Charles Percy Snow (1905–80) concerning the yawning gap between modern-day Art and Science will be finally bridged or filled. And, terraforming will take on a whole new virtual and geophysical “persona”.

5. EARTH’S ANTHROPOSTROME

When it has progressed to the Nth degree, macro-engineering is art and science applied to the Earth-crust and its superincumbent air as well as maintenance of all useful anthropogenic structures or buildings in harmony with the planet’s various sub-crust geological event-processes. In essence, we (in part following Yves Klein) are interested in understanding how humans may succeed in nesting below a purely anthropic wraparound shell and atop an increasingly reworked Earth-crust. Separating the two solid material surfaces—one a ceiling the other a floor—will be the conditioned air; due to the air’s viscosity it is dragged by our rotating planet but there will never again be storms such as tornadoes or hurricanes. Maximum air velocity might be only <5 on the Beaufort Wind Scale. Though lucid humans will dwell at the bottom of an Yves Klein-inspired Air Bag, the smart sack will merely be a traveling vehicle, useful as a protection from the omni-present hazards of both interplanetary and interstellar space.

The ultimate doomsday for the Earth-biosphere would come as a result of the absence of natural sunlight impinging our beloved homeland; it is rather macabre to contemplate all the geologic-biotic event-processes after such an interruption. Nevertheless, a reasonable robust supercomputer model of the termination of incoming solar sunlight proved that the simulated Earth-atmosphere could retain its basic characteristics for nearly 50 days following onset of total darkness (Hunt 1976). When the final segment of Anthropic Rock (Underwood, 2001) roofing is set into place—we ought to commence construction first at the Poles, then extend a solid material cover progressively over the Earth-crust/air by moving deliberately towards the Equator—only twilight will prevail worldwide since we have planned to introduce attenuated sunlight via fiber optic cables embedded in its concrete cover (Bentur, 2002).

Since the commencement of the Anthropocene more than 8,000 years ago (Crutzen, 2002), which humans have the powerful desire to spread to other Solar System planets by means of Terraforming, industrialized people have dug the most voluminous pit in the planet’s crust at Utah’s Bingham Canyon (LeCain, 2001) as well as the world’s most voluminous landfill. New York City’s Fresh Kills landfill (Walsh and LuFleur, 1995) currently is an object of macro-engineering’s scrutiny—the highest place between Maine and Florida on the USA’s East Coast—and is subject to the predicted impending 21st Century global sea level rise (Gornitz, Couch and Hartig, 2002). In our view, the Earth’s Anthropocene is best typified globally by the Fresh Kills “Global Standard Stratotype-section and Point” (Whitfield, 2004) at Staten Island within New York City composed of a monumental subsurface formation of Anthropic Rock and other aggregated artificial and natural discarded things that are emplaced beneath (any world) urban region (Kazuo, 2001). If all materials that normally flow into a major city were spread evenly on its ground, the topographic elevation of the land would steadily increase by $\sim 1\,\mathrm{m/century}$. Both landfills and buried cities will undergo profound and prolonged Anthropic Metasomatosis—a Earth-crust event-process whereby one solid substance is transmogrified into another solid substance by action of reactive fluids (Schuiling, 1998).

Yves Klein first imagined global human control of the Earth-crust (Noever and Perrin, 2004; Wilkinson, 2005) and the planet's complement of air coincident with the Space Age's start, when two divisions of humanity (the USA-western Europe and the USSR-eastern Europe), in their respective geographical regions, exemplified our geopolitically polarized world. These spatially large regions—all regions are "geomers", a three-dimensional segment of the Earth-crust and air, as the geographer Hans Carol (1915–71) defined that term in 1961 (Carol, 1961)—were a geopolitically significant part of the spreading Anthropostrome (Passerini, 1984). Pietro Passerini defined "Anthropostrome" as "... the huge ensemble of artifacts in which humans are so intimately mired that to a non-human onlooker they would appear almost indistinct from their inorganic framework." (Obviously, Passerini's "Anthropostrome" is almost indistinguishable from seamless machine-man interaction, the indissoluble combination of "biology, society and machines" George Bugliarello has dubbed "biosoma" (Bugliarello, 2000) The seismic response of cities has been characterized "... as a periodic distribution of simple oscillators" (Boutin and Roussilon, 2004); the total response of an urban "geomer" must include measurements from the substratum (Earth-crust) and the "surstratum" (the city). Yves Klein desired to make the Earth-crust a city before removing all the usual traces of civilization from the land's surface; Klein's vision is one of a fusion of crust and civilization, wherein geomer "surstratum" and "substratum" are one in the same.

Extrapolated from the futuristic scenario outlined by Yves Klein, and carried further here by us, humans will, literally, disappear from most of the Earth-crust's surface when the ESAC yet-to-be-inflated-with-air fabric still covers our world. We envision a geomer "Lazarus Effect"—the disappearance and apparent extinction of a dominant species (*Homo sapiens sapiens*) that later will reappear unscathed in the Earth-crust's geomer fossil records (e.g., Fara, 2001). The advent of a ubiquitous application of Molecular Nanotechnology will, subsequently, make more complicated the accurate decipherment of the formation and development of the Earth-crust and the planet's breathable air; as we stated in Section II, "infrastructure" may become equivalent with everything directly accessed by intelligent creatures living in this planet—it will connote the widest possible layer of life in terms of verticality. Project Mohole (1958–66) never put a drill bit into the Earth's Mohorovicic Discontinuity before the USA program to bore through it was cancelled (see *http://www.nas.edu/history/mohole/).* The boundary geological layer, perhaps a very thick "contact", sits between the crust and the underlying mantle and, of course, is named in honor of Andrija Mohorovičić (1857–1936). The geophysical characteristics of this stratum are variously given, but we hold the generality that the boundary marks a sharp seismic-velocity discontinuity that is so distinctive that it causes geo-engineers to name a volume of Earth-material "crust" and another, different volume of Earth-material "mantle". The Mohorovicic Discontinuity, evidently, indicates the depth in our planet's solid part at which Primary-wave velocities change strikingly from 6.7 to 7.2 km/s (in the lower crust) to 7.6 to 8.6 km/s or average 8.1 km/s (at the top of the mantle); its occurrence depth ranges from ~5 km beneath the ocean (~72% of Earth's present-day surface) to ~35 km below the continents, although it may occur at ~60 km beneath some high mountain ranges. It is often alleged the Mohorovicic Discontinuity "probably" represents a chemical change from basaltic to peridotitic or dunitic Earth-materials below, rather than a phase change by seismic velocities alone. It is variously estimated to be between ~0.2 km and ~3 km thick: in other words, Mohorovicic's Discontinuity solid liquid /gaseous volume could range from a minimum of $\sim 1.0112 \cdot 10^8 \text{ km}^3$ to a maximum of $\sim 1.5319 \cdot 10^9 \text{ km}^3$. For comparison, the ocean's volume is $\sim 1350 \cdot 10^6 \text{ km}^3$. (We wonder if the Mohorovicic Discontinuity stratum is really a wrung-out "healing" rock zone, emptied of water that is now present in the ocean and atmosphere because Earth has self-expanded in the past; see Herndon, 2005, 2006.) The Mohorovicic Discontinuity is the last global shell composed of solid material encompassing the Earth; the final global shell covering the solid planet is the atmosphere (Kerr, 2005).

When the Earth's enclosing smart wraparound "air-bag" compliant structure is fully inflated (globalized), counter-weighted by emplaced loose rock—which could shift catastrophically if disturbed by small asteroid impact—or regular, segmented concrete slabs (Abraham, 2003), the ESAC will have supportive roots in the form of deeply penetrating Earth-crust mine and access shafts and extensive, networked vehicular/pedestrian transportation tunnels (Edwards, 1965) and subterranean ore removal sites. Any loose materials stored, even if deposited in regular partitioned compartments on ESAC's outer surface, are tantamount to a potential colluvium threat, particularly if violent impacts by small asteroids occur.

6. GLOBALIZED COMPLIANT STRUCTURES: CIVILIZATIONS AS COSMIC AIR-BAG BIVOUACS

Supposing Earth is never disassembled to provide construction materials for a Dyson Shell (Dyson, 1960; Bradbury, 2001), we assume Earth will follow its natural course within the Solar System and the Milky Way Galaxy far into the geological and cosmological future. As shown by recent studies in the nascent field of *physical eschatology* (i.e., Adams and Laughlin, 1997), viable habitats for life and intelligence are limited by various astrophysical process-events on several timescales. For instance, toward the end of our Sun's lifetime, thermal pulsations on the asymptotic giant branch, will cause evaporation of Mercury, Venus, and probably Earth as well (Rybicki and Denis, 2001). There are ways in which our far-future descendents could influence this adverse course of process-events (Korycansky et al., 2001). In addition, Viorel Badescu and Richard B. Cathcart have, elsewhere and with true

"Stapledonian thinking", assumed that a technically advanced human civilization will eventually seek to drag the Earth by using a self-propelling Sun to do so (Badescu and Cathcart, 2000, 2005; see also Chapter 12 in this book).

Why would humans contemplate becoming so influential? Well, people do like to go places to satisfy their curiosity. And, people do prefer to avoid oncoming, predicted trouble if it is at all technically possible. We are witnessing a revival of what can be termed "neocatastrophism" (in order to distinguish from the 19th century catastrophism of Cuvier, Orbigny, and their students, which is often deemed discredited), encompassing a wider awareness of the risk lurking in our natural surroundings (Huggett, 1997; Schindewolf, 1962). After a revolution in Earth Sciences brought about by the discovery of the terminal Cretaceous process-event *circa* 1980, as well as spectator observations of the collision of the Comet SL-9 with Jupiter in 1994, the awareness of the sudden, catastrophic crises has strongly resurfaced as mainstream in much of the Earth and planetary sciences. In addition, we have recently become more and more cognizant of the radiation dangers due to supernovae and, especially, gamma-ray bursters (Scalo and Wheeler, 2002; Thomas et al., 2005) which could cause lethal consequences for (unprotected.) eukaryotic life forms up to a frightening $\sim$13 kpc from the burst location.

In this "neocatastrophic" context, it is only natural to conclude that the rational justification for various radical macro-projects is strengthened (cf., Pavlov et al., 2005). Such macro-projects will, after all, have as a main purpose protection and continuation of the unobstructed functioning of the terrestrial system. It is quite possible that Nth degree macro-engineering represents a generic outcome of the technological evolution of any intelligent and tool-making species in the Milky Way Galaxy. We may go even further in our imagination, and envisage that social and technological progress will enable our species to shift from a single Fullerian "Spaceship Earth" (ESAC) to a "Spaceship Solar System". Such a transition can only be fully described as panspermic and similar to migration by spacecraft as summarized in Paul Gilster's *Centauri Dreams: Imagining and Planning Interstellar Exploration* (Gilster, 2004). Badescu and Cathcart foresee a situation when our Sun becomes unstable and it will behoove Earthlings to exchange it for a longer-lived Sun. In essence, Yves Klein's architecture for Earth becomes a maternal womb-like wraparound musical-luministic *Gesamtkuntwerke*. Evidently, there will be a period when the ESAC must depart the discarded Sun's vicinity, endure a determinable flight trajectory and flight schedule to arrive near a selected vigorous destination star, before orbiting the new Sun. The envisioned Earth Shell Air-bag Civilization, a cosmic settlement-bivouac, ought to be up to that traveling task; in a more than metaphorical sense, we foresee a journeying ESAC as a very big piece of moving or orbiting "inscribed matter" (Rose and Wright, 2004), communicating to other intelligent creatures our active presence in this Universe.

REFERENCES

Abraham FF (2003) How fast can cracks move? A research adventure in materials failure using millions of atoms and big computers. Adv Phys 52:727–790

Adams FC, Laughlin G (1997) A dying Universe: the long-term fate and evolution of astrophysical objects. Rev Mod Phys 69:337–372

Addington M, Schodek D (2004) Smart materials and technologies for the architecture and design professions. Architectural Press, Oxford, p 208

Allen JP et al. (2003) The legacy of biosphere 2 for the study of biospherics and closed ecological systems. Adv Space Res 31:1629–1639

Allenby B (2004) Infrastructure in the anthropocene: example of information and communication technology. ASCE J Infrastructure Systems 10:79–86

Almquist E (2003) History of industrial gases. Kluwer, New York, pp 64–82

Anon. (2004) Good vibrations. Technol Rev 107:18

Ausubel JH (2004) Will the rest of the world live like America?. Technol Soc 26:343–360

Badescu V, Cathcart RB (2000) Stellar engines for Kardashev's type II civilization. J Br Interplanet Soc 53:297–306

Badescu V, Cathcart RB (2006a) Environmental thermodynamic limitations on global human population. Int J Global Energy Issues 25:129–140

Badescu V, Cathcart RB (2006b) Use of Class A and Class C stellar engines to control Sun movement in the Galaxy. Acta Astronautica 58:119–129

Bejan A (2004) Designed porous media: maximal heat transfer density at decreasing length scales. Int J Heat Mass Transfer 47:3073–3083

Bentur A (2002) Cementitious materials—nine millennia and a new century: past, present, and future. ASCE J Mater Civil Eng 14:2–22

Bolonkin A (2002) Optimal inflatable space towers of high height. 34th COSPAR Scientific Assembly, The Second World Space Congress, held 10–19 October, 2002 in Houston, TX, USA, p 2228

Bolonkin A (2004) Kinetic space towers and launchers. J Br Interplanet Soc 57:33–39

Bourdon D (1969) Christo. Art Artists 7:50

Boutin C, Roussilon P (2004) Assessment of the urbanization effect on seismic response. Bull Seismol Soc Am 94:251–268

Bradbury RJ (2001) Dyson shells: a retrospective. In: Kingsley SA, Bhathal R (eds) The Search for Extraterrestrial Intelligence (SETI) in the Optical Spectrum III. Proceedings of SPIE 4273, pp 56–62

Brin GD (1998) The transparent society: Will technology force us to choose between privacy and freedom? Perseus Books, New York

Bugliarello G (2000) The Biosoma: The synthesis of biology, machines, and society. Bull Sci Technol Soc 20:452–464

Butler EJ, Hoyle F (1979) On the effects of a sudden change in the albedo of the Earth. Astrophys Space Sci 60:505–511

Carol H (1961) Geography of the future. Professional Geographer XII: 14–18

Cartwright J, Huuse M (2005) 3D seismic technology: the geological 'Hubble'. Basin Res 17:1–20

Cathcart RB (1998) Taming Mars with a tent and a tunnel. Spec Sci Technol 21:117–131

Cathcart RB, Badescu V (2004) Architectural ecology: a tentative Sahara restoration. Int J Environ Stud 61:145–160

Caumon G et al. (2004) Building and editing a sealed geological model. Math Geol 36:405–442

Cockell CS (2002) 'Radio-agriculture'—ground and space-based determination of agricultural productivity. J Br Interplanet Soc 55:362–365

Cockell CS, Blaustein AR (2002) 'Ultraviolet spring' and the ecological consequences of catastrophic impacts. Ecol Lett 7:77–81

Collar JI (1996) Biological effects of stellar collapse neutrinos. Phys Rev Lett 76:999–1002

Crutzen P (2002) Geology of mankind. Nature 415:23

Daston L (1981) Biographies of scientific objects. University of Chicago Press, Chicago

Davies M (1981) A wall for all seasons. RIBA J 88:55–57

Dick SJ (2003) Cultural evolution, the postbiological universe and SETI. Int J Astrobiol 2:65–74
Dickinson RM, Grey J (1999) Lasers that beam power to Earth. Aerospace Am 37:5054
Dyson FJ (1960) Search for artificial stellar sources of infrared radiation. Science 131:1667–1668
Edwards LK (1965) High-speed tube transportation. Sci Am 213 (August issue):30–40
Ellis DH et al. (2003) Motorized migrations: the future or mere fantasy? Bioscience 53:260–264
Elvidge CD et al. (2004) U.S. constructed area approaches the size of Ohio. EOS Trans Am Geophys Union 85:233–240
Erren TC, Reiter RJ, Piekarski C (2003) Light, timing of biological rhythms, and chronodisruption in man. Naturwissenschaften 90:485–494
Fara E (2001) What are Lazarus taxa? Geol J 36:291–303
Fischer KW et al. (2004) Atmospheric laser communication. Bull Am Meteorol Soc 85:725–732
Fletcher L (2003) The costs and benefits of space weather. Contemp Phys 44:451–453
Gallupe RB et al. (1992) Electronic brainstorming and group size. Acad Manage 35:350–369
Gilster P (2004) Centauri dreams: Imagining and planning interstellar exploration. Springer, New York
Gitelson II, Lisovsky GM, MacElroy RD (2003) Manmade closed ecological systems. Taylor & Francis, New York
Godwin J (1996) Arktos: The polar myth. Adventures Unlimited, Kempton, Illinois
Goodman C (1987) Digital Visions: Computers and Art, Harry N. Abrams, New York, p 42
Gornitz V, Couch S, Hartig EK (2002) Impacts of sea level rise in New York City metropolitan area. Global Planet Change 32:61–88
Grigsby DG (2003) Geometry/labor = volume/mass? October 106:3–34
Hecht H (2001) Regularities of the physical world and the absence of their internalization. Behav Brain Sci 24:608–617
Herndon JM (2005) Whole-Earth decompression dynamics, Ar-Xiv preprint. http://www.arxiv.org/astro-ph/0507001
Herndon JM (2006) Solar system processes underlying planetary formation, geodynamics, and the georeactor, earth, moon and planets (in press). Ar-Xiv preprint http://www.arxiv.org/astro-ph/0602232
Hoffman RN (2002) Controlling the global weather. Bull Am Meteorol Soc 83:241–248
Huggett R (1997) Catastrophism: Asteroids, comets and other dynamic events in earth history. Verso, London
Hunt BG (1976) On the death of the atmosphere. J Geophys Res 81:3677–3687
Jaenicke R (2005) Abundance of cellular material and proteins in the atmosphere. Science 308:73
Jenkins CHM (2005) Compliant structures in nature and engineering. WIT Press, London
Kakalios J (2005) Resource letter GP-1: Granular physics or nonlinear dynamics in a sandbox. Am J Phys 73:8–22
Kalu A (2002) A contribution towards establishing a more comfortable space weather to cope with increased human space passengers for ISS Shuttles. In: 34th COSPAR Scientific Assembly, The Second World Space Congress, held 10–19 October, 2002 in Houston, TX, USA, COSPAR, USA, p 2110
Kazuo K (2001) Geologic interpretation of artificial strata in urbanized areas. J Geosci Osaka City University 44:121–135
Kerr RA (2005) Pursued for 40 years, the moho evades ocean drillers once again. Science 307:1707
Knapman J (2005) Dynamically supported launcher. J Br Interplanet Soc 58:90–102
Knott M (1996) Sky-high tower of power may ride the waves. New Scientist 149:23
Komerath NM, Wanis SS (2004) Radio waves for space-based construction. AIP Conf Proc 699:992–999
Korycansky DG, Laughlin G, Adams FC (2001) Astronomical engineering: a strategy for modifying planetary orbits. Astrophys Space Sci 275:349–366
Langlois AJ (2004) The elusive ontology of human rights. Global Soc 18:243–261
LeCain TJ (2001) The biggest mine. Invention Technol 16:10–19
Leitner B (1971) Sound architecture. Artforum 9:44–49
List R (2004) Weather modification—A scenario for the future. Bull Am Meteorol Soc 85:51–63
van Loon AJ (2004) From speculation to model: the challenge of launching new ideas in the Earth sciences. Earth Sci Rev 65:305–313

Lorenzini E, Sanmartin J (2004) Virtual-reality therapy. Sci Am 291:58–65
Lynch Z (2004) Neurotechnology and society (2010–2160). In: Roco MC, Montemagno CD (eds) The Coevolution of Human Potential and Converging Technologies. The New York Academy of Sciences, New York, pp 229–233
Macer DR (2002) Finite or infinite mind? A proposal for an integrative mental mapping project. Eubios J Int Bioethics 12:203–206
MacManus-Driscoll JL et al. (2004) Strongly enhanced current densities in superconducting coated conductors of $YBa_2Cu_3O_7$-x+ $BaZrO_3$. Nat Mater 3:439–443
Malozemoff AP et al. (2003) HTS wire: status and prospects. Physica C 386:424–430
Malozemoff AP et al. (2005) High-temperature cuprate superconductors get to work. Phys Today 58:42
Marques MAL et al. (2004) On the breaking of carbon nanotubes under tension. NANO Lett 4:811–815
Mason BG, Pyle DM, Oppenheimer C (2004) The size and frequency of the largest explosive eruptions on Earth. Bull Volcanol 66:735–748
May S (2003) Olafur eliasson: The weather project. Tate Publishing, London
Mehl RA et al. (2003) Generation of a bacterium with a 21 amino acid genetic code. J Am Chem Soc 125:935–939
Melnikov VM, Koshelev VA (1998) Large Space Structures Formed by Centrifugal Forces. Gordon & Breach, The Netherlands, pp 121–132
Metzger W (1930) Optische Untersuchungen am Ganzfeld, II. Zur Phanomenologie des homogenen Ganzfelds. Psychologische Forsch 13:6–29
Meyer WB (2002) Why indoor climates change: a case study. Climatic Change 55:395–407
Minkel JR (2004) Let the Sun shine in, Discover (July issue): 66–67
Moss RA (1999) Use of a superconductor cable to levitate an Earth tethered platform. J Astronaut Sci 37:465–475
Mouratidis J (1985) The origin of nudity in Greek athletics. J Sport Med 12:213–232
Naam R (2005) More than human: Embracing the promise of biological enhancement. Broadway Books, New York
Newman WR (2004) Promethean ambitions: alchemy and the quest to perfect nature. University of Chicago Press, Chicago
Noever P, Perrin F (2004) Yves klein: Air architecture. Hatje Cantz Verlag, Ostfildern, Germany
Nordell B (2003) Thermal pollution causes global warming. Global Planet Change 38:305–312
Palle E et al. (2004) Changes in Earth's reflectance over the past two decades. Science 304:1299–1301 (28 May)
Passerini P (1984) The ascent of the anthropostrome: a point of view on the man-made environment. Environ Geol Water Sci 6:211–221
Pattantyus-Abraham M (2004) What determines the nocturnal cooling timescale at 2 m?. Geophys Res Lett 31:L05109
Pavlov AA et al. (2005) Catastrophic ozone loss during passage of the solar system through an interstellar cloud. Geophys Res Lett 32:L01815
Phoenix C, Drexler E (2004) Safe exponential manufacturing. Nanotechnology 15:869–872
Polli A (2005) Atmospherics/weather works: a spatialized meteorological data sonification project. Leonardo 38:31–36
Reeser A, Schafer A (2003) Fresh kills landfill to landscape competition. Praxis: J Writing Building 4:18–63
Rose C, Wright G (2004) Inscribed matter as an energy-efficient means of communications with an extraterrestrial civilization. Nature 431:47–49
Roy KI, KennedyRG III, Fields DE (2004) Shell worlds: an approach to making large moons and small planets habitable. AIP Conf Proc 699:1075–1084
Rybicki KR, Denis C (2001) On the final destiny of the Earth and the Solar system. Icarus 151:130–137
Scalo J, Wheeler JC (2002) Astrophysical and astrobiological implications of gamma-ray burst properties, Astrophysical Journal 566:723–737
Schindewolf O (1962) Neokatastrophismus? Dtsch Geologische Gesellschaft Z Jahrgang 114:430–445

Schuelke M et al. (2004) Myostatin mutation associated with gross muscle hypertrophy in a child. N Engl J Med 350:2682–2688
Schuiling RD (1998) Geochemical engineering: taking stock. J Geochem Explor 62:1–28
Schwartz EI (2004) The sound war. Technol Rev 107:50–54
Seeman NC (2004) Nanotechnology and the double helix. Sci Am 290:64–75
Silberglitt R, Ettedgui E, Hove A (2002) Strengthening the grid: Effect of high-temperature superconducting power technologies on reliability, power transfer capacity, and energy use, RAND, Santa Monica
Smith RA (2004) Recasting concrete. Wall Street J CCXLIV:B1 & B4. (21 July)
Stasenko A (1999) The new Earth. Quantum 9:16–20
Steffen W et al. (2004) Abrupt changes: the Achilles heel of the Earth system. Environment 46:8–20
Sterling B (1999) Homo sapiens declared extinct. Nature 402:125–126
Stone R (2004) Iceland's doomsday scenario?. Science 306:1278–1281
Streicher JJ et al. (2004) Modeling the anatomical distribution of sunlight. Photochem Photobiol 79:40–47
Sturm BL (2005) Pulse of an ocean: sonification of ocean buoy data. Leonardo 38:143–149
Swart R et al. (2004) A good climate for clean air: linkages between climate change and air pollution. Climatic Change 66:263–269
Tan A, Lyatsky W, Xu S (2004) Distribution of areas of continents and islands. Math Spectrum 36:2–4
Task Committee on Outdoor Human Comfort of the Aerodynamics Committee of the American Society of Civil Engineers (2004) Outdoor human comfort and its assessment. ASCE Press, Reston
Tatem AJ et al. (2004) Momentous sprint at the 2156 Olympics? Nature 431:525
Taylor RLS (2001) The Mars atmosphere problem: paraterraforming—the worldhouse solution. J Br Interplanet Soc 54:236–249
Teller E et al. (1996) Completely automated nuclear reactors for long-term operation II: Toward a concept-level point-design of a high-temperature, gas-cooled central power station system. Lawrence Livermore National Laboratory, Livermore
Thomas et al. BC (2005) Terrestrial ozone depletion due to a Milky Way gamma-ray burst. Astrophys J 622:L153–L156
Trenberth KE, Smith L (2005) The mass of the atmosphere: a constraint on global analyses. J Climate 18:864–875
Turner WR, Nakamura T, Dinetti M (2004) Global urbanization and the separation of humans from nature. Bioscience 54:585–590
Underwood JR Jr (2001) Anthropic rocks as a fourth basic class. Environ Eng Geosci VII:104–110
Walsh DC, LaFleur RG (1995) Landfills in New York City: 1844–1994. Ground Water 33:556–560
Waltham C, Bendall S, Kotlicki A (2003) Bernoulli levitation. Am J Phys 71:176–179
Wells M (2002) An underground utopia. Futurist 36:33–36
White RM (2001) Climate systems engineering. Bridge 31:13–17
Whitfield J (2004) Time lords. Nature 429:124–125
Wilkinson BH (2005) Humans as geologic agents: a deep-time perspective. Geology 33:161–164
Willis KJ, Gillson L, Brncic TM (2004) How 'virgin' is virgin rainforest? Science 304:402–403
Wright PM (1998) The Earth gives up its heat. Renewable Energy World 1:21–25
Yandle B (1999) Grasping for the heavens: 3-D property rights and the global commons. Duke Environ Law Policy Forum X:13–44
Yoshida S (2004) The ICECUBE neutrino telescope. Mod Phys Lett A 19:1099–1106
Zhang M et al. (2004) Multifunctional carbon nanotube yarns by downsizing an ancient technology. Science 306:1358–1361
Zimmer D (2004) Dark science. Can Geographic 124:30

CHAPTER 10

CABLE ANTI-GRAVITATOR, ELECTROSTATIC LEVITATION AND ARTIFICIAL GRAVITY

ALEXANDER A. BOLONKIN
C&R Co,1310 Avenue R, #6-F, Brooklyn, NY 11229, USA
http://Bolonkin.narod.ru, E-mail: aBolonkin@gmail.com

Abstract: This Chapter proposes two new revolutionary ideas: (1) Simple cable anti-gravitator and (2) Electrostatic levitation and artificial gravity. Cable anti-gravitator provides a repel (opposed to gravity) force between two bodies by a rotational closed-loop cable. Electrostatic levitation allows people, cars, and tracks to fly over the Earth's surface. Artificial gravity allows astronauts to work near space ship without leash. Multiple related macro-engineering projects are herein proposed. These macro-projects are: High Tower; Space Elevator; Non-Rocket start from satellites and asteroids; Levitation Highway; Levitation Region in City; Vertical Takeoff and Landing aircraft; Flying, Walking, and Jumping man and High Altitude Crane

Keywords: anti-gravitator, levitation, artificial gravity

1. INTRODUCTION

This Chapter consists of an introduction and two main sections. In Section 2 it is described a method (and associated) devices that provide a repulsive (repel, push, opposed to gravitation) force between given bodies. The basic concept is that a strong, heavy cable is projected upwards using a motorized wheel on the ground. The upward momentum of the cable is transferred to the apparatus by means of a pulley/roller mechanism, which sends the cable back down to the motor. The momentum transferred from the cable to the apparatus produces a push force which can suspend the apparatus in the air or lift it. There is an equal and opposite force on the motorized wheel on the ground. The push force can be large (up to tens of tons-force) and operates over long distances (up to hundreds of kilometers in space). This force produces large accelerations and velocities of

V. Badescu et al. (eds.), Macro-Engineering: A Challenge for the Future, 175–213.

given bodies (vehicles). This device is called a kinetic (mechanical) anti-gravitator (kinetic repulsator or repellor) because gravitation attracts any two bodies, whereas the proposed device repels any two bodies.

The proposed method can be applied to: VTOL (vertical takeoff and landing) aircraft, non-rocket space apparatus, flying, walking, and jumping (pogo-stick) man (vehicles), long arm (long hand), high altitude crane, high tower (up to 200 km in height) and for construction of Space Elevator without rockets (see also Chapter 8).

In Section 3, first one reminds the necessary conditions to allow people and vehicles to levitate above the Earth, by using the electrostatic repulsive force. The author shows that by using small electrically charged balls, people and cars can take flight in the Earth atmosphere. Also, a levitated train can attain high speeds. It is also shown how this method may be used for creating artificial gravity (attraction force) into and out of space ships, space hotels, asteroids and small planets.

2. CABLE ANTI-GRAVITATOR

2.1 Introduction

At present, one knows several methods of flight on Earth and in space. Earth atmospheric devices (aircraft, helicopters) push down the air; space apparatus pushes away a rocket gas. These methods have many disadvantages. For example, we need an atmosphere of sufficient density or costly rocket fuel. The device cannot be immobile and suspended for an extended period of time at a given altitude over the surface of a planet that has gravity but no atmosphere.

The winged and rocket methods of flight have reached the peak of their development. In the last 30 years there has been no increase in flight speed, no wide application of vertical take-off and landing (VTOL) aircraft, no significant decrease in space delivery cost, no common space tourism and no non-vehicle individual flights in the atmosphere. Space launches are very expensive. The aviation, space, and energy industries need revolutionary ideas which will significantly improve the capability of air and space vehicles. Apart from rockets, another method for reaching space is the space elevator. It is very complex, expensive and technologically impossible at the present time.

Here, a new method in proposed (Bolonkin, 1965, 2002a, 2004a,b, 2005c, 2006). This method produces a push (repulsive, repel, opposed to gravitation) force between given bodies, for example, between the ground and a flying vehicle. The push (mechanical) force opposes the gravitational force between these bodies (for example, the ground and a flying vehicle). This force is created by a linear thin cable moved between the given bodies. If there is no roller and air friction and the distance between the given bodies is not changed, the suggested pusher does not require energy (except for the initial start and wheel friction). When the distance is increased, the energy is spent, when the distance is decreased, energy is gained. At first sight this push force may be surprising because the bodies are connected only by the flexible thin long cable. But there is no violation of the physics laws – simply, a momentum is transferred between the bodies through the moving cable. When this momentum per unit time (i.e. a force) exceeds the gravity force, the bodies will move away from one other; when the momentum per unit time is smaller than the gravity force, the bodies will be drawn together.

The proposed method makes it easy to achieve a large push (repulsive, repel) force over long distances, to accelerate a vehicle in flight or keep it at a given distance and to equalize the gravity force. The present section shows a number of useful applications of this method.

The proposed method is different from the 2000-km long space launcher (cable into tube) placed at an altitude of 80 km, proposed by K. Lofstrom (2002). The differences are shortly presented in Section 2.6.

2.2 Description of the Anti-gravitator

Some variants of the installation are shown in Figs. 1 to 6. The installation includes (see notations in Figs. 1 to 3): a linear closed-loop cable, top and bottom rollers, any conventional engine, and a load. Details of the top roller are shown in Fig. 1b, the bottom (lower) driver roller is shown in Fig. 3. Fig. 1a shows a double closed-loop cable moved in opposed direction. Fig. 2 shows a single closed-loop cable. The small rollers (Fig. 3) press on the cable and together with the large roller and engine move the cable. The possible cable cross-section areas are shown in Fig. 1c. Fig. 4 shows the anti-gravitator in a slope position.

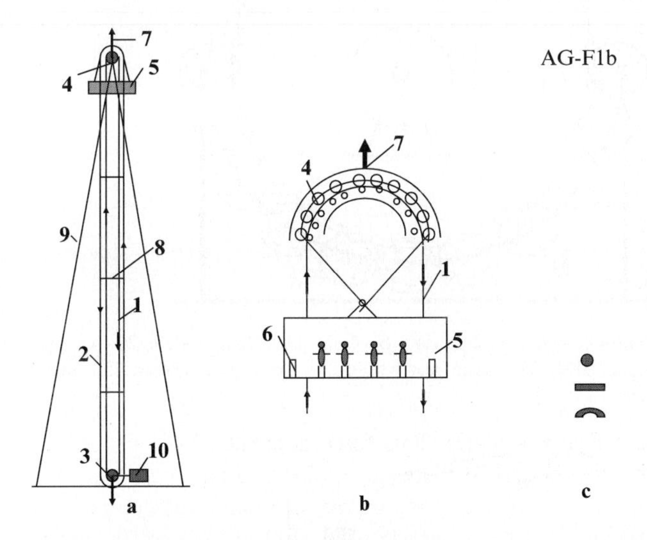

Figure 1. Push devices (cable anti-gravitators) with closed-loop cables: **a** – double cables are moved in opposite directions and located in one plane, **b** – top roller, **c** – forms of cable cross-sections. Notations are: 1 – one closed-loop cable; 2 – the second closed-loop cable; 3 – lower rollers; 4 – top rollers; 5 – suspended object; 6 – load; 7 – push (lift) force; 8 – spreader, 9 – braces, 10 – engine

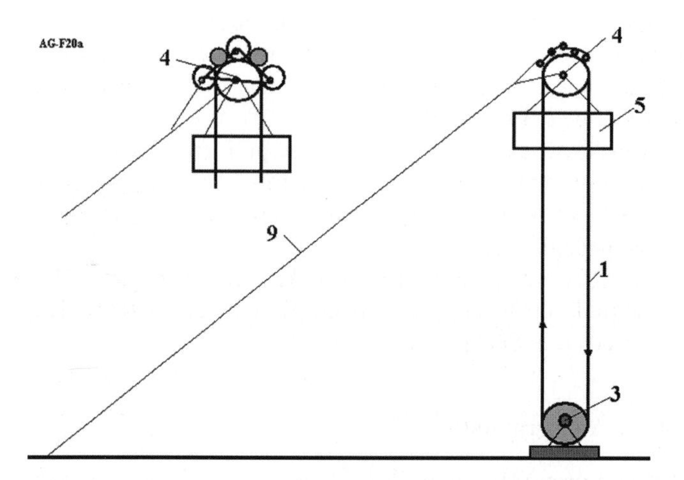

Figure 2. Cable anti-gravitator with single closed-loop cable. (Notations are same as in Fig. 1)

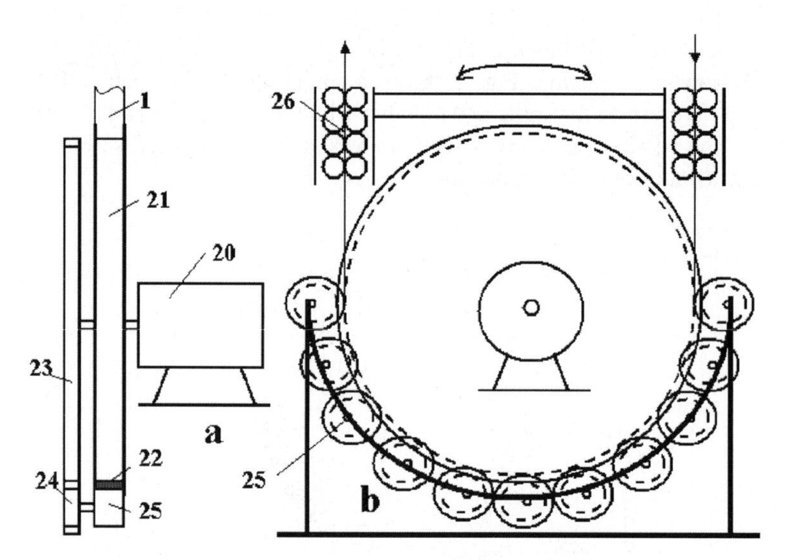

Figure 3. Drive roller of cable anti-gravitator. **a** – side view, **b** – front view. Notations are: 20 – engine; 21 – drive roller; 22 (1) – flexible cable; 23 – large gear wheel; 24 – small gear wheels; 25 – press drive rollers, 26 – directive rollers

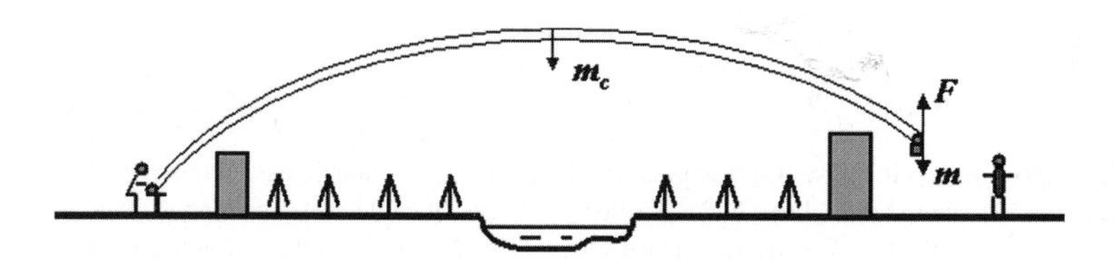

Figure 4. Anti-gravitator in slope position (super-hand). 52-video camera

The spool mechanisms are shown in Figs. 5 and 6. They allow reeling and unreeling of the left and right cable branches with different speeds as well as changing the length of the closed-loop cable without stopping of the facility. The spool may be one of two variants: a mobile spool (Fig. 5) or a motionless spool (Fig. 6).

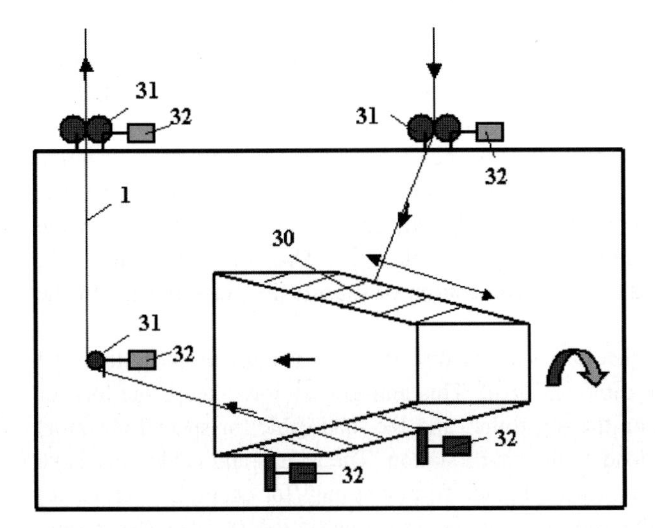

Figure 5. Revolving spool. Notation are: 30 – cable spool; 31 – directive rollers; 32 – spool engine. The left and right cables may have different speeds

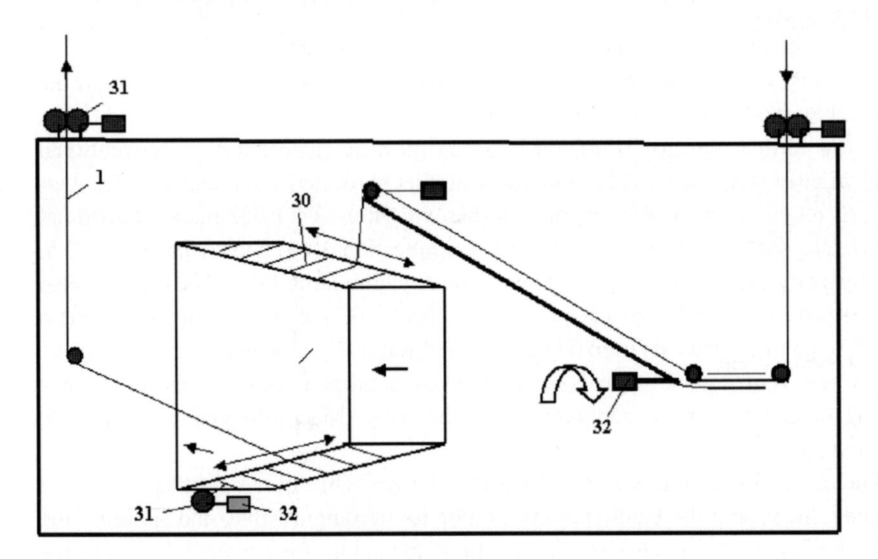

Figure 6. Motionless spool (the lever rotates around the spool). Notation are: 30 – cable spool; 31 – directive rollers, 32 – motor. The left and right cables can have different speeds

The installation works in the following way: the engine rotates the lower driver roller and continuously sends the closed-loop cable upwards at high speed. The cable reaches a top roller (which may be placed at high altitude), turns back and moves to the lower driver roller. When the cable turns back, it creates a push (repulsive, repel, reflective, centrifugal, momentum) force. This repulsive force can be calculated using centrifugal theory and can also be calculated using momentum or reflection theories (see the section 2.3 below).

The cable turns 180 degrees around pulleys. That turn produces a centrifugal force which supports or moves the load. However, Newton's laws say that for every action there is an equal and opposite reaction. In this case, the action comes from the wheel as this is what is pushing the cable and producing the net negative gravity field direction force on the cable. To do that, the wheel moves (it is pressed) by the cable producing the reaction, which adds up to (similarly to how the cable loads) an equal load in the positive gravity direction. This means the cable will push the lower wheel back toward the source of the gravity (in this case to the ground).

The repulsive force points in a vertical direction and it must be more than the gravitational force of the cable and load. This anti-gravity force keeps the load or space station suspended on the top roller; and the load cable (or special elevator) allows the delivery of a load to the space station. The rollers and cable may have high speed and stress. They must be made from a strong (for example, composite) material. In this case, the rollers have the same permitted (safety) stress (and permitted rotational speed) as the cable. The safety speed of the cable or roller is the speed permitted (admitted) by the maximum material strength divided by an assuring safety factor.

The moment of friction in the top roller can be compensated by guy lines as in Fig. 1a, or by the second closed-loop cable rotating in an opposed direction to the first cable and located in one plane (Fig. 1a).

A low altitude (up to 10 km) pusher may have its cable made of conventional steel wire (or steel fiber). This cable has a smaller permitted maximum speed and air drag. It requires less power for rotation than a light weight cable made of artificial fibers (see section 2.5 in this chapter for applications). As shown in section 2.5, the currently produced artificial fibers allow reaching altitudes of 100 km (see also Macro-projects 1 and 2 in Chapter 8, Section 2). If higher altitude is desired a multi-stage tower must be used (Chapter 8, Section 2, Fig. 8-4 and Macro-project 3), and when very high altitude is needed (geosynchronous orbit or more), a very strong cable made from nanotubes must be used (Macro-project 4, Chapter 8, Section 2).

The closed-loop cable may be of variable length. This allows starting from zero altitude, increasing the load (station) altitude to a required value, and spooling the cable for repair. If one changes the length of the cable (for example, by using the spool) the cable will lift the load to high altitude or into outer space.

The devices for this action are shown in Figs. 5 and 6. The proposed spools allow reeling and unreeling the left and right branches of the cable with different speeds to change the length of the cable.

The advantages of the proposed method are the same as for the centrifugal launcher (Bolonkin, 2003a). The suggested design has approximately half the construction cost of the circle launcher because it uses vertical cables (not circular cables as in Bolonkin, 2002a, 2004a). It also has approximately half the production delivery cost (up to \$2–4 per kg), because it has half the air drag and fuel consumption.

Experiments designed by author and performed by Mr. Gregory Lishanski in 2002–2003 show the revolving straight closed-loop cable is stabled in the vertical and horizontal positions.

2.3 Theory of the Cable Anti-gravitator

The theory of the anti-gravitator is briefly presented in this section. Computation procedures are described for several design solutions of practical interest.

2.3.1 Push force for immobile body

The relationships to be used in case of immobile bodies are different from those related to mobile bodies. The last category of bodies is considered in section 2.3.1.2.

2.3.1.1 Repulsive (repel, push) force in space without gravitation We may find the push force of the cable anti-gravitator from centrifugal theory (Fig. 7).

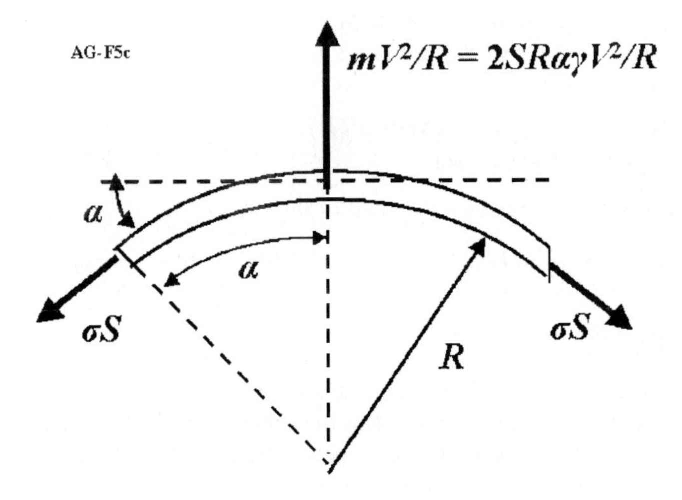

Figure 7. Forces for the rotational circular cable

We first take a small part of the rotational circular cable and write the equilibrium relationships for centrifugal force and tensile stress

$$\frac{2SR\alpha\gamma V^2}{R} = 2S\sigma \sin\alpha\,, \tag{1}$$

where α is the rim angle of the circular cable, V is cable speed, S is cable cross-section area, R is radius of rotational cable and σ is cable tensile stress.

When $\alpha \to 0$ the relationship between maximum rotational speed V and tensile strength σ of a closed-loop (curved) cable is given by:

$$V = \sqrt{\frac{\sigma}{\gamma}} = \sqrt{k}\,, \tag{2}$$

$$F = 2\sigma S, \tag{3}$$

where F is the repulsive (lift) force, $k = \sigma/\gamma$ is the relative cable stress. The more convenient quantity $K = 10^{-7}k$ is used for graphs. Eq. (2) may be found in the literature related to circular centrifugal cable launcher (Bolonkin, 2002a). Results of computation showing the dependence of V versus k for values of K ranging from 0 to 10 are presented in the same reference (see also Chapter 8, Section 2, Figs. 8-7 and 8-8 in this book). For example, the cable has the cross-section area $S = 1\,\text{mm}^2$, stress $\sigma = 100\,\text{kg/mm}^2$. Two cables produce a force 200 kg.

We can find the lift force using reflection theory (see textbooks on theoretical mechanics). Writing the momentum of the reflected mass per unit time gives

$$\begin{aligned} &F = mV - (-mV) = 2mV, \quad \text{where} \quad m = \gamma SV, \\ &\text{or} \quad \text{finally} \quad F = 2\gamma SV^2. \end{aligned} \tag{4}$$

Here m is the cable mass reflected per unit time. If Eq. (2) is substituted into Eq. (4), the same final expression for the repulsive (lift) force $F = 2\sigma S$ is obtained.

2.3.1.2 Repulsive force in constant gravity field In a constant gravity field without air drag, the repulsive force of the proposed device equals the centrifugal force F minus the cable weight W

$$\begin{aligned} F_g &= F - W = F - 2\gamma gSH = 2\gamma S(V^2 - gH) \\ &= 2S(\sigma - \gamma gH) = 2S\gamma\,(k - gH) \end{aligned}, \tag{5}$$

where H is the height of the kinetic device (top end of the cable).

2.3.1.3 Repulsive force for a mobile body For a mobile body the repulsive force is given by

$$F = 2\gamma S\,(V \pm v)^2 \tag{6}$$

where v is the speed of the body. The minus sign in Eq. (6) is used when the cable length is increased, while the plus sign is used when the cable length is decreased. From Eq. (6) it follows that the maximum body speed obtained from the cable cannot exceed the cable speed. Equation (6) is used for launching and landing of flight apparatus.

2.3.1.4 Restore force When the cable is deviated from a vertical position in the gravity field, the restore force is (m_c is the mass of the cable)

$$F_r = F - \frac{gm_c}{2}. \tag{7}$$

2.3.1.5 Air drag of the cable A consistent theory to allow computation of cable drag is still missing. This is mainly a consequence of the fact that no reliable experimental data are available concerning the air drag of a very long cable. However, the air drag of a double *subsonic cable* may be estimated by using the drag equations for plates (Anderson, 1989). Notations are the same as in Eqs. (36)–(40) of Chapter 8 of the present book (the Reynolds number is included):

$$\begin{aligned} D_L &= 0.5 \cdot 0.664\rho^{0.5}\ \mu^{0.5}V^{1.5}L^{0.8}s, \\ D_T &= 0.5 \cdot 0.0592\rho^{0.8}\ \mu^{0.2}V^{1.8}L^{0.8}s, \\ D &= 0.5(D_L + D_T). \end{aligned} \tag{8}$$

Here are: D_L – laminar drag, D_T – turbulent drag, D – total drag, L – cable length, s – cable perimeter, ρ – air density, μ – air viscosity. The cable has only one side, as opposed to a plate which has two sides; that way the multiplier 0.5 is included in Eq. (8). Results of computation are presented in Fig. 8.

The power P of cable air drag D is given by

$$P = DV = 0.5(D_T V + D_L V) = 0.5(P_T + P_L), \quad V = \sqrt{\sigma/\gamma}. \tag{9}$$

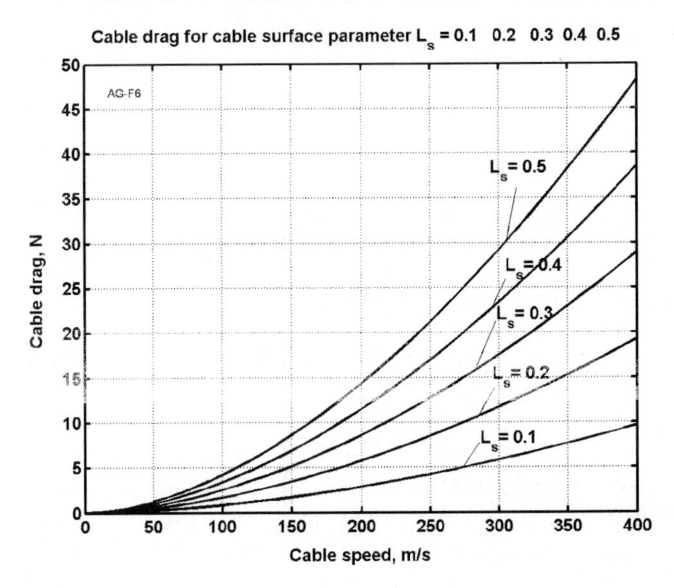

Figure 8. Air cable drag versus cable speed for the cable surface parameter $L_s = L^{0.8}s = 0.1$–0.5

The power of turbulent drag P_T and of laminar drag P_L, respectively, is

$$(10)\qquad \begin{aligned} P_L &= 0.5\cdot 0.664\rho^{0.5}\mu^{0.5}\left(\frac{\sigma}{\gamma}\right)^{1.25} L^{0.8}s,\\ P_T &= 0.5\cdot 0.0592\rho^{0.8}\mu^{0.2}\left(\frac{\sigma}{\gamma}\right)^{1.4} L^{0.8}s, \end{aligned}$$

where the total cable perimeter s of the round cables is (n is number of cable branches):

$$(11)\qquad s = 2\sqrt{\frac{\pi nF}{\sigma-\gamma gH}}.$$

Results of computation for the power P are presented in Fig. 9.

Most of the engine power (80–90%) overcomes the turbulent cable drag. In outer space there is no air, thus no air drag, and one may use a very long cable.

If the altitude H is small (up to 5–6 km), one may ignore the factor γgH. In this case, the cable depends on the relation $(\sigma^{0.9}/\gamma^{1.4})$. A cable with low tensile stress σ and high density γ (for example, a conventional steel cable) requires less power, because the safe maximum cable speed is small ($V \approx 250$–350 m/s). However, the required cable weight increases 10–15 times. The round and single closed-loop cable ($n = 2$) requires minimum power. The plate and semi-circular cables (Fig. 1c) require more power, but they may be more suitable for a drive mechanism.

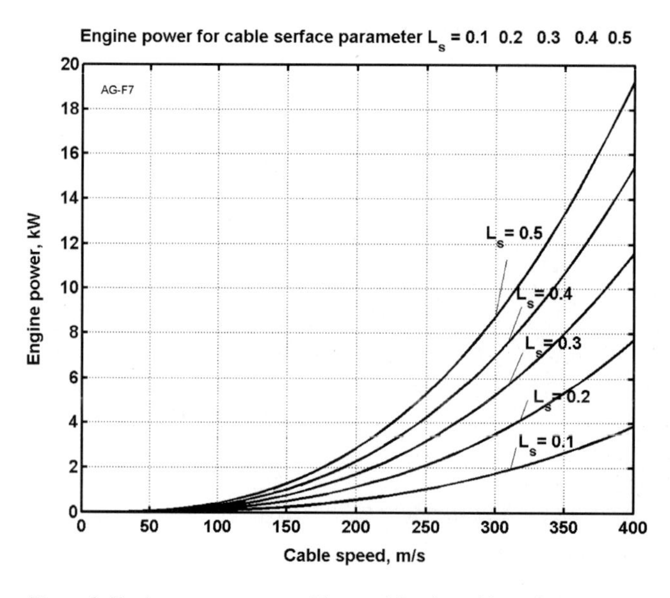

Figure 9. Engine power versus cable speed for the cable surface parameter $L_s = L^{0.8}s = 0.1$–0.5

If the space ship uses asteroid rubble (or ship's garbage) to create thrust via a repulsive engine, the space ship speed is (in an immovable coordinate system with the origin located in the system's center of gravity).

$$(12) \qquad V_s = V_3 \frac{m_3}{m_s},$$

where V_s, V_3 are the additional speeds of ship and garbage, respectively, while m_s, m_3 are masses of space ship and garbage respectively.

2.4 Advantages of the Method

The advantages of the proposed method are obvious. We can fly out of the Earth-atmosphere without using rockets. The facility is very simple (only an engine and a thin cable) and very light, especially when the cable is made of artificial fibers such as nanotubes. When the cable is located outside the atmosphere, it does not require power for rotation, except for a small amount of power to compensate friction losses in the rollers. We can create a huge force (up to tens or hundreds of tons-force) and operate at long distances (tens of kilometers in the Earth's atmosphere and hundreds of kilometers in space).

The cable anti-gravitator may have applications in many fields of civilian and military life: flights on Earth, in space, to planets and asteroids, launching space ships, lifting loads at high altitude, problems of communication, observation, search, and rescues operations. The same idea may be used to transfer mechanical energy over long distances (Bolonkin, 2004b), for moving aircraft in the air (Bolonkin, 2003b) or ground vehicles using an engine located at a large distance, to build simple air bridges over ocean straits and wide rivers or canals, and mountains, instead of tunnels (Bolonkin, 2003c). Some of these applications are considered below.

2.5 Potential Macro-projects

Note about material. Conventional steel cables have a maximum tensile stress of $\sigma = 300\,\text{kg/mm}^2$ and density of $\gamma = 7900\,\text{kg/m}^3$ while fiber steel cables have a tensile strength of about $\sigma = 2000\,\text{kg/mm}^2$. At present, industry widely produces cheap artificial fibers with a maximum tensile stress of $\sigma = 500\text{–}620\,\text{kg/mm}^2$ and density (Carbon Fiber, 1995) $\gamma = 800\text{–}1800\,\text{kg/m}^3$. Whiskers have $\sigma = 2000\text{–}8000\,\text{kg/mm}^2$ and density $\gamma = 2000\text{–}4000\,\text{kg/m}^3$, and nanotubes, created in scientific laboratories, have $\sigma = 20{,}000\,\text{kg/mm}^2 = 2 \cdot 10^{11}\,\text{N/m}^2$ and density $\gamma = 800\text{–}1800\,\text{kg/m}^3$ (Dresselhous, 2000). Theory predicts that nanotubes may have $\sigma = 100{,}000\,\text{kg/mm}^2$ and density $\gamma = 800\text{–}1800\,\text{kg/m}^3$. We will consider a double closed-loop cable in the macro-projects below. We will also use the conventional steel cable that has confirmed tensile stress of $\sigma = 50\text{–}100\,\text{kg/mm}^2$ or the conventional fibers with a maximum confirmed strength of $\sigma = 200\,\text{kg/mm}^2$ and density $\gamma = 1800\,\text{kg/m}^3$. This means the safety factor is 3–6 or 2.5–3.1. The use of whiskers or nanotubes dramatically improves the parameters of the cable anti-gravitator for long distance uses.

2.5.1 Vertical Takeoff and Landing (VTOL) conventional aircraft

Let us estimate the parameters of the cable anti-gravitator for conventional aircraft (Fig. 10). Assume the mass of the aircraft is $M = 20$ tons, the takeoff speed is less than or equal to $V_m = 80\,\mathrm{m/s}$ (most aircraft have takeoff speeds of 50–70 m/s), safety stress of the artificial fiber cable is $\sigma = 200\,\mathrm{kg/mm}^2$, its density is $\gamma = 1800\,\mathrm{kg/m}^3$ and the start and landing acceleration is $a = 3g \approx 30\,\mathrm{m/s}^2$. This acceleration is reasonably low for ordinary (non-trained) people (with training, people can withstand $8g$ permanently and $16g$ for a short time). The required cable length is $L = V_m{}^2/2a = 80^2/60 = 107\,\mathrm{m}$. The total cross-section area of all artificial cable is $S = Ma/(g\sigma) = 20000 \cdot 30/(10 \cdot 200) = 300\,\mathrm{mm}^2$, and one branch of the four-cable system has a diameter of $d = 10\,\mathrm{mm}$. Cable mass is $m = SL\gamma = 300 \cdot 10^{-6} \cdot 107 \cdot 1800 = 58\,\mathrm{kg}$. The cable speed is $V_c = (200 \times 10^7/1800)^{0.5} = 1054\,\mathrm{m/s}$ (see Eq. 2). If the roller is made of the same composite material as the cable, both of them will keep the same speed. Time of acceleration is $t = V/a = 80/30 = 2.7$ seconds. The engine power equals the kinetic energy divided by time $P = MV^2/2t = 20000 \cdot 80^2/(2 \cdot 2.7) = 23704\,\mathrm{kW}$. This value is two to three times more than the power of a typical aircraft engine. However, the start engine may be located on the ground and may have any power wanted. If we want to use an aircraft engine (for cable drive), we must decrease the start acceleration by 2–3 times or use a flywheel. In landing, this energy will return to the fly wheel because the distance between the ground and the apparatus is reduced.

2.5.2 Non-rocket space apparatus

The results used in previous section will apply for a space ship starting from an asteroid or planet without atmosphere. However, if the final speed is high, we must use Eq. (6).

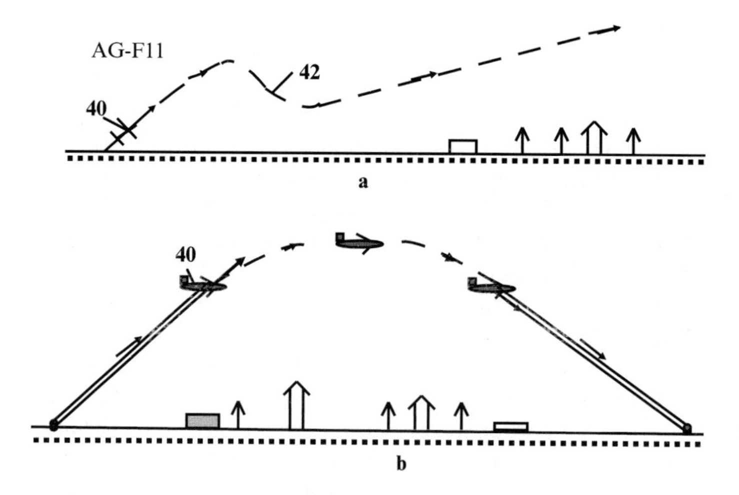

Figure 10. Slope takeoff (a) and landing (b) of conventional aircraft using the cable anti-gravitator. Notations are: 40 – aircraft, 42 – aircraft start trajectory

2.5.3 *Flying man*

Taking on altitude $H = 100\,\text{m}$, the maximum load is $M = 200\,\text{kg}$ (this is enough for Man, another passenger, an engine and a parachute for safety) (Fig. 11a). The steel cable has tensile stress $\sigma = 100\,\text{kg/mm}^2$ and density $\gamma = 7900\,\text{kg/m}^3$. The required total cross-section area of all cables is $S = Mg/\sigma = 2\,\text{mm}^2$, cable diameter is $d = 1.6\,\text{mm}$, the perimeter of the four cables is $s = 10\,\text{mm}$. The cable mass is $m = SL\gamma = 2 \cdot 10^{-6} \cdot 100 \cdot 7800 = 1.56\,\text{kg}$, and cable speed is $V = \sqrt{\sigma/\gamma} = \sqrt{10^9/7900} = 356\,\text{m/s}$. Area parameter is $L_s = L^{0.8} s = 100^{0.8} \cdot 0.01 = 0.4$. The cable drag is $D = 31\,\text{N}$ (Fig. 8 or Eq. 8–10), and the required engine power is $P = 11\,\text{kW}$ (Fig. 9). The cable can be made of transparent fibers and it will be invisible from a long distance.

2.5.4 *Walking man or vehicle*

The lower rollers can be made separately and have separate controls. This allows The man to walk, run, and move with high speed (Fig. 11b). For example, if the "flying Man" described above takes one step (length 100 m) in 2 seconds, his speed will be 180 km/hour.

2.5.5 *Jumping (pogo-stick) man*

Assume the short cable anti-gravitator gives a man of mass $M = 100\,\text{kg}$ the speed $V = 70\,\text{m/s}$ with acceleration $a = 3g = 30\,\text{m/s}^2$ (Fig. 12a). The cable length is $L = V^2/2a = 70^2/2 \cdot 30 = 82\,\text{m}$. The time of acceleration is $t = V/a = 70/30 = 2.33$ seconds. The total cross-section areas of all cables is $S = Ma/\sigma = 100 \cdot 30/(200 \cdot 10^7) = 1.5\,\text{mm}^2$, and the cable mass is $m = SL\gamma = 1.5 \cdot 10^{-6} \cdot 82 \cdot 1800 = 230\,\text{g}$. The jump distance at an angle $\alpha = 45°$ without air drag (it is small at this speed) is $J = V^2/g = 70^2/10 = 490\,\text{m}$, the altitude is $H = V^2 \sin\alpha/2g = 70^2 \sin 45°/20 = 173\,\text{m}$, jump time is about 10 seconds. The

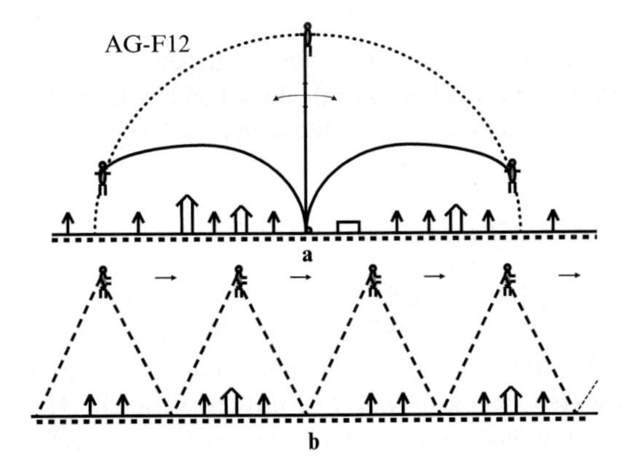

Figure 11. **a)** Flying man using the cable anti-gravitator. **b)** Legged (walking) man using two cable anti-gravitators

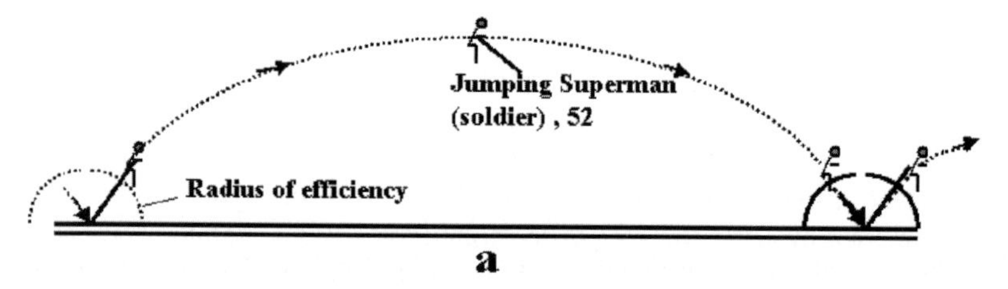

Figure 12. a) man using the jump cable anti-gravitator (50)

required starting thrust is 300 kg-force, and the start (jump) power is about $P = E/t = mV^2/2t = 100 \cdot 70^2/(2 \cdot 2.33) = 105\,\text{kW}$, but the start energy will be restored in landing except for the air drag loss of 10–20%. If we have an energy accumulator, a permanent power of 5 –10 kW will be enough for this device.

2.5.6 *Jumping vehicle*

Assume the cable anti-gravitator gives a vehicle of mass $M = 1000\,\text{kg}$ the speed $V = 200\,\text{m/s}$ with acceleration $a = 8g = 80\,\text{m/s}^2$ (which is acceptable for trained soldiers, for example). The cable length is $L = V^2/2a = 200^2/2/80 = 250\,\text{m}$. The time of acceleration is $t = V/a = 2.5$ seconds. The jump distance at an angle of 45° without air drag is about 4 km, the altitude is 1.4 km, and the jump time is about 20 seconds. The cross-section area of all the cables is $S = Ma/\sigma = 1000 \cdot 80/200/10^7 = 40\,\text{mm}^2$. Cable mass is $m = SL\gamma = 40 \cdot 10^{-6} \cdot 250 \cdot 1800 = 18\,\text{kg}$. The required start thrust is 8 tons-force and the jump power is about $P = 8000\,\text{kW}$, but the start energy will be restored in landing except for the air drag loss of 10–20%. If we have an energy accumulator, an engine with 500–800 kW power will be enough for this device. The vehicle may have a small wing (surface area $2\,\text{m}^2$) and glide from an altitude of 1.4 km for the distance of 14–17 km to the selected place for the next jump.

2.5.7 *Long arm (long hand)*

The proposed method allows us to create a “long arm” which suspends a video camera or weapon aloft (Fig. 4). Assume the load mass of the long hand is $M = 2\,\text{kg}$ and the hand has a length of 1 km. The hand uses a steel cable with $\sigma = 100\,\text{kg/mm}^2$ and $\gamma = 7.9\,\text{g/cc}$. Maximum speed is

$$V = \sqrt{\sigma/\gamma} = \sqrt{10^9/7900} = 356\,\text{m/s}. \tag{19}$$

The cross-section area is $S = M/\sigma = 2/100 = 0.02\,\text{mm}^2$, $d = 0.08$, $s = 1\,\text{mm}$, and the cable mass is $m = SL\gamma = 0.02 \cdot 10^{-6} \cdot 1000 \cdot 7900 = 158\,\text{g}$. The cross-section area parameter is $L_s = L^{0.8}s = 1000^{0.8} \cdot 0.001 = 0.25$. The cable drag is $D = 20\,\text{N}$ (Fig. 6 or Eqs. 8–10) and the required engine power $P = 6.8\,\text{kW}$ (Fig. 9). The operator (e.g. a soldier) can observe regions within a 1 km radius and immediately apply the weapon if necessary. The radius may be increased up to 10 km. If using a more powerful cable anti-gravitator that can hold a load of 200 kg with a net and catcher installed at the end of the cable, the operator can catch the soldier and deliver him or her to another place. This may be very useful for rescue and anti-terrorist operations in urban regions.

2.5.8 High altitude crane

The construction of skyscrapers needs high cranes. Consider the design of a crane of $L = H = 500\,\text{m}$ height using the proposed method. Take the useful load as 1 ton and the steel cable as having safe tensile stress of $\sigma = 50\,\text{kg/mm}^2$ and cable density of $\gamma = 7.9\,\text{g/cc}$. The total cross-section cable area is (Eq. 5) $S = F/(\sigma - \gamma g H) = 22\,\text{mm}^2$. The cable mass is $m = S\gamma H = 2 \cdot 10^{-6} \cdot 500 \cdot 7900 = 87\,\text{kg}$, and safe cable speed is $V = (\sigma/\gamma)^{0.5} = 250\,\text{m/s}$. If the installation has four cables of diameter $d = 2.6\,\text{mm}$ each, the total perimeter of the four cable is $s = 4\pi d = 33.2\,\text{mm}$, the parameter $L_s = L^{0.8} s = 500^{0.8}\ 0.0332 = 4.8$, the cable air drag is $D = 200\,\text{N}$ (Fig. 6, Eq. 8) and the required power to support cable rotation is $P = 50\,\text{kW}$ (Fig. 7, Eq. 10). This is the highest (500 m) and the lightest (87 kg) crane in the world, having a load capability of 1 ton.

2.5.9 High tower

Consider the design of a tower of height $L = H = 4\,\text{km}$ using the proposed method. Take the useful load as 30 tons and the steel cable as having a safe tensile stress of $\sigma = 50\,\text{kg/mm}^2$ and cable density of $\gamma = 7.9\,\text{g/cc}$. The total cross-section (all branches) of the cable area is (Eq. 4) $S = F/(\sigma - \gamma g H) = 1630\,\text{mm}^2$. Cable mass is $m = S\gamma H = 51.5$ tons, and safe cable speed is $V = (\sigma/\gamma)^{0.5} = 250\,\text{m/s}$. If the installation has four cables, the diameter of one cable is $d = 23\,\text{mm}$, the total perimeter of the four cable is $s = 4\pi d = 0.289\,\text{m}$, the parameter $L_s = L^{0.8} s = 4000^{0.8} \cdot 0.289 = 220$, the cable air drag is $D = 9500\,\text{N}$, and the required power to support cable rotation is $P = 2.3\,\text{MW}$. This is highest (4 km) and the lightest (52 tons) tower in the world, which has a load capability of 30 tons at the top.

Note that the computations of a similar tower in Chapter 8 (Section 2) of this book assumes an artificial fiber cable, having a safe tensile stress $\sigma = 200\,\text{kg/mm}^2$ and density $\gamma = 1800\,\text{kg/m}^3$. The results in Chapter 8 showed the required speed $V = 1000\,\text{m/s}$ and power $P = 11\,\text{MW}$. In this chapter the only change was the cable material (i.e. steel rope instead of artificial fiber). As a result, the required support power decreases by approximately 5 times, and the total production cost for delivery of one tourist decreases from \$3.35 to \$1.4. This happens because the required speed decreases from 1000 m/s to 250 m/sec, decreasing the required power and fuel consumption. The power depends on speed following a (nearly) cubic law (i.e. $\sim V^{2.8}$; see Eqs. 8–10).

This shows that the parameters of the considered example are very far from optimal. On the other hand, the cable mass increases from 1.13 tons up to 51.5 tons.

2.5.10 The construction of the space elevator without rockets

The space elevator may be a promising breakthrough in space. However, the space elevator needs an equalizer load weighting hundreds of tons, and thousands of tons of cable. Also, delivery using rockets into geosynchronous orbit is very expensive. When nanotubes will become cheaper, the proposed cable anti-gravitator will allow the construction of the space elevator without rockets. The author shows in Chapter 8 (Section 2) of this book that if $K = \sigma/\gamma/10^7 > 5$ the cable space tower (cable mast) can be up to 150,000 km high or more. This is enough to reach geosynchronous orbit (i.e. 37,000 km distance from Earth).

2.6 Discussion

Lofstrom (2002) published a description of a space launcher facility. The device proposed in this chapter has the following technical and physical differences as reported to Loftstrom's installation.

The Loftstrom facility has a the 2000-km long launch path located at an altitude of 80 km, which accelerates the space vehicle to space speed. The Loftstrom space launcher is *non-connected plates enclosed in an immobile tube*. The *plates* are made from rubber-iron material and are moved using an electromagnetic linear engine. The *plates* are turned by electromagnets.

The method proposed in this article is based on a kinetic device which creates a push (repulsive, repel) force between two given bodies (for example, between a planet and the apparatus). This force supports a body at a given altitude. The body is connected to the cable by rollers that slide along the cable. The cable may be made of artificial fiber and moved by the rollers and any engine. The cable anti-gravitator supports any body at altitude (for example, towers) and may also be used to launch vehicles. The Loftstrom device is only a space launcher and cannot permanently support a body at altitude or towers.

The cable anti-gravitator creates a permanent controlled force. If the distance between bodies does not change, the cable anti-gravitator requires only a small amount of energy to compensate for the friction in the rollers and air.

The cable anti-gravitator may be applied in many fields and may be very useful in the future. The method does not need special technology. Current cheap and widely produced materials such as steel or artificial fiber can be used to operate up to an altitude of 75–225 km. The cable anti-gravitator is simple and does not need complex equipment. If the installation is designed correctly, Mr. Gregary Lishanski's (2002–2003) experiments showed that the high speed closed-loop straight-line cable is stable in any direction including the horizontal position.

Other applications of this idea may be found in papers published by the author from 1965 to 2006.

3. ELECTROSTATIC LEVITATION AND ARTIFICIAL GRAVITY

3.1 Creating Repulsive Forces

Only two methods are known to create repulsive forces: magnetism and electrostatics. Magnetism is well understood and the usage of superconductive magnets for levitating a train is a rather standard envisaged application. However, repulsive magnets have only a short-range force. They work well for ground trains but are not appropriate for air flight. Electrostatic flight needs powerful electric fields and powerful electric charges. The Earth's electric field is very weak and cannot be used for levitation. The main innovations presented in this chapter are methods for creating powerful static electrical fields in the atmosphere and powerful, stable electrical charges of small size which allow levitation (flight) of people, cars, and vehicles in the air. We also show how this method can be utilized into and out of a space ship (space hotel) or on an asteroid surface for creating artificial gravity.

Magnetic levitation has been widely discussed in the literature for a long time. However, there are few comprehensive scientific works related to electrostatic levitation. Electrostatic charges may have a high voltage and may create corona discharges, breakthrough and relaxation.

Several important innovations in this field were first proposed in Bolonkin (1982) and some practical applications were given in Bolonkin (1983). These ideas were published in Bolonkin (1990, p. 79). In a series of papers (Bolonkin, 2005a, 2005b, 2005e) the theory was developed and more details are given. Some macro-projects are also presented, to allow estimation of the parameters of the proposed flight systems.

3.2 Brief Description of Innovation

It is known that similar electric charges repel and electric charges of opposite sign attract each other. A large electric charge (for example, positive) located at altitude induces the unlike (negative) electric charge at the Earth's surface (Fig. 13) because

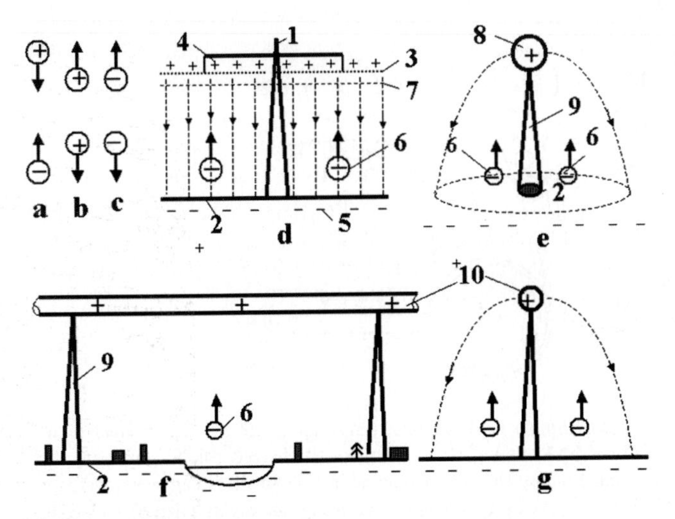

Figure 13. Explanation of electrostatic levitation: a) Attraction of unlike charges; b,c) Repulsion of like charges; d) Creation of the homogeneous electric field (highway); e) Electrical field from a large spherical charge; f,g) Electrical field from a tube (highway) (side and front views). Notations are: 1 – column, 2 – Earth (or other) surface charged by induction, 3 – net, 4 – upper charges, 5 – lower charges, 6 – levitation apparatus, 7 – safety net, 8 – charged air balloon, 9 – column, 10 – charged tube

the Earth is an electrical conductor. Between the upper and lower charges there is an electric field. If a small negative electric charge is placed in this electric field, this charge will be repelled by the charges of the same sign (on the Earth's surface) and attracted by the upper charge (which has an opposite sign) (Fig. 13d). That is the electrostatic lift force. Most lift forces are determined by the Earth's charges, because the small charges are conventionally located near the Earth's surface. As shown below, these small charges may be connected to a given body and have enough force to lift and support them in the air.

The upper charge may be located on a column as shown in Fig. 13 or a tethered air balloon (if we want to create levitation in a small town) (Fig. 13e), or an air tube (if we want to build a big highway), or a tube suspended on columns (Fig. 13f,g). In particular, the charges may be at two identically charged plates, used for a non-contact train (Fig. 15a).

A lifting charge may use charged balls. If a thin film ball with maximum electrical intensity of below $3 \cdot 10^6$ V/m is used, the ball will have a radius of about 1 m (for a man's mass, of about 100 kg). For a 1 ton car, the ball will have a radius of about 3 m (see the computation below and Fig. 14g,h,i). If a higher electric intensity is used, the balls may be small and located underneath clothes (for computation see

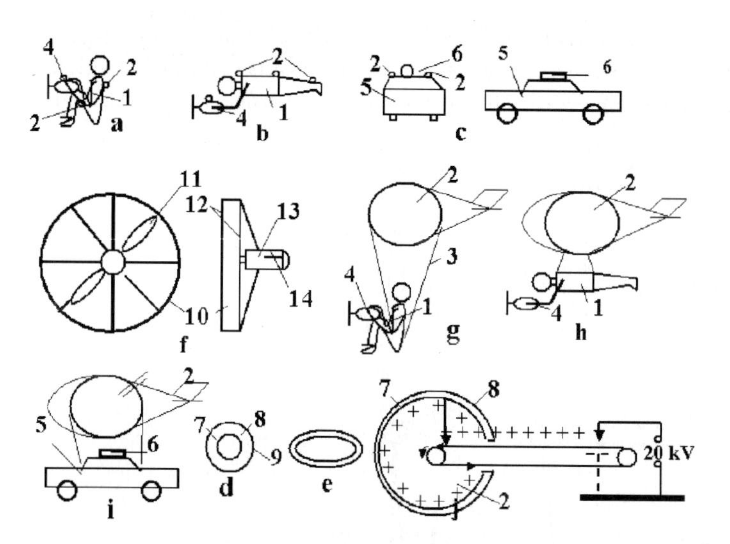

Figure 14. Levitation apparatus: a,b) Single levitated man (mass up to 100 kg) using small highly charged balls 2. a) Sitting position; b) Reclining position; c) Small charged ball for levitating car; d) Small highly charged ball; e) Small highly charged cylindrical belt; f) Small air engine (forward and side views); g) Single levitated man (mass up to 100 kg) using a big non-highly charged ball which doesn't have an ionized zone (sitting position); h) The some man in a reclining position; i) Large charged ball to levitate a car which doesn't have an ionized zone; j) Van de Graaff generator. Notations: 1 – man; 2 – charged lifting ball; 4 – handheld air engine; 5 – car; 6 – engine (turbo-rocket or other); 7 – conducting layer; 8 – insulator (dielectric); 9 – strong cover from artificial fibers or whiskers; 10 – lagging; 11 – air propeller; 12 – preventive nets; 13 – engine; 14 – control knobs

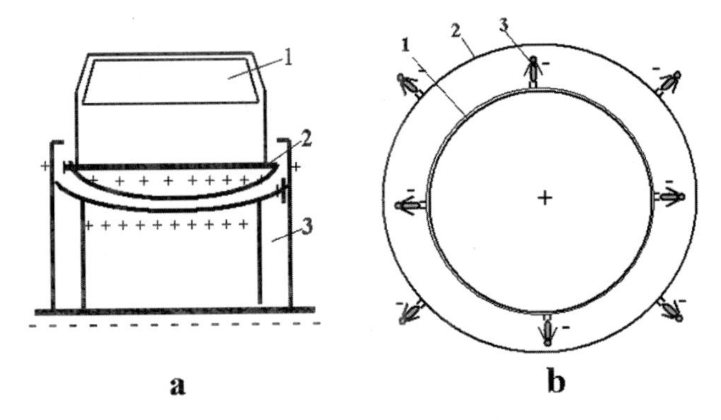

Figure 15. Levitated train on Earth and artificial gravity into and on space ships and asteroids. a) Levitated train; b) Artificial gravity on a space ship. Notations: a) 1 – train; 2 – charged plates; 3 – insulated column; b) 1 – charged space body; 2 – space ship; 3 – man

below and Fig. 14 a,b,c). This method may be also used for a levitated train and artificial gravity in space ships, hotels, and asteroids (Fig. 15a,b).

The proposed method has important advantages in comparison to conventional vehicles (see Figs. 13 and 14). For example, no very expensive highways are necessary. Rivers, lakes, forests, and buildings are not obstacles for this method.

In given regions (Figs. 13 and 14) people (and cars) can move at high speeds (people up to 70 km/hour and cars up to 200–400 km/hour) in any direction using simple equipment (small balls under their clothing and small engines (Fig. 14a,b,c)). They can perform vertical takeoffs and landings. People can reduce their weight and move at high speed, jump a long distance, and lift heavy weights. Building high altitude homes will be easier.

A space ship (hotel) definitely needs artificial gravity. Any slight carelessness in space can result in the unleashed astronaut, instruments or devices drifting away from the space ship. Presently, they are connected to the space ship by cables.

To produce artificial gravity in space one may use the rotation of the space ship and magnetism. Both methods have drawbacks. The rotation creates artificial gravity only inside the space ship (hotel). Observation of space from a rotating ship is very difficult. The magnetic force is only effective over a very short distance. The magnets stick together and a person has to expend a large effort to move.

If there is a charge inside the space ship and small electric charges of opposite sign are attached to objects elsewhere, then they will fall back to the ship if they are dropped.

The same situation occurs for astronauts on asteroids or small planets which have a weak gravitational field. If one charges the asteroid and astronauts with unlike electric charges, the astronauts will always return to the asteroid during walking and jumping.

A number of problems should be solved before the method is implemented. For example, a high electrical intensity is needed if one wants small charged balls to be used. This problem is treated below, among other related aspects.

3.3 Theory of Electrostatic Lift Force and Results

The theory of electrostatic lift force is briefly presented below. Some results of computation are also given.

3.3.1 Brief information about electric charges, electric fields, and electric corona

Electric charge creates an electric field. Every point this field has a vector called electrical intensity, of magnitude, E, measured in Volts/meter. If the unlike charges (or non-insulated electrodes under voltage) are located in air at ground level and the electrical intensity is less than $E_c = (3\,\text{to}\,4)\cdot 10^6\,\text{V/m}$, the discharge current will be very small. If $E > E_c = 3\cdot 10^6\,\text{V/m}$ and we have a closed-loop high-voltage circuit (for non-insulated electrodes), and electric current appears. The current increases by following an exponential law when the voltage is increased. In a homogeneous electric field (as between plates), the increasing voltage makes a spark (flashover, breakthrough, lighting). A non-homogeneous electric field (as between a sphere and plate or an open sphere) makes an electric corona. Electrons break away from the metal negative electrode and ionize the air. Positive ions hit the non-insulated positive electrode and knock out electrons. These unlike ions can cause a particle blockade (discharge) of the main charge. The efficiency of ionization by positive ions is much less than for electrons of the same energy. Most ionization occurs as a result of secondary electrons released at the negative electrode by positive ion bombardment. These electrons produce ionization as they move from the strong field at the electrode out into the weak field. This, however, leaves a positive-ion space charge, which slows down the incoming ions. That has the effect of diminishing the secondary electron yield. Because the positive ion mobility is low, there is a time lag before the high field conditions can be restored. For this reason the discharge is somewhat unstable.

The air contains a small amount of free electrons. These electrons can also create an electric corona around the positive non-insulated electrode, but under higher voltage than the negative electrode. The effect here is to enable the free electrons to ionize by collision in the high field surrounding this electrode. One electron can produce an avalanche in such a field, because each ionization event releases an additional electron, which may contribute to further ionization. To sustain the discharge, it is necessary to collect the positive ions and to produce the primary electrons far enough from the positive electrode to permit the avalanche to develop. The positive ions are collected at the negative electrode, and it is their low mobility that limits the current in the discharge. The primary electrons are thought to be produced by photo-ionization.

The particular characteristics of the discharge are determined by the shape of the electrodes, the polarity, the size of gap (ball), and the gas or gas mixture and its pressure. In high voltage electric lines the corona discharge that surrounds a high-potential power transmission line represents power loss and limits the maximum potential which can be used.

The proposed method is very different from the conventional techniques described in textbooks. The charges are isolated here by using an insulator (dielectric). They cannot emit electrons in the air. There is not a closed circuit. This resulting device is rather similar to the single polarity electrets, where positive charges are inserted into an insulator (Kestelman, 2000). Electrets have typical surface charges of about $\sigma = 10^{-8}\,\text{C/cm}^2$, PETP up to $1.4 \cdot 10^{-7}\,\text{C/cm}^2$ (Kestelman, 2000, p. 17), and TSD with plasticized PVB up to $1.5 \cdot 10^{-5}\,\text{C/cm}^2$ (Kestelman, 2000, p. 253). This means the electrical intensity near their surface reaches ($E = 2\pi k\sigma$, $k = 9 \cdot 10^9$) $6 \cdot 10^6\,\text{V/m}$, $80 \cdot 10^6\,\text{V/m}$, and $8500 \cdot 10^6\,\text{V/m}$ respectively. The charges are not blockaded and the discharge (half-life time) continues from 100 days up to several years. In humid air the electrets lose part of their properties, but in dry air they regain them.

Natural Earth radioactivity and cosmic rays create about 1.5–10.4 ions in $1\,\text{cm}^3$ every second (see Kikoin, 1976, p. 1004). These ions gradually recombine back into conventional molecules.

In a vacuum the discharge mechanism is different. In non-insulated negative metal electrodes, the electrons may be extracted from the conducting electrode by the strong electric field. The critical surface electric intensity, E_0, is about $100 \cdot 10^6\,\text{V/m}$ at the non-insulated negative electrode. This intensity is about 1000 times stronger at the positive electrode because the ions are very difficult to tear away from the solid material. Conducting sharp edges increase the electric intensity. That is why it is better to charge the planet or asteroid surface with positive charges. A very sharp spike allows the electrical energy of the charged ball to be regained.

3.3.2 *Size of corona (ionized sphere)*

The size of the corona may be found as a spherical area where the electrical intensity is more than safe air intensity:

$$E \geq E_c, \quad \text{or} \quad \frac{kq}{R_c^2} \geq E_c, \quad \text{where} \quad q = \frac{E_a a^2}{k},$$
$$\text{finally} \quad R_c \leq \sqrt{\frac{kE_a a^2}{kE_c}}, \quad \bar{R}_c = \tfrac{R_c}{a} \leq \sqrt{\frac{E_a}{E_c}}, \tag{20}$$

where E – electrical intensity of the charge; E_c – electrical intensity at the beginning of the corona, $E_c \approx 3 \cdot 10^6$; E_a – electrical intensity at the ball surface; a – ball radius; R_c – radius of corona; $k = 9 \cdot 10^9$.

To find the admissible (safe electrical intensity), E_a, for a negatively charged ball in an insulated cover from the point of rupture (spark) into a neutral environment the following relationships may be used. One uses the notation: U – ball voltage; U_i – safe voltage of ball insulator; E_i – safe electrical intensity of ball insulator; δ – thickness of the ball cover; ε – dielectric constant. Then, from the inequality $U < U_1$, where $U = (kq/\varepsilon)\,[1/a - 1/(a+\delta)]$ and $U_i = \varepsilon E_i \delta$, one derives the condition $(kq/\varepsilon)[1/a - 1/(a+\delta)] < \varepsilon E_i\delta$. Note that $q = a^2E_a/k$. Finally, one obtains $\bar{\delta} = \delta/a > E_a/(\varepsilon E_i) - 1$.

$$\text{(21)} \qquad \text{If } \bar{\delta} = 0, \quad \text{then} \quad E_a < \varepsilon E_i.$$

The last inequalities in Eqs. (20) and (21) are the required results.

3.3.2.1 Numerical example The ball is covered by Mylar with $E_i = 160\,\text{MV/m}$, $\varepsilon = 3$ (see Table 1). Then $E_a = 3 \cdot 160 = 480$ MV/m, and the relative radius of the ionized sphere (Eq. (1) is $(480/3)^{0.5} = 12.6$. If $a = 0.05$ m, the real radius is $R_c = 12.6 \cdot 0.05 = 0.63$ m.

3.3.3 Size of corona (cylindrical cable or belt)

The radius of the corona (ionized cylinder) can be found using the same method. From inequality $E \geq E_c$ where $E = 2k\tau/R_c$, $\tau = aE_a/2k$, we find $R_c \leq aE_a/E_c$ or

$$\text{(22)} \qquad \bar{R}_c = R_c/a \leq E_a/E_c,$$

where τ is the linear charge.

To find, by using the same method (Eq. 21), the admissible (safe) quantity, E_a, for a negatively charged cable (belt, tube) in an insulated cover from the point of rupture into a neutral environment, the following relationship may be used:

$$\text{(23)} \qquad \text{From} \quad U < U_i, \quad U = 2k\tau \ln\left(\frac{a+\delta}{a}\right), \quad U_i = \varepsilon E_i \delta,$$

$$\text{(24)} \qquad \text{we get} \quad 2k\tau \ln\left(\frac{a+\delta}{a}\right) < \varepsilon E_i \delta, \quad \text{or} \quad \tau < \frac{\varepsilon E_i a\delta/a}{2k\ln(1+\delta/a)},$$

Table 1. Properties of some insulators. *For room temperature, 500–700 MV/m, ** 400–500 MV/m

Insulator	Resistivity (Ohm-m)	Dielectric strength E_i (MV/m)	Dielectric constant, ε	Tensile strength (kg/mm^2, $\sigma \times 10^7$N/m^2)
Lexan	10^{17}–10^{19}	320–640	3	5.5
Kapton H	10^{19}–10^{20}	120–320	3	3.45
Kel-F	10^{17}–10^{19}	80–240	2–3	3.45
Mylar	10^{15}–10^{16}	160–640	3	13.8
Parylene	10^{17}–10^{20}	240–400	2–3	6.9
Polyethylene	10^{18}–$5 \cdot 10^{18}$	40–680*	2	2.8–4.1
Poly(tetra-fluoraethylene)	10^{15}–$5 \cdot 10^{19}$	40–280**	2	2.8–3.5
Air (1 atm, 1 mm gap)	-	4	1	0
Vacuum ($1.3 \cdot 10^{-3}$ Pa, 1 mm gap)	-	80–120	1	0

(Reference: Encyclopedia of Science & Technology, 1997, McGraw-Hill, New York, Vol. 6, pp. 229, 231)

In the limit $\delta/a \to 0$ we have:

$$(25) \qquad \bar{\tau} = \frac{\tau}{a} < \frac{\varepsilon E_i}{2k}.$$

From $E_a = k(2\tau)/a$ we get:

$$(26) \qquad \frac{\tau}{a} = \frac{E_a}{2k}, \quad \frac{E_a}{2k} < \frac{\varepsilon E_i}{2k}.$$

and finally:

$$(27) \qquad E_a < \varepsilon E_i,$$

where the last inequality is the required result.

3.3.4 Discharging by corona

The results below show how the large number ($\sim 10^9$ part/m^3) of charged particles influences the main charge. If 1 m^3 of air contains d positive-charged (electrons or ions) particles and the charge density is constant, the charge, q, of a sphere with radius r is

$$(28) \qquad q = \frac{4}{3}\pi r^3 ed,$$

where $e = 1.6 \cdot 10^{-19}$ C – charge of one particle (electron or single charged ion); d – particle density. On the other side, the main charge, q_o, will be partially blockaded until the intensity at radius r becomes E_c. Consequently, the relation is

$$(29) \qquad q_0 - q = \frac{E_c}{k} r^2, \quad \text{or} \quad \frac{4}{3}\pi edr^3 + \frac{E_c}{k} r^2 - q_0 = 0$$

where $k = 9 \cdot 10^9$. The Eq. (29) has only one real root. Results of this computation are presented in Figs. 16 and 17, which show that a large particle density decreases the main charge. However, experiments are necessary to confirm these theoretical findings.

3.3.5 Data about the ball material

The properties of electrical insulation vary depending on the impurities in the material, temperature, thickness, etc. and they are different for the same material with a different dielectric. For example, the resistivity of fused quartz is 10^{15} Ohm · cm for $T = 20\,°$C, the resistivity of the quartz fused (from crystal) reaches about 10^{24} Ohm · cm for $T \approx 20°$C (see p. 231 and p. 329, Fig. 20.2 in Kikoin, 1976). Properties of some materials are presented in Table 1.

For small balls, the tensile stress is important for reducing the weight (one reminds that positive charges tear the ball). The author suggests that an artificial fiber with a maximum tensile stress at 500–620 kg/mm^2 (fiber) or up to 2000 kg/mm^2 (whiskers) is more suitable. These fibers can also be used to strengthen balls insulated by a dielectric (used, for example, as an additional cover, Fig. 14d).

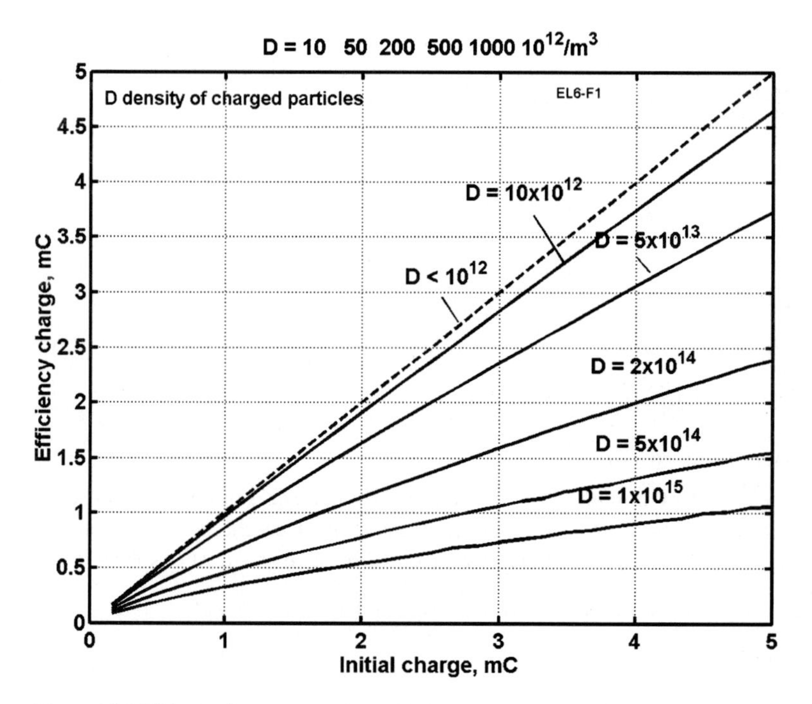

Figure 16. Efficiency charge versus the main charge and density of charged particles in the environment (ionized zone)

3.3.6 *The half-lifetime of the charge*

3.3.6.1 Spherical ball Let us consider a complex condition; where the unlike charges are separated only by an insulator (charged spherical condenser). A number of notations are adopted now: R – insulator resistance; ρ – specific resistance of insulator; δ – thickness of insulator; i – electric current intensity; U – electric voltage; E – electrical field intensity; q – electric charge; a – internal radius of the charged ball; C – capacity of the ball; t - time; t_h – electric charge half-lifetime; and $k = 9 \cdot 10^9$.

From the known relation of an electric circuit, $Ri - U = 0$ (where $i = dq/dt$) and the common formulas of the electric condenser:

(30) $$R = \rho\frac{\delta}{4\pi a^2}, \quad U = -\frac{q}{C}, \quad C = \frac{a}{k},$$

(31) we get equation $$R\frac{dq}{dt} + \frac{q}{C} = 0.$$

(32) Its solution is $$q = q_0 \exp\left(-\frac{4\pi ak}{\rho\delta}t\right),$$

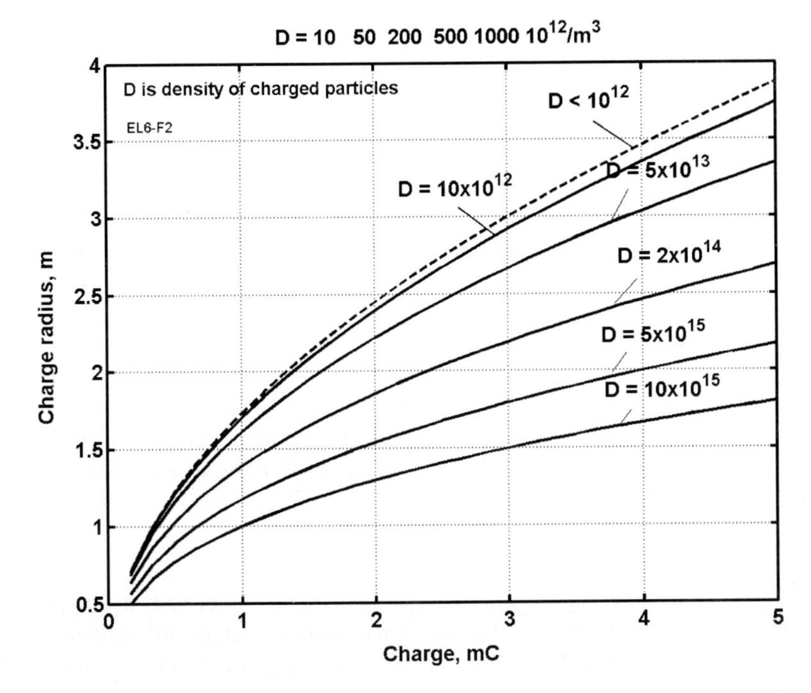

Figure 17. Critical radius of main charge versus the main charge and density of charged particles in the environment (ionized zone)

To evaluate the half lifetime, first one uses in Eq. (32) the condition $q/q_0 = 1/2$, which yields:

$$(33) \qquad -\frac{4\pi a k}{\rho\delta} t_h = \ln\frac{1}{2} = -0.693 \approx -0.7,$$

From Eq. (33) the half-life time of the charge (spherical ball) is obtained as:

$$(34) \qquad t_h = 0.693 \frac{\rho\delta}{4\pi k a},$$

3.3.6.1.1 Example

Let us take typical data: $\rho = 10^{19}\,\Omega$-m, $\delta/a = 0.2$. Then $t_h = 1.24 \cdot 10^6$ seconds $= 144$ days.

3.3.6.2 Cylindrical tube The computation is similar to that for tubes (1 m charged cylindrical condenser). From known relationships of electricity:

$$(35) \qquad q = q_0 \exp\left(-\frac{1}{RC}t\right), \quad C = \frac{1}{k\ln(1+\delta/a)},$$

and the from the relations $R = \rho\delta/(2\pi a)$ and $-0.693 = -t_h/(RC)$ one derives the estimated half lifetime:

$$(36) \qquad t_h = \frac{0.693\rho\delta}{2\pi k a \ln(1+\delta/a)},$$

In the special case when $\delta/a \to 0$ we obtain:

$$(37) \qquad t_h \approx 0.7\frac{\rho}{2\pi k}.$$

3.3.7 *Rupture (breakthrough) of insulator*

The breakthrough can only occur when the charge contacts an unlike charge or conducting material. The voltage between the charges must be less than $U = \delta U_r$, where U_r is the breakthrough voltage for a given insulator and δ is the thickness of the insulator. For a good insulator this is up to $U_r \approx 700$ million V/m, and for thin mica up to $U_r = 1000$ million V/m.

3.3.8 *Levitation between flat net and ground surface*

Levitation between flat net and ground surface is the simplest from both the point of view of utilization and computation. The top of the column contains the insulated metallic net with a high voltage (Fig. 13d) (it may be a direct current electricity line). This induces the opposite charge in the Earth and powers the static electric field. The man (car) has charged balls or a balloon with like charges to the Earth's charge. These balls repel from the Earth's surface (charges) and support the man (car) in the air.

We adopt the following notations: E_o – electrical field intensity between the net and the Earth's surface; E_a – electrical field intensity at the ball's (balloon) surface from the internal ball (balloon) charge; a – internal radius of ball (balloon); M – mass of the flight vehicle (man, car); $g = 9.81$ m/s^2 – Earth's gravity; U – voltage between the net and Earth; h – altitude of the net.

The lifting force, L, and the radius, a, of a small lifting balloon with charge q can be computed by using the relations:

$$(38)-(40) \quad L = qE_0, \quad E_a = k\frac{q}{a^2}, \quad U = E_0 h,$$

For example, the required radius is given by:

$$a = \sqrt{k\frac{q}{E_a}} = \sqrt{\frac{kMg}{E_a E_0}}.$$

3.3.8.1 *Example*

Assume $E_o = E_a = 2.9 \cdot 10^6$ V/m, the man's mass is $M = 100$ kg, and the car's mass is 1000 kg. The radius of a single support balloon will be $a \approx 1.1$ m for the man and $a \approx 3.3$ m for the car, respectively. Note that this voltage is lower than the discharge voltage for a non-insulated conductor and the ionized zone is absent. We may change the single balloon to some small highly charged balls or a belt with an ionized zone.

The flying vehicles may be protected from contact with the top net by a dielectric (insulator) safety net located below the top net (Fig. 13d, mark 7).

3.3.9 *Electrostatic levitation of a train*

Two identically charged closed plates of area S have a repelling force L:

$$L = 2\pi k \sigma_c^2 S, \tag{41}$$

where σ_c is surface charge density. For example, two 1 m^2 plates with identical charge $\sigma_c = 2 \cdot 10^{-4}$ C/m^2 will have a specific lift force of $L = 2260$ N/m^2 = 226 kgf/m^2. Conventional electrets have $\sigma_c = 10^{-4} - 1.4 \cdot 10^{-3}$ C/m^2 charge and can be used for a non-contact train (Fig. 15a).

3.3.10 *Top tube highway*

The following notation is adopted in this section: τ – the linear charge of 1 meter of tube; a – radius of tube cross section; E_a – electric field intensity at tube surface; E_0 – electric field intensity at the Earth's surface at a point under the tube (for other points $E = E_0 \cos^3 \alpha$ where α is the angle between a vertical line from the tube center and a line to a given point; electric lines are perpendicular to the Earth's surface, there is no lateral acceleration); h – altitude; $k = 9 \cdot 10^9$ – coefficient; C_1 – capacity of 1 meter of tube; U – voltage; W – electrical energy of 1 meter of tube; F_h – electric force of 1 meter of tube between the tube and the Earth's surface; F_a– radial tensile force of 1 meter of tube.

The parameters of a tube highway may be calculated as follows. From the relations $\tau = aE_a/(2k)$ and $E_0 = 4k\tau/h$ one derives

$$\frac{E_0}{E_a} = \frac{a}{h} \tag{42}$$

The next relations are also used:

$$C_1 \approx \frac{1}{2k \ln(2h/a)}, \quad U = \frac{\tau}{C_1}, \quad W = \frac{\tau^2}{2C_1}. \tag{43}$$

Equations (42) and (43) allow obtaining the following formulas for forces (under the assumption $\tau = \text{const}$)

$$F_h = \frac{\partial W}{\partial h} = \frac{2k\tau^2}{h}, \quad F_a = \frac{\partial W}{\partial a} = -\frac{2k\tau^2}{a}. \tag{44}$$

Let us evaluate the thickness and mass of the top tube (with a thin cover). Additional notation is first adopted: σ – safety tensile stress of tube cover; δ – thickness of the tube cover; M_1 – mass of 1 m of tube cover; γ – density of tube cover. From $F_a = -2\sigma\delta$ and Eq. (44) one obtains:

$$(45) \qquad \delta = \frac{k\tau^2}{a\sigma}$$

and the required mass is given by:

$$(46) \qquad M_1 = 2\pi\gamma a\delta.$$

The lift force of the tube as an *air balloon filled by helium* can be computed by using the relation

$$(47) \qquad F_L = (\rho - \rho_g)\pi a^2 \bar{\rho}(h) g,$$

where $\rho = 1.225$ kg/m^3 – air density; ρ_g – filling gas density (for helium $\rho_g = 0.1785$ kg/m^3); a – radius of tube; $\bar{\rho}(h)$ – relative air density (it depends on altitude; for $h = 0$ km, $\bar{\rho}(h) = 1$, for $h = 1$ km, $\bar{\rho}(h) = 0.908$).

Note that E_c decreases proportional to the atmospheric density. Unfortunately, in many case the attractive electric force F_h is larger than the air lift force F_L. See the example in Macro-project 2.

3.3.11 Spherical main ball on mast and air balloon

The parameters of charges of the main ball and spherical balloon may be easily calculated. The notation in this section is: q – the charge of air balloon (sphere); a – radius of air balloon; E_a – electric field intensity at the balloon's surface; E_o – electric field intensity at the Earth's surface at a point under the balloon (for other points $E = E_0 \cos^3 \alpha$ where α is the angle between a vertical line from the balloon center and a line to a given point); h – altitude; $k = 9 \times 10^9$ – coefficient; C – electrical capacity of balloon; U – voltage; W – electrical energy of the balloon; F_h – electrical force between the balloon and the Earth's surface. From $E_a = kq/a^2$ and $E_0 = 2kq/h^2$ one finds:

$$(48) \qquad \frac{E_0}{E_a} = 2\left(\frac{a}{h}\right)^2$$

The balloon electrical capacity is given by:

$$(49) \qquad C = \left[k\left(\frac{1}{a} - \frac{1}{2h - a}\right)\right]^{-1} \approx \frac{a}{2k}$$

The voltage and the electric energy of the balloon may be calculated by using the following relations:

$$(50) \qquad U \approx \frac{kq}{a}, \quad W = \frac{q^2}{2C}$$

Finally, the electrical force between the balloon and the Earth's surface is obtained from:

$$F_h = \frac{\partial W}{\partial h} = -\frac{kq^2}{4h^2}, \quad (51)$$

The thickness and mass of the top atmospheric air balloon as a spherical capacitor 1 m radius (with a thin cover) may be found from known equations of electrostatic theory (for force, electrical energy and electric capacity, respectively):

$$F_a = \frac{\partial W}{\partial a}, \quad W = \frac{q^2}{2C}, \quad c = \frac{a}{k}. \quad (52)$$

For $q = a^2 E_a / k = \text{const}$ we have the electrical force

$$F_a = -\frac{kq^2}{2a^2} = -\frac{(aE_a)^2}{2k}. \quad (53)$$

The balloon internal pressure under like charges is defined by

$$p = -\frac{F_a}{4\pi a^2} = \frac{E_a^2}{8\pi k} \quad (54)-(55)$$

Now, the additional notation is: σ – safe tensile stress of the balloon cover; δ – thickness of the balloon cover; M_b – mass of the balloon cover; γ – density of the balloon cover. By using the mechanical equilibrium condition (i.e. $\pi a^2 p = 2\pi a\delta\sigma$), we find the thickness of the balloon cover and balloon mass

$$\delta = \frac{ap}{2\sigma} = \frac{aE_a^2}{8\pi k}, \quad M_b = 4\pi a^2 \gamma\delta = \frac{a^3 E_a^2 \delta}{4k\sigma}. \quad (56)$$

The lift force of the air balloon filled by helium may be computed by using the Eq. (47), i.e.:

$$F_L = \frac{4}{3}(\rho - \rho_g)\pi a^3 \bar{\rho}(h) g, \quad (57)$$

Here the notations of section 3.3.10 apply. As stated in section 3.3.10, in many cases the attracted electric force F_h is more than the air lift force F_L. See the computation in Macro-project 3.

3.3.12 Small spherical lifting balls

Assume the electrical intensity of the main top charge is significantly larger than the lifting charges. The parameters of large spherical balls with thin covers may be computed by using the equations above. The parameters of small balls with thick covers may be computed by using the relations given in this section. First, the notation used here is: L– total lift force; n – number of balls; E_a – electric field intensity at the ball surface from electrical charge of the ball q; a – internal radius of the ball. The lift force is:

$$(58) \qquad L = nqE_0.$$

The following obvious relations also apply:

$$(59) \qquad q = CU, \quad C = \frac{r}{k}, \quad U = aE_a, \quad q = \frac{a^2 E_a}{k}.$$

Equations (58) and (59) allow to showing the lift force as a function of various parameters:

$$(60) \qquad L = n\frac{a^2 E_a E_0}{k}.$$

To derive the thickness of ball (thick) cover and the mass of the ball additional notation is needed: $R = a + \delta$ – external radius of ball; σ – safe tensile stress of the ball material; δ – thickness of the ball cover; M_b – mass of the ball cover; γ – density of the ball cover.

First, from the mechanical equilibrium condition (i.e. $p_b = E_a^2/(8\pi k)$ one finds:

$$(61) \qquad \sigma = \frac{p_b}{(R/a)^2 - 1} = \frac{E_a^2}{8\pi k\left[(R/a)^2 - 1\right]}$$

For given R, Eq. (61) allows computation of a. Finally, the cover thickness is obtained from $\delta = R - a$. The mass of the ball may be computed as a function of a as follows:

$$(62) \qquad M_b = \frac{4}{3}\pi\gamma a^3\left(\frac{R^3}{a^3} - 1\right).$$

Results are shown in Figs. 18 to 21. Notice that the lifting balls have a large ratio of lift force/ball mass, about 10,000–20,000.

3.3.13 *Long cylindrical lifting belt*

Let use denote: l – length of the belt: δ – thickness of the belt; γ – density of belt cover; a – internal radius of the belt cross-section area; σ – safe tensile stress of the belt cover; M – mass of the belt; q – charge of the belt; τ – electric charge per unit length of belt; E_a– electrical intensity of the belt surface; F_a – electrostatic

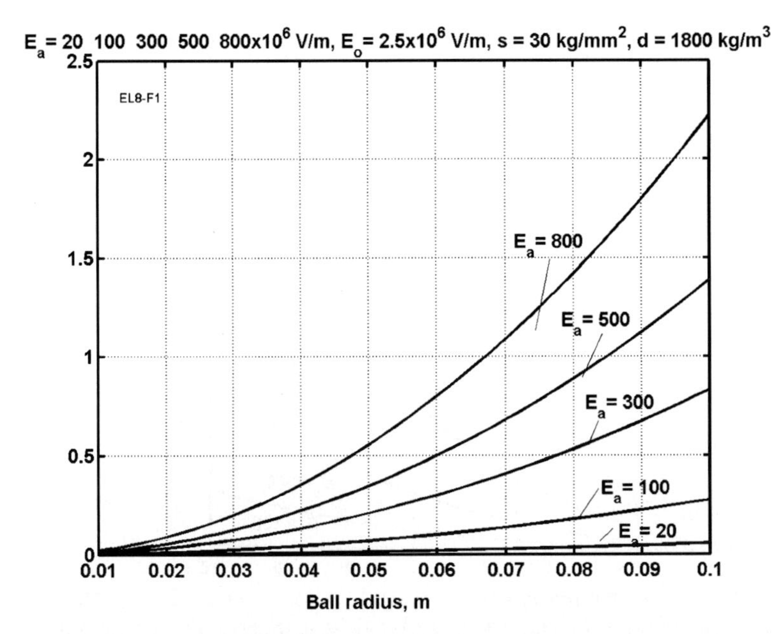

Figure 18. Electrostatic lift force (kN) of small lifting ball versus radius of ball for the electrical intensity of the ball's surface E_a= (20–800)$\cdot 10^6$ V/m, general electrical intensity $E_o = 2,5 \cdot 10^6$ V/m, safe tensile stress of the ball cover 30 kg/mm^2, specific density of the ball cover 1800 kg/m^3

force in the tube (see Eq. 44). The following assumption is adopted in this section: $a << l$. The lift force may be computed by:

$$(63) \qquad L = E_0 \tau l,$$

where τ still has to be computed. The electrostatic force F_a enters the following two relations:

$$(64) \qquad F_a = \frac{2k\tau^2}{a} \quad , \quad 2\sigma\delta = F_a.$$

The second relation in Eq. (64) is a consequence of mechanical equilibrium. Equations (64) allow evaluation of τ as follows:

$$(65) \qquad \tau = \sqrt{\frac{a\delta\sigma}{k}}.$$

Now, the maximum charge, the mass per unit length of belt, the total belt mass and the electrical intensity of belt surface are given, respectively, by:

$$(66) \qquad q = \tau l, \quad M_1 = 2\pi\gamma a\delta, \quad M = M_1 l, \quad E_a = k\frac{2\tau}{a}.$$

Results are presented in Figs. 22 and 23. See also the example in Macro-project 3.

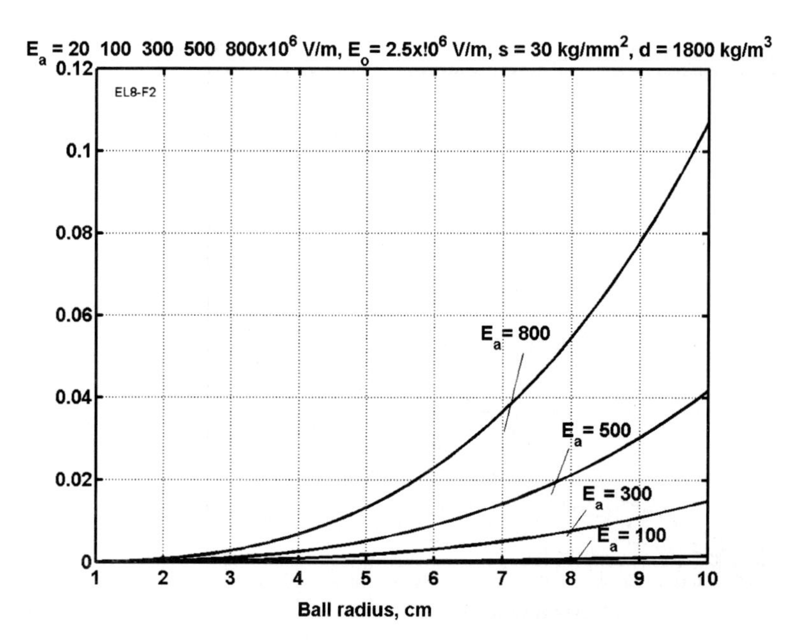

Figure 19. Mass (kg) of small lifting ball versus radius of ball for the electrical intensity of the ball's surface $E_a = (100\text{–}800) \cdot 10^6$ V/m, general electrical intensity $E_o = 2.5 \cdot 10^6$ V/m, safe tensile stress of the ball cover 30 kg/mm^2, specific density of the ball cover 1800 kg/m^3

3.4 Aerodynamics of the Levitated Vehicles

The drag, D, (required thrust, T) and the required power, W, of the levitated person, car or vehicle, may be computed by using, respectively:

$$(67, 68) \qquad T = D = C_D \frac{\rho V^2}{2} S, \qquad W = \frac{VD}{\eta}.$$

where C_D – aerodynamic drag coefficient (for a sitting person $C_D \approx 0.5$; for a lying man $C_D \approx 0.3$; for a car $C_D \approx 0.25$; for a sphere $C_D \approx 0.1 - 0.2$ (depending on the size and speed); for a dirigible $C_D \approx 0.06 - 0.1$) $\rho = 1.225$ kg/m^3 – standard air density; V – speed; S – vehicle cross section area; W – required power; η – propeller efficiency, $\eta = 0.7 - 0.8$.

For example, a flying person ($S = 0.3$ m^2) has $D = 5.5$ N for speed $V = 10$ m/s(36 km/hour). One only needs a small motor, $W = 0.073$ kW.

3.5 Control and Stability

Control is accomplished using the direction (and magnitude) of motor thrust (and variable torque) and the charging and discharging of lifting charges. The levitated vehicle will be stable in a vertical position if its center of gravity is lower than the

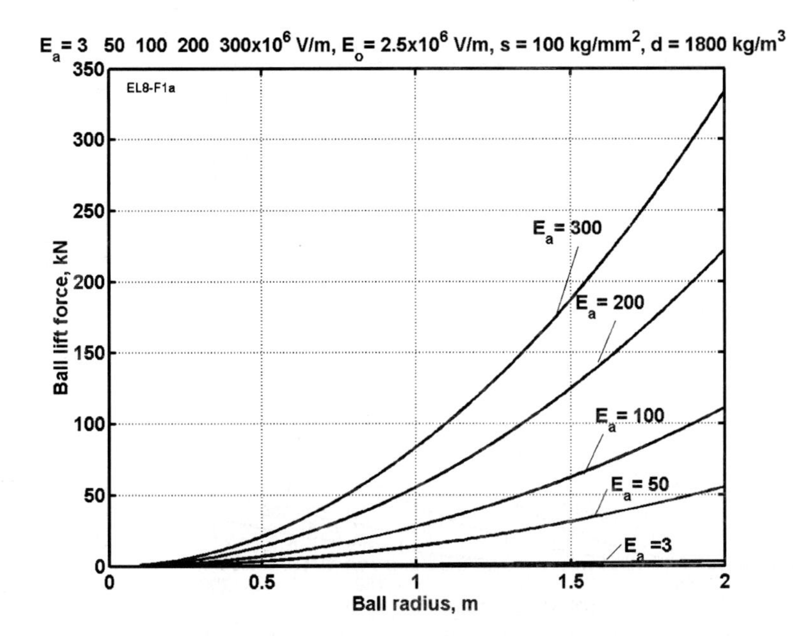

Figure 20. Electrostatic lift force (kN) of the lifting ball versus ball radius for the electrical intensity of the ball's surface E_a= (3–300)·10^6 V/m, general electrical intensity $E_o = 2.5 \cdot 10^6$ V/m, safe tensile stress of ball cover 100 kg/mm^2, specific density of ball cover 1800 kg/m^3

center of levitation (lift) force. The dipole moment of the particular vehicle's design may give additional stability. Note that electric lines are vertical at the Earth's surface (Figs. 13), which means that the lift force is vertical.

3.6 Flight in Thunderstorms

Thunderstorms produce an electric field of about 300,000 to 1,000,000 V/m. This field can be used for levitation.

3.7 Charging

In the author's opinion, the easiest method of charging and maintaining the charge is by using a Van de Graaff electrostatic generator. Other high voltage generation devices can be used.

3.8 Safety

The interaction between static electric fields an human body is not fully understood. People in an electric field of about 300,000 to 1,000,000 V/m during a thunderstorm or under high voltage electrical lines feel normal. The inside space of conventional

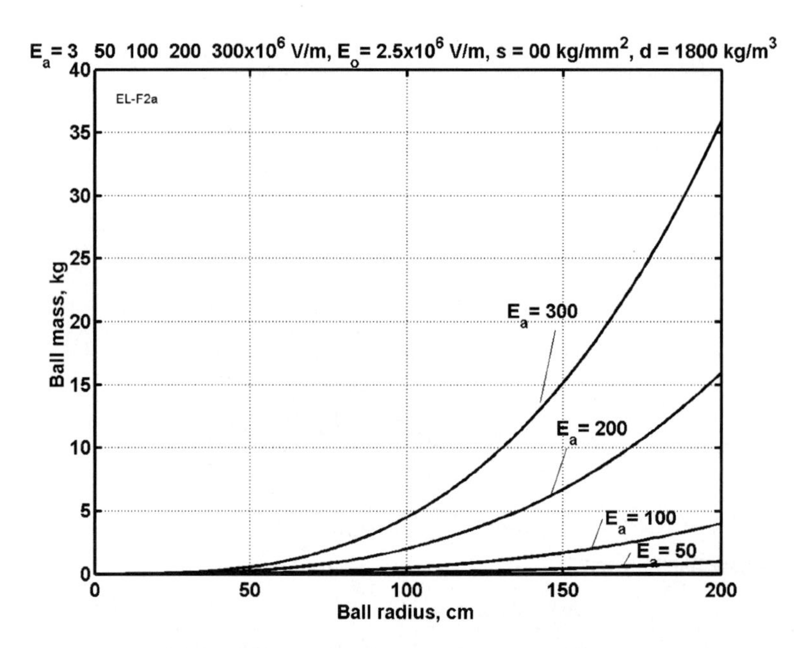

Figure 21. Mass (kg) of the lifting ball versus ball radius for the electrical intensity of the ball's surface E_a= (3–300)·10^6 V/m, general electrical intensity $E_o = 2.5 \cdot 10^6$ V/m, safe tensile stress of the ball cover 100 kg/mm^2, specific density of the ball cover 1800 kg/m^3

metallic cars does not have an electric field. People may wear clothes armored by conductive filaments as a defense against electric fields.

3.9 Charged Ball as an Accumulator of Energy

The energy required to charge a ball (or, in other words, the energy accumulated in the ball) may be easily evaluated. First, the following notation is adopted: M – mass of the ball; σ – safe tensile stress of the ball cover; γ – specific density of the ball cover; $R = a + \delta$ – external radius of the ball. The following relations apply for the energy accumulated W and the ball electrical capacity, respectively:

$$W = \frac{1}{2}\frac{q^2}{C}, \quad C = \frac{a}{R}. \tag{69}$$

The electric field intensity is given by:

$$E_a = k\frac{q}{a^2} \tag{70}$$

By using Eqs. (69) and (70) one finally finds the accumulated energy:

$$W = \frac{1}{2}\frac{a^3 E_a^2}{k}. \tag{71}$$

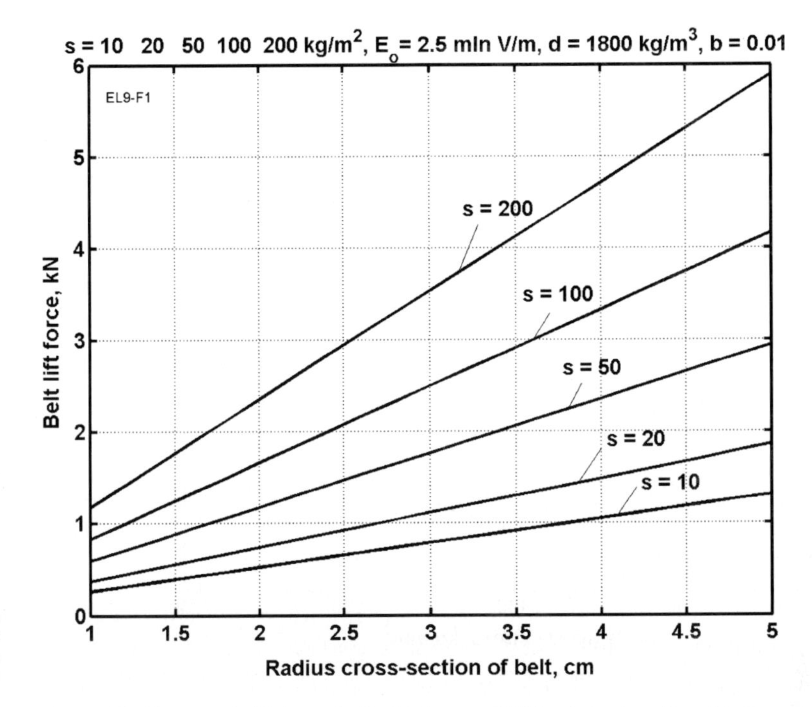

Figure 22. Electrostatic lift force (kN) of a 1 m small lifting belt via radius of ball cross-section area for general electrical intensity $E_o = 2.5 \cdot 10^6$ V/m, safe tensile stress of the ball cover 10–200 kg/mm^2, $\sigma = (10\text{–}200) \cdot 10^7$ N/m^2, density of ball cover $\gamma = 1800$ kg/m^3

The ball mass at safe stress levels with repelling charges can be calculated using Eq. (60) re-written under the form:

$$\left(\frac{R}{a}\right) = \sqrt{\frac{E_a^2}{8\pi k\sigma} + 1}. \tag{72}$$

Finally, the ball mass M and the ratio W/M are given, respectively, by:

$$M = \frac{4\pi\gamma a^3}{3}\left[\left(\frac{R}{a}\right)^3 - 1\right], \quad \frac{W}{M} = \frac{3E_a^2}{8\pi\gamma k\left[(R/a)^3 - 1\right]}. \tag{73}$$

The accumulated relative energy for $\sigma = 200$ kg/mm^2 may be close to conventional power and last much longer than electrical energy in a typical condenser. This electrical energy may be reclaimed (by using a sharp spike) or used for launching or accelerating space vehicles if one uses two like charges (balls) and allow them to repel each other. This method of transforming electrical energy into thrust may be more useful than the thrust from a conventional electric space engine because one can create a significantly large thrust by utilizing asteroids.

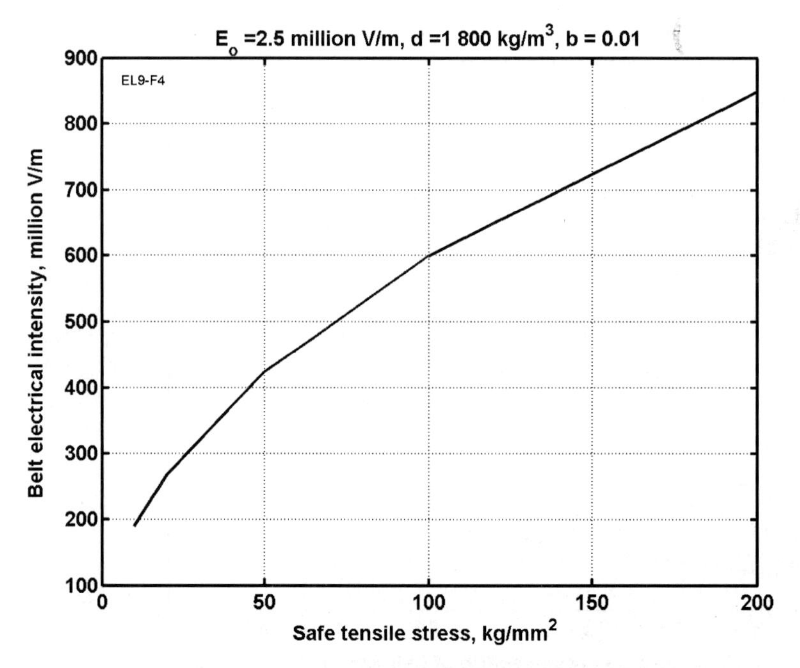

Figure 23. Belt electrical intensity via safe tensile stress for general electrical intensity $E_o = 2.5 \cdot 10^6$ V/m, safe tensile stress of the ball cover $10 - 200$ kg/mm^2, $\sigma = (10 - 200) \cdot 10^7$ N/m^2; γ – specific density of ball cover 1800 kg/m^3, relative thickness of belt cover, $\delta = 0.01a$

3.10 Macro-projects

Let us estimate the main design parameters for some possible macro-engineering applications. This will give perspective to our findings. Note that the parameters are not optimized. Indeed, the purpose here is just to show the method may be implemented by using current technology.

3.10.1 Levitation highway

The height of the top net is 20 m (Fig. 13d). The electrical field intensity is $E_o = 2.5 \cdot 10^6$ V $< E_c = (3-4) \cdot 10^6$V. The voltage between the top net and the ground is $U = 50 \cdot 10^6$V. The width of each side of the road is 20 m. We first find the size of the lifting ball for a person (100 kg), car (1000 kg), or track (10,000 kg). Below R_c is the radius of the ionized zone.

3.10.1.1 Flying person One adopts the following initial data: mass (of a single man) $M = 100$ kg, $\varepsilon = 3$, $E_i = 200 \cdot 10^6$ V/m, g ≈ 10 m/s^2). We find the ball radius and corona radius (for $E_a \leq \varepsilon E_i = 3 \cdot 200 \cdot 10^6$ V/m):

$$(74)-(75) \qquad a = \sqrt{\frac{kMg}{E_0 E_a}} \approx 0.08 \text{ m}, \quad R_c = \sqrt{\frac{E_a}{E_c}} \approx 1 \text{ m}.$$

Notice that the radius of a single ball supporting the man is only 8 cm. In case of using two balls $a = 5 - 6$ cm and $R_c = 0.75$ m. Even smaller balls may be used, of course. If the man uses a 1 m cylindrical belt, the radius of the belt cross-section area is 1.1 cm, $\sigma = 100$ kg/mm^2, $E_a = 600 \cdot 10^6$ V/m (Figs. 22 and 23). The belt may be more comfortable for some people.

3.10.1.2 Flying car With similar calculations one can find that a car of mass $M = 1000$ kg will be levitated using a single charged ball with $a = 23$ cm and $R_c = 3.2$ m (or two balls with $a = 16$ cm and $R_c = 2.3$ m).

3.10.1.3 Flying truck The truck mass is $M = 10,000$ kg. It will be levitated by using a single charged ball with $a = 70$ cm and $R_c = 10$ m (or two balls with $a = 0.5$ m and $R_c = 7$ m).

3.10.2 *Levitating tube highway*

Assume the levitation highway has the design of Fig. 13c,d where the top net is changed to a tube. Take the data $E_0 = 2.5 \cdot 10^6$V $< E_c = (3\text{–}4) \cdot 10^6$ V, $E_a = 2 \cdot 10^8$ V/m, $h = 20$ m. This means the electrical field intensity, E_o, at ground level is the same as in the previous case. The required radius, a, of the top tube is

$$(76) \qquad \frac{a}{h} = \frac{E_0}{2E_a} = 0.00625, \quad \text{and} \quad a = 0.00624h = 0.125 \text{ m}.$$

The diameter of the top tube is 0.25 m, the top ionized zone has a radius of $R_s = aE_a/E_0 = 10$ m.

3.10.3 *Charged ball located on a high mast or tower*

Assume there is a mast (tower) 500 m high with a ball of radius $a = 32$ m at its top charged up to $E_a = 3 \cdot 10^8$ V/m. The electric charge and the electric field intensity are given, respectively, by

$$(77) \qquad q = \frac{a^2 E_a}{k} = 34\ C, \quad E_0 = k\frac{2q}{h^2} = 2.45 \times 10^6 \text{ V / m}.$$

This electrical field intensity at ground level is enough strong that within a radius of approximately 1 km, people, cars and other loads can levitate.

3.10.4 *Levitation in low cumulonimbus and thunderstorm clouds*

In these clouds the electrical intensity at ground level is about $E_o = 3 \cdot 10^5 - 10^6$ V/m. A person can use an appropriate number of balls and levitate.

3.10.5 *Artificial gravity on space ships or asteroids*

Assume the space ship is a sphere with an inner radius of $a = 10$ m and external radius of 13 m. We can create an electrical field intensity $E_o = 2.5 \cdot 10^6$ V/m without an ionized zone. The electrical charge is $q = a^2 E_o/k = 2.8 \cdot 10^{-2}$ C. For a man weighing 100 kg (g = 10 m/s^2, force $F = 1000$ N), it is sufficient to have a charge of $q = F/E_o = 4 \cdot 10^{-4}$ C and a small ball with $a = 0.1$ m and $E_a = qk/a^2 = 3.6 \cdot 10^8$ V/m. In outer space at the ship's surface, the artificial gravity will be $(10/13)^2 = 0.6 = 60\%$ of g.

3.10.6 Charged ball as an accumulator of energy and rocket engine

Computations show that the ratio of accumulated energy per ball mass W/M, calculated from safe tensile stress, does not depend on E_a. A ball cover with a tensile stress of $\sigma = 200$ kg/mm^2 reaches 2.2 MJ/kg. This is close to the energy of conventional powder (3 MJ/kg). If whiskers or nanotubes are used, the ratio stored energy per mass will be close to than of liquid rocket fuel.

Two like charged balls repel one another and can give significant acceleration for a space vehicle, VTOL aircraft, or weapon.

3.11 Discussion

Electrostatic levitation could create a revolution in transportation, construction, entertainment, aviation, space flights and the energy industry.

The proposed method needs development and testing. The experimental procedure is not expensive. This experiment may be carried out in any high voltage electric laboratory. The proposed levitation theory is based on the well-known electrostatic theory. Problems may arise with discharging, blockage of the charge by the ionized zone, breakdown, and half-lifetime of the discharge. However, a careful choice of suitable electrical materials and electric field intensity may solve all these aspects. Most of these problems do not occur in a vacuum (for example, in outer space).

Another problem is the effect of the strong electrostatic field on a living organism. Only experiments using test animals may give an answer to this question. Any case, there are protective methods (such as conducting clothes or vehicles made of metal or covered by conducting paint) which might offer a defense against the electric field.

ACKNOWLEDGMENTS

The content of this chapter was presented as papers by A. Bolonkin "Cable anti-gravitator", AIAA-2005-4505, and "Problems of Electrostatic Levitation and Artificial Gravity", AIAA-2005-4465, at 41 Joint Propulsion Conferences, 10-13 July, 2005, Tucson, Arizona, USA.

REFERENCES

Anderson JD (1989) Hypersonic and high temperature gas dynamics. McGraw-Hill Book Co, New York

Bolonkin AA (1965) Theory of flight vehicles with control radial force. Collection Researches of Flight Dynamics, 'Mashinostroenie' Publishers, Moscow, 1965, pp 79–118 (in Russian)

Bolonkin AA (1982) Installation for creating open electrostatic field. Patent applications #3467270/21, 116676, 9 July 1982. USSR Patent office

Bolonkin AA (1983) (Electrostatic) Method for tensing of films. Patent application #3646689/10, 138085, 28 September 1983. USSR Patent office
Bolonkin AA (1990) Aviation, Motor, and Space Design, Collection. Emerging Technology in the Soviet Union, Delphic Ass, 1990, USA, pp 32–80
Bolonkin AA (2002a) Transport system for delivery of tourists at altitude 140 km. IAC-02-IAA.1.3.03. 53rd International Astronautical Congress. The World Space Congress – 2002, Houston, Texas, USA, 10–19 October 2002
Bolonkin AA (2003a) Centrifugal keeper for space stations and satellites. J Br Interplanet Soc 56(9/10):314–327
Bolonkin AA (2003b) Air cable transport. J Aircraft 40(2):265–269
Bolonkin AA (2003c) Air cable transport and bridges, TN 7567. International Air & Space Symposium – the Next 100 Years, Dayton, Ohio, USA, 14–17 July 2003
Bolonkin AA (2004a) Kinetic space towers and launchers. J Br Interplanet Soc 57:33–39
Bolonkin AA (2004b) High efficiency transfer of mechanical energy, AIAA-2004–5660. International Energy Conversion Engineering Conference, Rhode Island, 16–19 August 2004
Bolonkin AA (2005a) Electrostatic utilization of asteroids for space flight, AiAA-2005–4032.41 Joint Propulsion Conferences, Tucson, Arizona, USA, 10–13 July 2005
Bolonkin AA (2005b) Electrostatic solar wind propulsion system, AIAA-2005–4225. 41 Joint Propulsion Conferences, Tucson, Arizona, USA, 10–13 July 2005
Bolonkin AA (2005c) Kinetic anti-gravitator, AIAA-2005–4505. 41 Joint Propulsion Conferences, Tucson, Arizona, 10–13 July 2005
Bolonkin AA (2005d) Sling rotary space launcher, AIAA-2005–4035. 41 Joint Propulsion Conferences, Tucson, Arizona, 10–13 July 2005
Bolonkin AA (2005e) Problems of electrostatic levitation and artificial gravity, AIAA-2005–4465. 41 Joint Propulsion Conference, Tucson, Arizona, USA, 10–13 July
Bolonkin AA (2006) "Electrostatic Linear Engine", AIAA-2006-5229, 42 Joint Propulsion Conference, Sacramento, USA, July 9–12, 2006
Carbon and High Performance Fibers (2000) Directory, NY, 1995
Dresselhous MS (2000) Carbon Nanotubes. Springer, New York
Kalashnikov CK (1985) Electricity, Nauka, Moscow (in Russian)
Kestelman VN, Pinchuk LS, Goldale VA (2000) Electrets in Engineering, Fundamentals and Applications. Klumer Academic Publisher, Dordrecht
Kikoin IK (ed) (1976) Tables of physical values, Directory, Atomisdat, Moscow (in Russian)
Koshkin NI, Shirkevich MG (1982) Directory of elementary physics, Nauka, Moscow (in Russian)
Lofstrom KH (2002) The launch loop: A low cost earth-to-high-orbit launch system. http://www.Launchloop.com/launchloop.pdf
Shortley G, Williams D (2000) Elements of Physics, 5th edn. Prentice Hall, Inc, New Jersey, USA

CHAPTER 11

PLANETARY MACRO-ENGINEERING USING ORBITING SOLAR REFLECTORS

COLIN R. McINNES
Department of Mechanical Engineering, University of Strathclyde, Glasgow G1 1XJ, UK

Abstract: The prospect of engineering a planetary climate raises a multitude of issues associated with climatology, engineering on truly macroscopic scales and indeed the ethics of such ventures. Depending on personal views, such large-scale engineering is either an obvious necessity for the deep future, or yet another example of human conceit. In this chapter a simple climate model will be used to assess the possibility of engineering the Earth's climate (geo-engineering) using large orbiting reflectors. Two particular cases will be considered: active cooling of the climate to mitigate against anthropogenic climate change due to a doubling of the carbon dioxide concentration in the Earth's atmosphere and active heating of the climate to mitigate against an advance of the polar ice sheets of a magnitude comparable to that induced by the Milankovitch cycles. These two cases will be used as representative scenarios to allow the scale of the engineering challenge to be determined. In addition, even more visionary applications of solar reflectors to slowly manipulate the Earth's orbit will be investigated. While, engineering on such scales appears formidable at present, emerging capabilities to process lunar and asteroid material will allow such ventures to be considered in the future. This chapter aims to provide a foretaste of such future possibilities

Keywords: geo-engineering, climate variability, solar reflectors, orbital mechanics

1. INTRODUCTION

1.1 Earth Climate Variability

Human civilisation has developed during a time when the climate has been in a relatively benign state, with modern technological civilisation quickly developing during a temperate period between extremes. These favourable circumstances have allowed rapid global population growth combined with relative affluence in the

V. Badescu et al. (eds.), Macro-Engineering: A Challenge for the Future, 215–250.

West. However, the recent controversy surrounding human driven climate change has brought into sharp focus the fact that the climate is not static. The popular view of the climate as being perpetually in equilibrium is only due to the narrow window of human history through which we view the past. Indeed the question arises as to what the optimum climate state is (it does not necessarily need to be that of the present era).

Natural climate variability has been a major driver in the long-term expansion of human activity. Historically, population growth, migration and the development of technology have been, at least partly, in response to the driver of climate variability and its influence on the complex web of human activities. However, as western society becomes more sophisticated and geographically integrated, future climate variability will likely have a more profound influence than has been the case in the past. For example, if a period similar to the 'little ice age' of 1645–1715 (Free and Robock, 1999) were to recur in western society, the resulting change in energy demand would have major consequences for energy prices and possibly economic stability.

Aside from such relatively recent and geographically localised events, there is of course significant long term variability in the Earth's climate. The advance and retreat of glaciers appears to be forced by the Milankovitch cycles which can trip the climate into periodic ice ages (Muller and MacDonald, 1997). These cycles are due to oscillations in the elements of the Earth's orbit about the Sun, combined with periodic changes to the orientation of the Earth's spin axis, both resulting in changes to the relative flux of energy received by the Earth at polar and equatorial latitudes. The key periodicities are oscillations in the Earth's orbit eccentricity with a period of order 10^5 years, oscillations in the tilt of the spin axis between 21.5° and 24.5° with a period of order 41,000 years and the precession of the spin axis, with a period of order 23,000 years. While the total change in insolation due to the Milankovitch cycles is small (eccentricity oscillations have the largest effect, resulting in a change in total insolation of less than 1%), the key effect appears to be the change in the distribution of heat input as a function of latitude, which can show large excursions. Indeed, it is the insolation at high latitudes which is the key controlling parameter, since this directly effects the growth and retreat of ice sheets.

In addition to these cyclic effects, at the end of the last ice age rapid large scale climatic events have been documented, such as so-called Dansgaard-Oeschger and Heinrich events (Rahmstorf, 2003). Indeed, it is now recognised that apart from ice ages, many rapid climatic changes have occurred in the past, possibly due to the bi-stable nature of thermohaline circulation loops which are responsible for significant heat transport through the Earth's oceans.

Aside from natural climate variability, anthropogenic climate change has become a major issue due to the industrial release of carbon dioxide and other emissions. There is strong empirical evidence that such emissions are leading to an enhanced greenhouse effect and a rise in mean global temperatures. While natural climate variability has been of a greater magnitude, it is the apparent rapid change in the global mean temperature which raises concerns, along with the possibility of other rapid climate shifts being triggered in turn. It is widely accepted that a doubling of carbon dioxide concentration from a pre-industrial value of 280 ppm to 560 ppm will lead to an increase in mean global temperature of order 1.75 K. Proxy records appear to show that mean global temperatures have already risen by 0.6 ± 0.2 K in the last century, although there is still debate as to the appropriate use of such data (Mann et al., 1998). As will be seen in Section 2.3, a doubling of carbon dioxide concentration would require a reduction in total solar insolation of order 1.8% to mitigate its effects.

Most attention and effort has been focused on emission reductions in order to minimise the peak concentration of carbon dioxide. While such approaches have dominated the debate on anthropogenic climate change, there is a growing discussion of engineering solutions which can be deployed in addition to emission reductions, or as an alternative method of mitigation should emission controls provide to be politically unobtainable. While even the discussion of such geo-engineering solutions can be controversial, it seems prudent to investigate geo-engineering both as an effective tool to mitigate the effects of anthropogenic climate change, and more importantly, as a tool to deal with the extremes of long term natural climate variability (Cicerone et al., 1992; Keith, 2000; Schneider, 2001). It should be noted given the global effects of geoengineering, serious political issues will arise concerning any future implementation.

Lastly, it should be noted that simple climate models can posses multiple, overlapping equilibria with the same forcing from solar insolation (Emanuel, 2002). These models help explain the fact that the climate can switch rapidly between different states and is sensitive to relatively small changes in solar insolation driven, for example, by the Milankovitch cycles. The intriguing possibility then arises that the complexity of the climate can be actively exploited, so that geo-engineering can be deployed in a more subtle and sophisticated manner than has been considered in the past. By exploiting the non-linearity of the climate, it may be that geo-engineering schemes can be considered which can actively control such transitions between climate states using only modest, localised engineering intervention. This is an exciting possibility for the future which may significantly reduce the scale of endeavour required compared to the macro-engineering discussed here.

1.2 Geo-engineering

1.2.1 Climate Response

The definition of geo-engineering is open to interpretation, since it can be argued that the industrial release of carbon dioxide and other emissions is a form of geo-engineering, given the likely climatic change which it will bring about. However, geo-engineering will be used here as a term to describe deliberate, active intervention to modify the climate in a controlled manner, with presumed beneficial effect. While geo-engineering has a long history in various guises (Schneider, 2001), it is only relatively recently that space-based geo-engineering using orbiting solar reflectors has been considered (although as early as 1929 Oberth discusses the use of reflectors for localised climate engineering (Oberth, 1972). While only large-scale ventures are investigated here, there are a range of other measures available including active carbon sequestration from fossil fuels, natural carbon sequestration through ocean fertilisation, modification of surface albedo through ice cover and land use and the control of atmospheric emissivity through long-lived radiatively active gases (including carbon dioxide) (Schneider, 2001). More ambitious and dramatic (and perhaps reckless) schemes which involve modification to large-scale ocean currents have also been proposed in the past (Higuchi, 1970).

Critics of geo-engineering have argued that a coarse modification of the total solar insolation will lead to a range of undesirable consequences. In particular, it has been argued that while carbon dioxide traps heat during both the day and night cycles, a reduction in the solar insolation will only be experienced during the day cycle. In addition, the averaged effect of such geo-engineering would be most pronounced at the equator, leading to a less distinctive diurnal cycle and lower temperature gradients with latitude. Remarkably however, a detailed numerical study of the effect of geo-engineering has demonstrated that in principle a reduction in solar insolation can have precisely the intended effect (Govindasamy and Caldeira, 2000; Govindasamy et al., 2002, and Chapter 6 of this book). The response of the terrestrial climate to a 1.8% reduction in solar insolation and a doubling of carbon dioxide concentration was investigated using high fidelity numerical simulation.

Contrary to expectations, the simulation shows that reducing the total solar insolation will indeed compensate for increased atmospheric carbon dioxide, significantly ameliorating increases in both mean global temperatures and indeed local variations. These simulations also show that sea ice coverage is maintained at current levels, even in the presence of such a large increase carbon dioxide content, leading to no mean increase in precipitation.

Other secondary effects of anthropogenic climate change, such as major disruption to ocean currents due to an increase in fresh water flows to arctic seas are also apparently avoided. While these results are encouraging, it is clear that much more detailed investigation is required to determine the potential long-term effects of geo-engineering.

1.2.2 Aerosol Deposition

Most proposals for large-scale geo-engineering involve modifying the total solar insolation through scattering sunlight back to space. A range of strategies have been proposed to mitigate against anthropocentric climate change, including large-scale deposition of scattering aerosols in the stratosphere on a global scale, either sulphur dioxide particles (Schneider, 2001) or di-electric aerosols (Teller et al., 1997, 2004). Early proposals by Budyko (as described by Schneider, 2001; Keith, 2000) estimated that approximately 10^7 tons per annum of sulphur dioxide is required to offset a doubling of carbon dioxide concentration, while Teller et al. estimated that approximately 10^7 tons per annum of $\sim$100 nm aerosols are required to increase the Earth's albedo by 1%. Teller et al. also demonstrated that by using optically resonant scattering particles the mass requirements for such a venture could be reduced to 10^5–10^6 tons per annum, with either mesh microstructures or 4 mm helium filled aluminium balloons sized to float to 25 km, with a long life in the stratosphere, but oxidizing rapidly in the troposphere.

Climate cooling following major volcanic eruptions, such as Mount Pinatubo in 1991 (Fig. 1), demonstrate that such active control is possible, although engineered ventures will be significantly better optimized. The Pinatubo eruption deposited an estimated $1.7 \cdot 10^7$ ton of sulphur dioxide into the atmosphere, leading to a mean hemispheric surface cooling of order 0.5 K (Hansen et al., 1992). The climate cooling from the Pinatubo eruption is likely to have been greater than the effective warming from anthropogenic carbon dioxide during 1991–1993.

However, there are concerns that large scale deposition of engineered aerosols may lead to changes in atmospheric chemistry, in particular enhancing ozone depletion. In addition, the population of scattering particles will need to be continuously replenished as particles are washed out by precipitation, although this does ensure that the process is reversible on a relatively short timescale.

Figure 1. Large-scale deposition of aerosols by Mount Pinatubo (credit: Dave Harlow, United States Geological Survey). [Please see this figure in the color section at p. 315]

1.2.3 Occulting Disks

Aside from aerosol deposition (albedo modification) and carbon sequestration (atmospheric emissivity modification), large-scale engineering using orbiting reflectors has been considered by various authors (Seifritz, 1989; Early, 1989; Hudson, 1991; Mautner, 1991; McInnes, 2002b) to manipulate the total solar insolation. These concepts centre on fabricating and deploying a large occulting disk (or more likely disks) to reduce the total solar insolation in order to mitigate against increased carbon dioxide emissions. For example, the use of vast numbers ($\sim 5 \cdot 10^4$) of $100\,\text{km}^2$, actively controlled occulting disks in Earth orbit has been considered, but would likely lead to an apparent flickering of the Sun ($\sim$2% amplitude) and would create a significant orbital debris hazard (Keith, 2000). In addition, various proposals for an artificial ring of passive scattering particles have been documented (Mautner, 1991; Pearson et al., 2002). A more effective scheme is to station a large occulting disk (or disks), typically with a mass of order $10^{10}\,\text{kg}$, close to the Sun-Earth L_1 (Lagrange) equilibrium point some $1.5 \cdot 10^6\,\text{km}$ sunwards of the Earth.

The equilibrium location at the L_1 point is unstable, necessitating the use of active control using solar radiation pressure. Indeed, the use of a large numbers of smaller occulting disks would mitigate against the potentially catastrophic effect of the loss a single large disk. Despite these difficulties, occulting disks near L_1 appear to be the least invasive geo-engineering tool, whose effect can be more easily controlled than other schemes. In addition, if a large number of actively controlled reflective disks are available, climate heating can also be considered, as will be discussed later in Section 4.

As has been noted elsewhere (Mautner and Parks, 1990), a key advantage of using large solar reflectors for planetary engineering is the vast energy leverage obtained in a relatively short time. The total accumulated solar energy intercepted by the reflector quickly grows beyond the energy required for its fabrication, leading to a highly efficient tool for climate engineering. While solar reflectors offer many advantages, there are challenges associated with the fabrication and active control of such large, gossamer structures. In addition, it is almost certain that such structures would be fabricated in-orbit, either using lunar material or material processed from a suitable near Earth asteroid. Therefore, a prerequisite for geo- and planetary engineering using solar reflectors is a capability to effectively and economically exploit the resources of the moon or asteroids. The issues associated with fabrication of large solar reflectors will be discussed later in Section 5.

1.2.4 Small Angle Scattering

Since light only has to be scattered through a small angle ($\sim$0.01 rad) at L_1 to miss the Earth (as opposed to large angle scattering in Earth orbit), Teller et al. make the intriguing suggestion that instead of a solid disk (or disks), processed metallic sheets providing small angle scattering could be utilised, leading to a deployed system mass as low as $3 \cdot 10^6\,\text{kg}$ (Teller et al., 1997). A metallic mesh with a thickness similar to the skin depth of the optical radiation is required, with a grid spacing of order one half of the wavelength of the radiation being scattered. This open mesh micro-structure leads to an extremely low areal density of order 10^{-3} g m^{-2}, and so provides a relatively modest deployed mass.

While the extremely low areal density provides significant mass advantages, issues arise concerning active control of such a highly gossamer structure, along with long term space environment effects due to sputtering of solar wind particles on its micro-structure. However, by reducing the deployed mass to as low as $3 \cdot 10^6$ kg, terrestrial fabrication can be contemplated if low cost launch services are available. Clearly such concepts appear to offer significant advantages in scale over truly macro-engineering concepts and should be explored greater in detail.

1.3 Planetary Engineering

While the scale of endeavour required to deploy geo-engineering schemes is impressive, on an even more ambitious scale the same technologies which can be envisaged to engineer the Earth's climate can be scaled to engineer the climates of Mars and possibly Venus. Such terraforming schemes (engineering an Earth-like climate) have long been discussed, although the concept became somewhat more mainstream with the work of Sagan and others (Sagan, 1961, 1973). Bioengineered schemes have been proposed, including the delivery of customized organisms to convert carbon dioxide to oxygen in the atmosphere of Venus, and darkening the polar caps of Mars to reduce their albedo, again using customized organisms.

1.3.1 Terraforming Mars

More direct terraforming schemes have been proposed for Mars, including the use of orbiting reflectors to increase the total solar insolation. Such reflectors can be used to increase the mean surface temperature through direct radiative forcing, or can attempt to catalyze favourable climatic transitions by preferentially heating the polar caps in an attempt to release frozen carbon dioxide. Zubrin and McKay (1997) estimate that a rise in polar temperature of order 4 K may be sufficient to drive a transition to a new climate state in which the polar caps have evaporated, delivering an atmosphere with a surface pressure of 50–100 mb, as the first step of a longer term terraforming process. It is estimated that sufficient carbon dioxide exists in both the polar caps and regolith to ultimately raise the surface pressure to 300–600 mb. The reflectors to produce such a transition are envisaged to be in static equilibria relative to the planet (Zubrin and McKay, 1997), or in circular displaced polar orbits (McInnes, 2002a). The reflector system proposed by Zubrin was to be used solely to heat the polar caps, while the much larger systems discussed by both Birch (1992) and Fogg (1992) were seen as providing an increase in total insolation, averaged over the planetary surface. These reflectors are envisaged as part of a larger terraforming effort, with the reflectors used to increase the total planetary insolation by as much as 30%, again discussed by Fogg (1992).

Aside from solar reflectors, halocarbons synthesised on Mars have been considered as a tool to quickly raise the surface temperature and so liberate trapped carbon dioxide (Gerstell et al., 2001; Badescu, 2005).

1.3.2 Terraforming Venus

Even more ambitious ventures have been discussed by Birch (1991) who investigates the use of a large occulting disk at Venus to cool the atmosphere to allow the precipitation of carbon dioxide, again as the first step of a longer term terraforming process. The surface temperature of Venus is of order 730 K, driven by an intense greenhouse effect. By lowering the atmospheric temperature to the critical temperature of carbon dioxide at 304 K, liquid rain can form which then precipitates out of the atmosphere to the planetary surface. If the temperature is further reduced to the triple point of carbon dioxide at 217 K, the liquid carbon dioxide oceans which have formed freeze, ultimately trapping the original atmospheric carbon dioxide in solid form on the surface.

2. CLIMATE DYNAMICS

2.1 Energy Balance Model

In order to determine the requirements for climate engineering, it is necessary to investigate the response of the Earth's climate to large-scale engineering intervention. In particular, the response of the global mean temperature to changes in total solar insolation is of interest, as are any non-linear effects which would pose a risk of unintended and potentially catastrophic modifications to the climate. By determining a relationship between global mean temperature and solar insolation, the requirements for solar reflectors can be obtained and the scale of engineering required can be determined. The Energy Balance Model (EBM) is a simple, yet elegant climate model which captures the essential large-scale features of the Earth's climate dynamics (McGuffie and Henderson-Sellers, 1997). This low order (so-called zero dimensional) model does not contain the sophistication of numerical General Circulation Models (GCM), but does allow insights into the dynamical processes at work. The EBM assumes that any change to the heat balance of the Earth is simply due to an inequality between absorbed heat flux Q_{IN} and emitted heat flux Q_{OUT}.

Therefore, assuming some mean specific heat capacity C per unit area and global mean temperature T, the dynamics of the Earth's climate can be written simply as

$$C\frac{dT}{dt} = Q_{IN} - Q_{OUT}, \tag{1a}$$

so that

$$C\frac{dT}{dt} = Q(1-\alpha) - \varepsilon\sigma T^4, \tag{1b}$$

where $Q(342.5\,\mathrm{Wm}^{-2})$ is the total solar insolation, $\alpha(\sim 0.3)$ is the mean planetary albedo, ε is the mean emissivity (~ 0.62) and $\sigma(= 5.67 \cdot 10^{-8}\,\mathrm{Wm}^{-2}\,\mathrm{K}^{-4})$ is the Stefan-Bolztman constant. It should be noted that the solar insolation is defined in terms of the solar constant $F(= 1370\,\mathrm{Wm}^{-2})$ as $Q = F/4$, since the Earth presents only a circular cross-sectional area to the incoming flux but radiates over its entire spherical surface area. Assuming that the climate is in equilibrium ($dT/dt = 0$), a naïve initial estimate of the mean planetary temperature can be obtain from Eq. (1b) as

$$\overline{T} = \left[\frac{Q(1-\alpha)}{\varepsilon\sigma}\right]^{1/4}, \tag{2}$$

which yields an estimated mean temperature of 14°C (287 K), in good agreement with the observed global mean temperature. In order to provide a more accurate EBM, an empirical relationship which models the emitted heat flux Q_{OUT} can be used. In addition, the mean planetary albedo can be modelled as a simple function of temperature in order to capture the change in albedo with temperature due to ice cover. This more complete and comprehensive EBM can then be written as

$$C\frac{dT}{dt} = Q(1-\alpha(T)) - A - BT, \tag{3}$$

where the coefficients for Q_{OUT} are determined from empirical observational data as $A = 204\,\text{Wm}^{-2}$ and $B = 2.17\,\text{Wm}^{-2}\,\text{K}^{-1}$ (McGuffie and Henderson-Sellers, 1997). The linear approximation used for Q_{OUT} in Eq. (3) can be obtained from a truncated expansion of Eq. (1b), although the coefficients will be somewhat different with $A = 314.9\,\text{Wm}^{-2}$ and $B = 4.61\,\text{Wm}^{-2}\,\text{K}^{-1}$. Again, assuming that the climate is in equilibrium ($dT/dt = 0$), a new estimate of the mean planetary temperature can be found from Eq. (3) as

$$\overline{T} = \frac{Q(1-\alpha) - A}{B}, \tag{4}$$

where α is again assumed to be of order 0.3. This new estimate of the global mean temperature is found to be 16.5°C (289.5 K), which is again in good agreement with the observed global mean temperature. In order to capture ice-albedo feedback effects (Budyko, 1969), a suitable functional form for the albedo is given by

$$\alpha(T) = \alpha_I - \frac{1}{2}(\alpha_I - \alpha_F)(1 + \tanh(\gamma T)), \tag{5}$$

which provides a continuous variation of the mean albedo from an ice free state ($\alpha_F = 0.3$) to an ice covered state ($\alpha_I = 0.6$), as shown in Fig. 2 for the choice of shaping parameter $\gamma = 0.5$. The key feature which the EBM now provides is a set of multiple equilibria for the mean temperature T. Using Eq. (3) and Eq. (5), these multiple equilibria can be identified, as shown in Fig. 3, where three possible equilibrium temperatures now exist.

It can be shown that the upper equilibrium temperature ($T_U = 16.5$°C (289.5 K)) and lower equilibrium temperature ($T_L = -30.9$°C (242.1 K)) are stable equilibria and are separated by an unstable intermediate state ($T_I = 0.7$°C (273.7 K)). The upper equilibrium temperature corresponds to the present day Earth, while the lower

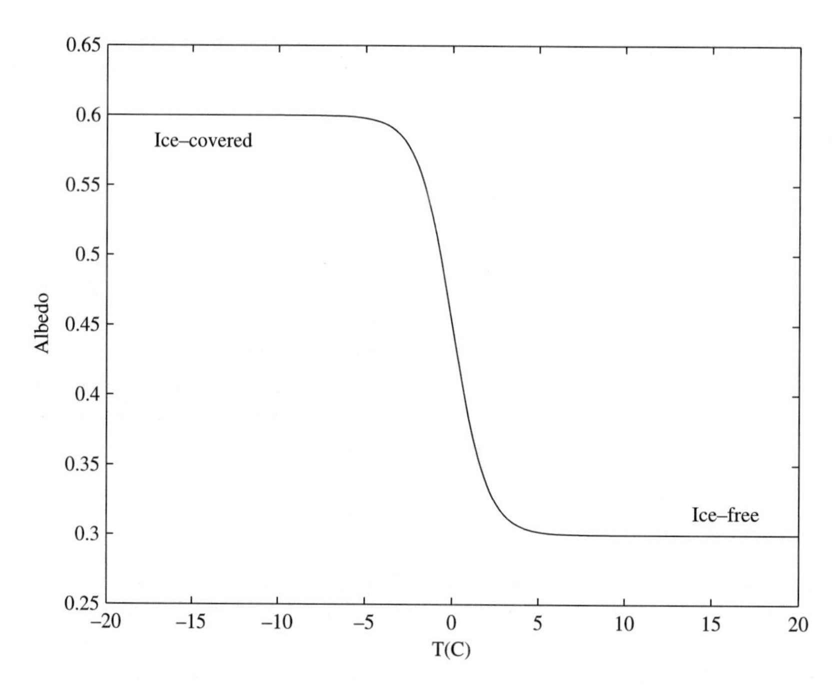

Figure 2. Albedo model with ice-covered (0.6) and ice-free (0.3) states

temperature corresponds to a hypothetical ice covered state. Some evidence exists that such a totally ice covered state has occurred in the past (Hoffman and Schrag, 2002). It should be noted that the coefficients A and B are unlikely to be valid at the lower temperature state. However, the model provides a useful qualitative view of multiple equilibria in the climate and the requirements for transitions between them, as will be discussed next. It will be seen that T_U and T_I are energetically relatively close, so that it does not require a significant reduction in solar insolation to drive a transition towards T_L.

2.2 Implications for Geo-engineering

In order to engineer the climate, Eq. (2) shows that there are three parameters which can in principle be manipulated to modify the mean global temperature. The mean albedo can be modified by engineering changes to ice and cloud cover and by depositing scattering particles into the stratosphere, as discussed in Section 1.2, while the mean emissivity can be modified by controlling the release of long-lived radiatively active gases (such as carbon dioxide). However, perhaps the most directly available parameter for manipulation is the solar insolation Q, again discussed in Section 1.3. By using orbiting solar reflectors, Q can either be enhanced or diminished, which in turn will displace the upper equilibrium temperature T_U. By engineering changes to the total solar insolation, invasive and potentially damaging schemes, such as manipulating the chemistry of the atmosphere can be avoided. However, as a cautionary note, the limits to solar insolation modification should first be defined.

Using the EBM with ice-albedo feedback, defined by Eq. (5), it is found that as the total solar insolation is decreased, the upper (stable) and intermediate (unstable) equilibrium temperatures merge, and finally the system quickly moves to the lower (stable) equilibrium temperature. This process can be seen in Fig. 3 for an insolation decrease of order 15%. In this new equilibrium state, a significant increase in solar insolation is then required to drive the mean temperature back to the upper temperate state. It is clear then that such a dramatic transition to the climate is not directly reversible, and in fact the climate exhibits a form of hysteresis. This hysteresis effect can be seen in Fig. 4, which shows the change in equilibrium temperature as the solar insolation is modified. Two rapid transitions are seen corresponding to the transition from T_U to T_L (transition A) with only a modest reduction in solar insolation of order 10% and the transition from T_L to T_U (transition B), but only if the solar insolation is raised to order 40% above its present value. Again, it should be noted that the coefficients in Eq. (3) are unlikely to be valid at the lower temperature state. However, the model does provide a useful qualitative view of multiple climate equilibria.

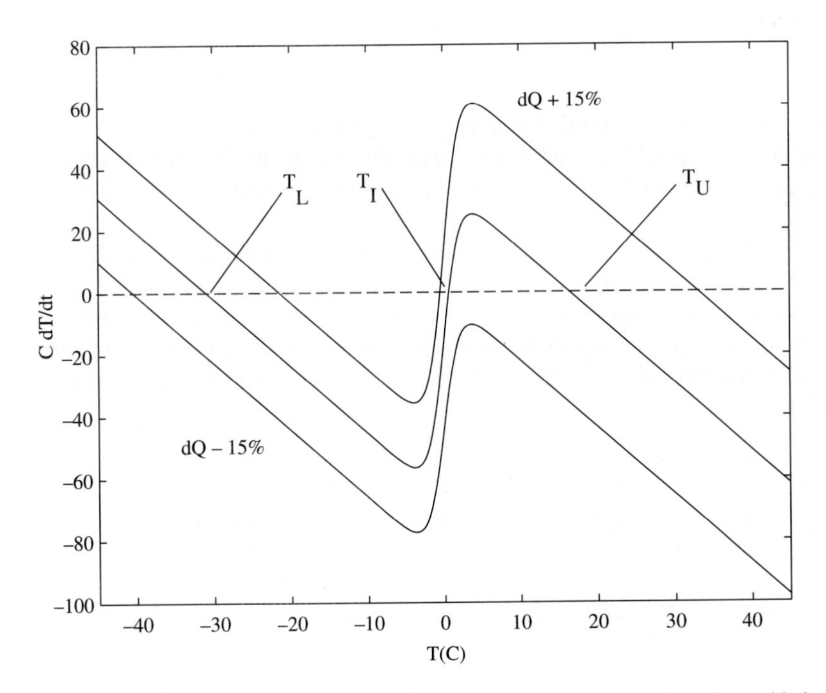

Figure 3. Equilibrium temperatures (stable warm state T_U and cool state T_L and unstable intermediate state T_I) for EMB with perturbations of ±15% to the solar insolation Q

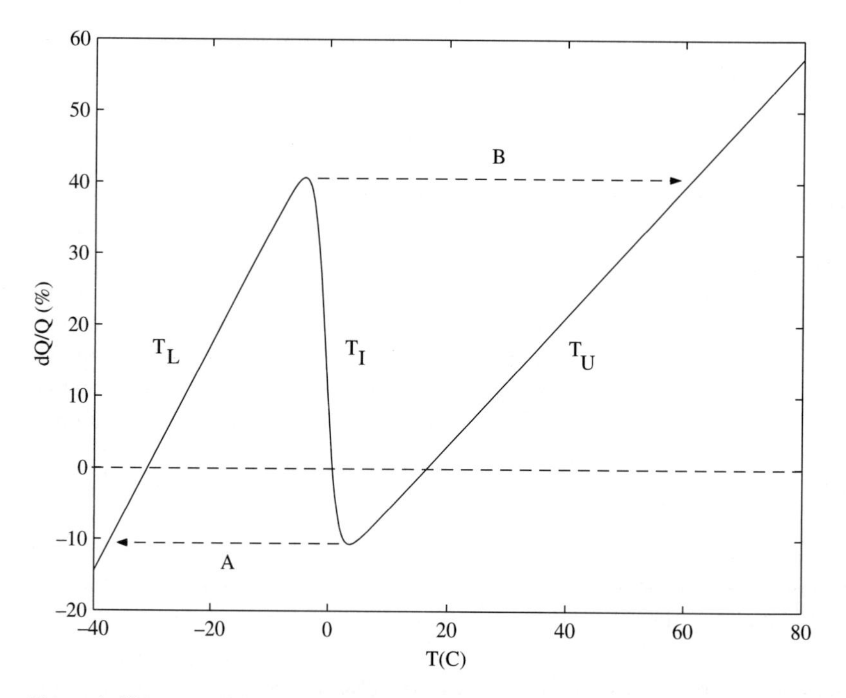

Figure 4. Climate transitions due to changes in total solar insolation (stable warm state T_U and cool state T_L and unstable intermediate state T_I)

A further insight into the implications of geo-engineering can be found by noting that the deterministic EBM defined by Eq. (3) should in principle include a stochastic forcing term to account for the short period dynamics which the simple EBM does not capture (short term weather fluctuations) (Berglund and Gentz, 2001). These stochastic effects will then result in fluctuations about the upper equilibrium temperature. The question then arises as to the possibility that such stochastic forcing could drive the mean temperature from T_U to T_L, even although the deterministic EBM does not allow such a transition. In order to assess the effect of such stochastic forcing, the EBM defined by Eq. (3) will be written in terms of an effective potential function as

$$C\frac{dT}{dt} = -\frac{dV(T)}{dT} + \rho\xi(t), \tag{6}$$

where $V(T)$ is an effective potential defined by

$$V(T) = AT + \frac{1}{2}BT^2 - Q\int_0^T (1 - \alpha(T'))dT', \tag{7}$$

and the term $\xi(t)$ is zero mean, uncorrelated white noise ($\langle\xi(t)\rangle = 0$, $\langle\xi(t), \xi(t')\rangle = \delta(t - t')$), with ρ the noise intensity. The functional form of Eq. (7) is shown

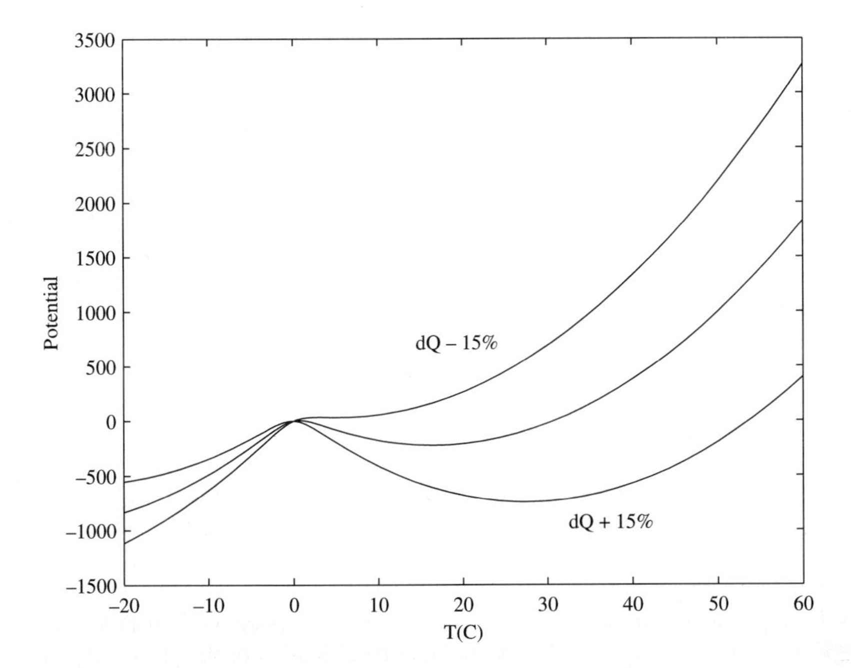

Figure 5. Effective potential with changes in total solar insolation of ±15%

in Fig. 5, where the stable local minimum in the potential formed by T_U can be seen, along with an unstable local maximum formed by T_I. It can also be seen that as the solar insolation is reduced, a single stable local minimum will be formed at T_L as T_U and T_I merge.

For a fixed solar insolation ($\delta Q = 0$), the deterministic EBM defined by Eq. (3) shows that the mean temperature will always remain at T_U, as long as the potential V_U is less than the potential at the unstable intermediate temperature V_I. However, with stochastic forcing, the mean temperature will experience excursions from T_U, with the magnitude of the excursion dependent on the intensity of the stochastic forcing, defined by the noise intensity ρ.

While the most likely outcome is that the mean global temperature will remain close to T_U, there is a finite probability that the stochastic forcing could drive the climate from T_U to T_L by escaping over the local maximum in the effective potential at T_I. Such processes have been proposed as an explanation for the periodicities of ice ages.

The Milankovitch cycles will have the effect of oscillating the height of the potential well at T_U, such that when the difference in potential with T_I reaches a minimum there is a greater chance of stochastic forcing driving the climate towards T_L in a stochastic resonance process (Benzi et al., 1983).

Given that the stochastic model allows a range of potential climates, the climate can be described by a probability density function. It can be shown using Eq. (6)

that the probability p of the mean global temperature being in the range $[T, T+dT]$ is given by

$$p(T) = \frac{1}{N}\exp\left[-\frac{2V(T)}{\rho^2}\right], \tag{8}$$

for some normalisation N, so that again the mean temperature is unlikely to depart from a neighbourhood ρ about a local minimum. While Eq. (8) provides a probability density function for the global mean temperature, the key issue is transitions between states. Therefore, the time τ to exit from the upper potential well (Kramer's time) can be found (Berglund and Gentz, 2001) and scales as

$$\tau \sim \exp\left[\frac{2(V_I - V_U)}{\rho^2}\right], \tag{9}$$

Therefore, if as a result of geo-engineering ventures the potential height $(V_I - V_U)$ decreases, there will be a growing probability of a catastrophic shift in the climate, even although the deterministic EBM model would suggest that no transition can take place. Importantly, the exit time from the local minimum of the effective potential at T_U scales exponentially with the height of the potential barrier so that any geo-engineering venture will likely lead to a modification of the probability of a climate transition, even although such a transition does not appear possible using a deterministic model.

2.3 Global Warming Mitigation

As discussed in Section 1.1, it is generally agreed that there has been an increase in the global mean temperature, with a rise of order $0.6 \pm 0.2\,\mathrm{K}$ during the last century. This rise in global mean temperature is specifically due to a significant increase in radiatively active gases in the atmosphere, principally carbon dioxide, leading to a reduction in the effective emissivity of the atmosphere. The reduction in emissivity can be expressed as a change in radiative forcing. For a doubling of the carbon dioxide content of the atmosphere it can be shown that the coefficient A is reduced by $4.17\,\mathrm{Wm}^{-2}$, resulting in a lower re-radiation of heat due to an enhanced greenhouse effect (Govindasamy and Caldeira, 2000). In principle, this change in radiative forcing can be offset by a reduction in the effective solar insolation through geo-engineering. Assuming that the climate system is in a neighbourhood of the upper equilibrium T_U, the required change in solar insolation by δQ to offset a change in radiative forcing by δA is given by

$$\delta Q = Q - \frac{B\overline{T} + (A - \delta A)}{1 - \alpha}. \tag{10}$$

Therefore, Eq. (10) shows that to maintain the mean temperature at T_U of 16.5°C (289.5 K) in the presence of a doubling of the carbon dioxide content of the

Earth's atmosphere, a reduction in solar insolation $\delta Q/Q$ of 1.74% is required for δA=4.17 Wm^{-2}. The requirements to engineer such a reduction in effective solar insolation will be determined later in Section 3. Again, it is noted that detailed numerical studies of the effect of geo-engineering have demonstrated that reducing solar insolation appears to compensate for increased atmospheric carbon dioxide, significantly ameliorating increases in both mean global temperatures and local variations (Govindasamy and Caldeira, 2000; Govindasamy et al., 2002).

Assuming that such a reduction in solar insolation δQ can be engineered, the response of the mean global temperature can be obtained from Eq. (3). If the change in insolation is small, so that the albedo is fixed, there is a change in mean global temperature such that

$$T(t) = \overline{T} + \frac{\delta Q}{B}(1-\alpha)(1-\exp(-t/\tau)), \quad (11)$$

where $\tau = C/B$ is the relaxation time of the system. By considering the heat capacity of the top 70 m of the oceans (mixing layer), it is found that $\tau \sim 3$ years, so that the response of the climate to changes in insolation is slow, but at an acceptable time scale given the nature of the venture. For $t >> \tau$ the resulting change in mean global temperature is of order $(1-\alpha)\delta Q/B$, so that a reduction in solar insolation $\delta Q/Q$ of 1.74% to offset a doubling of the carbon dioxide content of the atmosphere is equivalent to reducing the mean global temperature by 1.92 K. The requirements to engineer such a reduction in solar insolation using a large occulting disk (or disks) will be considered later in Section 3.

2.4 Ice Age Mitigation

The requirements for Earth cooling to mitigate against an anthropogenic increase in the carbon content of the atmosphere can be stated with relative precision. However, the requirements for Earth heating to mitigate against the natural variability which gives rise to periodic ice ages are more difficult to define. As has been discussed in Section 1.1, long period oscillations of the Earth's orbital elements and spin axis lead to a redistribution of heat between the equatorial and polar regions. While the change in total insolation is in fact rather small, the phasing between different oscillations may lead to an onset of glaciation (cool summers in the Northern hemisphere for example). Other approaches postulate the stochastic resonance effect discussed in Section 2.2 as the mechanism to explain why a relatively modest oscillation in total insolation can lead to abrupt switches between different climate states.

Later in Section 4.4, a more detailed 1-dimensional EBM which includes latitude dependence will be used to assess the effect of an increase in solar insolation. Such a model is required since the zero-dimensional EBM used earlier cannot capture the high latitude cooling and heating which controls the onset, or termination, of ice cap growth. It will be shown that an increase in total insolation of order

3.6% (equivalent to increasing the mean global temperature by 4 K) can dissipate the present day polar caps. This will be used as a bench mark to determine the reflector requirements for Earth climate heating. This requirement is similar to the 4% increase in insolation defined by Teller et al. (2004) to mitigate against possible climate cooling towards an ice age (circa 3000) when the mid-21st century pulse of carbon is sunk into the ocean, increasing the effective emissivity of the atmosphere.

3. EARTH CLIMATE COOLING

3.1 Requirements

The concept of using a large occulting disk near the Sun-Earth L_1 point to reduce solar insolation has been discussed by various authors, as noted in Section 1.3.

In this section it will be shown that there is in fact a minimum occulting disk mass which can be obtained if the disk is positioned at an optimum location along the Sun–Earth line, sunward of the classical L_1 Lagrange point. This optimum location is found from an analysis of the three-body mechanics of the problem (McInnes et al., 1994). The location of the disk can be optimized since the solar radiation pressure exerted on the disk will modify the location of the classical L_1 Lagrange point. As the disk mass falls, the L_1 point will be displaced sunwards due to the solar radiation pressure induced force exerted on the disk increasing (McInnes, 2002b). However, as the disk moves sunward the required disk area to maintain the necessary reduction in solar insolation at the Earth will grow, leading to an increase in disk mass. These two processes must then be balanced in order to minimize the total disk mass through an optimum choice of disk location.

3.2 Occulter Orbit

Now that the required change in solar insolation has been obtained from Section 2.3, a suitable occulting disk can be sized. For a disk of radius R_S at some distance r_S from the Earth, the disk will subtend a solid angle Ω_S of $\pi R_S^2/r_S^2$, as shown in Fig. 6. Similarly the Sun, of radius R_O at distance r_O from the Earth, will subtend a solid angle Ω_O of $\pi R_O^2/r_O^2$, so that the disk will partially occult the Sun and reduce the insolation at the Earth by a factor Ω_S/Ω_O. It should be noted that by partly occulting the solar disk the solar flux F is reduced such that $\delta Q = \delta F/4$, however the relative change in insolation $\delta Q/Q$ is identical to the relative change in flux $\delta F/F$. Therefore, the reduction in insolation produced by the occulting disk is defined by

$$\frac{\delta Q}{Q} = \left[\frac{R_S}{R_O}\right]^2 \left[\frac{r_O}{r_S}\right]^2. \tag{12}$$

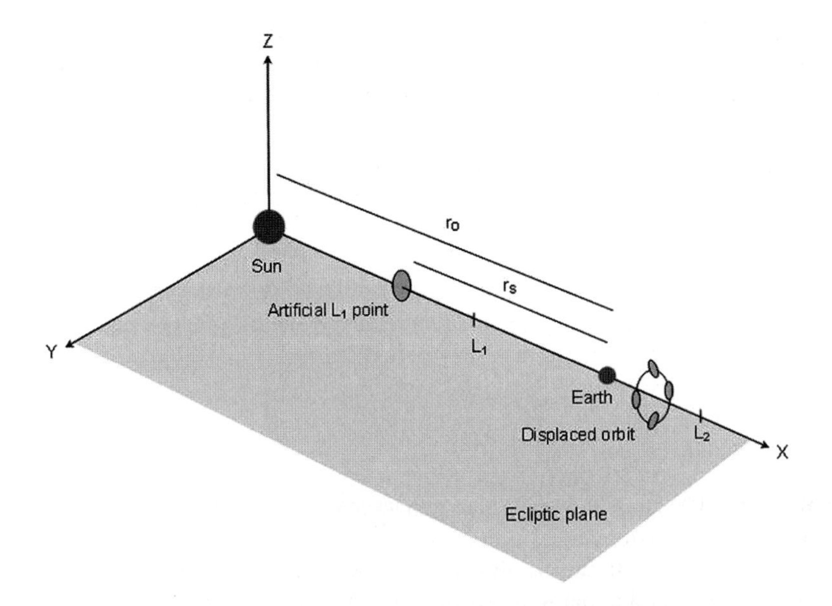

Figure 6. Occulting solar disk and reflectors stationed along the Sun-Earth line

Using Eq. (12), the required disk radius may now be obtained as a function of its distance r_S from the Earth to provide the required change in solar insolation of $\delta Q/Q$ as

$$(13) \qquad R_S = R_O \left[\frac{r_S}{r_O}\right]\left[\frac{\delta Q}{Q}\right]^{1/2}.$$

However, before the occulting disk is sized, the optimum distance of the disk from the Earth can be determined to minimise the total disk mass. As discussed in Section 3.1, the disk will be located near the interior Sun-Earth Lagrange point. However, due to the solar radiation pressure acting on the disk, the Lagrange point will be displaced sunwards. A trade-off therefore exists between lowering the disk mass and displacing the Lagrange point sunward, and ultimately increasing the disk mass due to the increased disk area to provide the required partial occultation of the Sun. This trade-off leads to an optimum disk location which will minimise the total disk mass.

The condition for equilibrium of the occulting disk in the Sun-Earth three-body problem can be determined from a simple force balance. Although the general solution for artificial equilibria for a reflector is known (McInnes et al., 1994), only a simple 1-dimensional problem need be considered here to locate the displaced L_1 Lagrange point. Since the mass of the Earth M_E is essentially negligible relative to the solar mass M_O, the centre-of-mass of the Sun-Earth system will be taken as being located at the centre-of-mass of the Sun, as shown in Fig. 6. This approximation has a negligible effect on the subsequent analysis. The condition for equilibrium may now be obtained by balancing the gravitational force from the Sun and the Earth, the centripetal force and the solar radiation pressure induced force acting on the occulting disk a_S such that

$$\frac{GM_E}{r_S^2} - \frac{GM_O}{(r_O - r_S)^2} + \omega^2 (r_O - r_S) + a_S = 0, \quad \omega = \sqrt{\frac{GM_O}{r_O^3}}, \tag{14}$$

where ω is the orbital angular velocity of the Earth relative to the Sun and G is the gravitational constant. The inverse square solar radiation pressure induced acceleration acting on the occulting disk may be written as

$$a_S = \frac{2\kappa P_E A_S}{M_S}\left[\frac{r_O}{r_O - r_S}\right]^2, \tag{15}$$

where $P_E (= 4.56 \cdot 10^{-6}\,\mathrm{Nm}^{-2})$ is the solar radiation pressure experienced by an absorbing surface at 1 astronomical unit (r_O) and κ is a function of the optical properties of the reflector. It can be shown (McInnes, 1999) that for a specular reflector with Lambertian thermal re-emission the function κ is given by

$$\kappa = \frac{1}{2}\left[(1+\eta) + \frac{2}{3}(1-\eta)\frac{\varepsilon_F - \varepsilon_B}{\varepsilon_F + \varepsilon_B}\right], \tag{16}$$

where η is the specular reflectivity, ε_F is the emissivity of the front (Sun facing) side of the disk and ε_B is the emissivity of the rear (Earth facing) side of the occulting disk, while the disk has an area A_S of πR_S^2. Using Eqs. (13) and (15), the disk mass M_S may now be written as

$$M_S(r_S) = 2\pi\kappa P_E R_O^2 \left[\frac{\delta Q}{Q}\right]\left[\frac{r_S}{r_O - r_S}\right]^2 \frac{1}{a_S(r_S)}, \tag{17}$$

where a_S is determined from Eq. (14). Since Eq. (17) is now a function of r_S only, the variation of the mass of the occulting disk with location along the Sun-Earth line can be investigated to attempt to minimise the disk mass.

3.3 Occulter Sizing

The mass of the occulting disk may now be determined for a required reduction in solar insolation. Assuming a reduction of $\delta Q/Q = 1.74\%$, as discussed in Section 2.3, the variation of the disk mass with equilibrium location is shown in Fig. 7, where it can be seen that there are two limiting conditions.

Firstly, as the shield is located closer to the classical interior L_1 Lagrange point, the disk mass grows and is unbound as the Lagrange point is approached. This growth in mass is required to reduce the solar radiation pressure induced force exerted on the disk, which would otherwise displace the equilibrium location

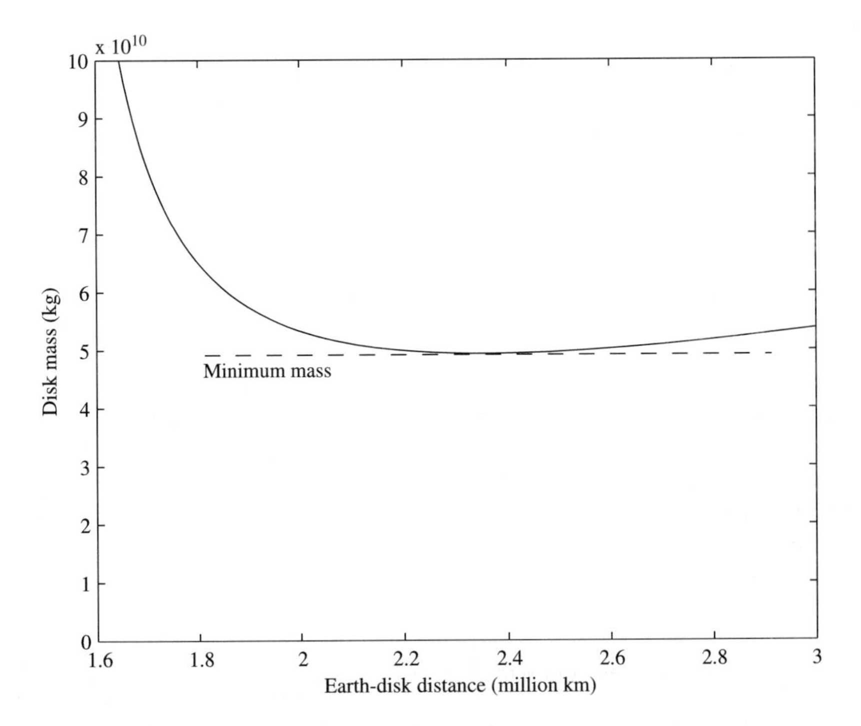

Figure 7. Optimum occulting disk location ($\kappa = 0.17$)

sunward. Similarly, as the location of the disk is moved sunwards, the required mass of the disk will fall due to the increased solar radiation pressure induced force required for equilibrium. However, as the disk is moved sunwards significantly from the classical L_1 point, its mass will start to grow as the disk area increases to maintain the required solid angle subtended at the Earth to reduce the total solar insolation. These two opposing processes lead to a minimum disk mass, as can be seen in Fig. 7.

The minimum disk mass can now be determined by finding the turning point of Eq. (17). It can be seen that there is a single location which will minimize the disk mass, independent of the required reduction in solar insolation or the disk optical properties. This location can be found by minimising the function

$$f(r_S) = \left[\frac{r_S}{r_S - r_O}\right]^2 \frac{1}{a_S(r_S)}, \tag{18}$$

where it is found that $f'(r_S) = 0$ when the disk location r_S is $2.36 \cdot 10^6$ km from the Earth. This optimum location is somewhat sunward of the classical interior Lagrange point at $1.50 \cdot 10^6$ km and is again independent of the disk properties, representing the true optimum location for an occulting disk. Assuming that the disk must provide a reduction in solar insolation of $\delta Q/Q = 1.74\,\%$, a disk with a radius of 1450 km and a total mass of $2.6 \cdot 10^{11}$ kg is required if $\kappa \sim 0.91$, representative of a reflecting metallic occulting disk. However, if $\eta \sim 0$ and $\varepsilon_F \sim 0$ then $\kappa \sim 0.17$, representative of a non-reflecting black occulting disk, resulting in substantial mass savings, although the optimum location of the reflector remains unchanged. In this case a total mass of $4.9 \cdot 10^{10}$ kg is required, again with a radius of 1450 km, as detailed in Tables 1 and 2.

For comparison, the mass of a range of terrestrial engineering ventures are listed in Table 3. It can be seen that the Chinese Three Gorges Dam requires approximately $6 \cdot 10^{10}$ kg of concrete, and so forms a structure with a comparable mass to the solar occulting disk. While the challenges posed are significant, it interesting to note that measured in terms of mass, large-scale geo-engineering represents a venture of comparable scale to current large-scale terrestrial engineering endeavours.

While the analysis above has considered a single occulting disk, a more likely scenario would be the use of a large number (swarm) of smaller elements with the same total area as that required for a single large disk. Such a swarm of disks may either be independent free-flying elements, or could be used to assemble a large occulter, as shown in Fig. 8. As will be discussed in Section 5 below, the mass requirements for a solid occulter (as opposed to a small angle scattering mesh)

Table 1. Occulting disk optical properties

Disk type	η	ε_F	ε_B	κ
A (reflecting)	0.82	0.06	0.06	0.91
B (non-reflecting)	0	0.01	0.5	0.17

Table 2. Occulting disk physical properties

Disk type	Disk mass (kg)	Disk radius (km)	Density (gm^{-2})
A (reflecting)	$2.6 \cdot 10^{11}$	$1.44 \cdot 10^3$	40.2
B (non- reflecting)	$4.9 \cdot 10^{10}$	$1.44 \cdot 10^3$	7.4
Teller et al. (1997)	$3 \cdot 10^6$	$6.20 \cdot 10^2$	$1.2 \cdot 10^{-3}$

Table 3. Mass comparison with terrestrial engineering ventures

Scale	Mass (kg)	Engineering venture
10^8	$6.5 \cdot 10^8$	'Knock Nevis' oil tanker (fully laden)
10^9	$6 \cdot 10^9$	Great pyramid of Giza
10^{10}	$6 \cdot 10^{10}$	Concrete used for Three Gorges dam
10^{11}	$2 \cdot 10^{11}$	Water stored in London's reservoirs
10^{12}	$7 \cdot 10^{12}$	World annual CO_2 emissions

Figure 8. Assembly of a large occulting disk from a swarm of discrete elements (credit: Dario Izzo, ESA/ACT). [Please see this figure in the color section at p. 316]

necessitate the use of lunar, or more likely near Earth asteroid material. A scenario can be envisaged whereby individual elements are fabricated from asteroid material and the total effective occulting area grows over a period of time, to match the required reduction in solar insolation to maintain a constant global mean temperature as carbon emissions rise.

4. EARTH CLIMATE WARMING

4.1 Requirements

Now that the requirements for a swarm of occulting disks for Earth climate cooling has been determined, the effect of such a system on climate warming will be assessed. As discussed in Section 1.1, it is clear that the Earth experiences significant natural climatic variability, with the implication that such variability will either have to be adapted to through changes to economic and social behaviour, or that it will require geo-engineering to mitigate. While future large scale glaciation is unlikely to be of concern for some time, Ruddiman argues that it is anthropogenic emission of carbon dioxide (and methane) from agriculture, deforestation and industrial activity which has prevented the advance of glaciers in recent times (Ruddiman, 2003). Ruddiman terms this the anthropogenic era to highlight the strong link between human activities and climate dynamics. Again, as noted in Section 1.1, relatively fast cooling events have occurred in the past, possibly connected with the bi-stable nature of thermohaline circulation loops. In this light, geo-engineering appears to be an essential tool for the future in order to manage extremes in natural climate variability and mitigate its effects on human activity.

In order to explore the possibility of climate heating, the same equivalent area of solar collector which was defined for climate cooling in Section 3.2 will be assumed, with type A (reflective) disks. This will be used as a reference area, with the assumption that the capability required to fabricate and deploy a system of occulting disks for climate cooling can be utilised for climate warming. Alternatively, it can be assumed that the occulting disks sunward of L_1 can be re-deployed and re-tasked for climate warming (if they are of type A). For a total reflective area A_R, and solar flux F, the increase in insolation can be estimated as

$$\delta Q = \frac{FA_R}{4\pi R_E{}^2} \cos\alpha, \tag{19}$$

where α is the aspect angle of the reflectors and R_E is the radius of the Earth. For the parameters defined in Section 2, and assuming α~45°, the increase in insolation is 12.4 Wm^{-2}, corresponding to $\delta Q/Q = 3.6$ %. Using the approximation that the change in mean global temperature is of order $(1-\alpha)\,\delta Q/B$, an increase in solar insolation $\delta Q/Q$ of 3.6% is equivalent to increasing the mean global temperature by 4 K. The result of such an increase in total isolation will be discussed later in Section 4.4.

4.2 Reflector Orbit

In order to re-direct solar radiation to increase the total planetary insolation, reflectors must be deployed on suitable orbits. One of the key issues is management of the momentum accumulated by the reflectors due to solar radiation pressure, which poses difficulties for reflectors deployed directly in Earth orbit. In addition, since the reflectors act as mirrors, they will project an image of the solar disk onto the Earth. At the Earth's distance from the Sun, the solar disk subtends an angle θ of order 0.01 rad, therefore a reflector at distance d from the surface of the Earth will project a spot of diameter D~$d\theta$, assuming a perfectly flat reflecting surface. In order that the spot size is less than the diameter of the Earth, the distance d must be less than $1.4 \cdot 10^6$ km, a distance comparable to that to the L_1 point. For a solid reflector, locations near the L_1 point are not attractive due to the long path length the reflected radiation must traverse and consequently the extremely high pointing accuracy required. In addition, in order to establish an artificial equilibrium point using solar radiation pressure, the aspect angle of the reflectors is rather large (McInnes et al., 1994). However, the re-directed radiation is incident on the day side of the Earth which avoids illuminating the night sky. Artificial equilibria between the Earth and the L_2 point have been considered by Zubrin, for heating the poles of Mars (Zubrin and McKay, 1997). However, it has been shown that when a reflector with non-ideal reflectivity is considered the required aspect angle of the reflector is such that the re-direct radiation is in fact largely unable to illuminate the poles (McInnes, 2002a).

As noted above, reflectors in Earth orbit will be strongly perturbed by solar radiation pressure. An attractive family of orbits for solar reflectors are displaced orbits which can be generated by orienting the reflector such that a component of the solar radiation pressure induced force exerted on the reflector is directed normal to the orbit plane (McInnes, 2002a). These orbits are therefore circular and near polar, but are displaced behind the Earth in the anti-Sun direction, as shown in Fig. 9 (and Fig. 6) for a reflector orbit radius ρ and displacement distance z. The momentum accumulated by the reflector due to solar radiation pressure is offset by the z-component of gravitational force. Using a 2-body analysis, the requirements for such displaced orbits can now be investigated. However, it is found that such orbits can also be established using a full 3-body analysis of the dynamics of the problem (Bookless and McInnes, 2004). Using cylindrical polar coordinates, the equations of motion of a reflector can be written as

$$\text{(20a)} \quad \ddot{\rho} - \rho\dot{\theta}^2 = -\frac{\rho}{r}\left(\frac{\mu}{r^2}\right) + a\cos^2\alpha\sin\alpha,$$

$$\text{(20b)} \quad \rho\ddot{\theta} + 2\dot{\rho}\dot{\theta} = 0,$$

$$\text{(20c)} \quad \ddot{z} = -\frac{z}{r}\left(\frac{\mu}{r^2}\right) + a\cos^3\alpha.$$

where μ is the product of the gravitational constant and the mass of the Earth. The reflector orientation is defined by a pitch angle α and the acceleration induced by solar radiation pressure a is defined by Eq. (15). It is assumed that a is constant (although the magnitude of the solar radiation pressure induced force scales as $cos^2\alpha$). For a circular displaced orbit it is required that $\ddot{\rho} = 0$ and $\ddot{z} = 0$. Therefore, defining the orbital angular velocity $\omega = \dot{\theta}$, while noting $\ddot{\rho} = 0$, it can be shown that the required reflector pitch angle and acceleration can be written as

$$\text{(21a)} \quad \tan\alpha = \frac{\rho}{z}\left[1 - \left[\frac{\omega}{\tilde{\omega}}\right]^2\right], \quad \tilde{\omega}^2 = \frac{\mu}{r^3}$$

$$\text{(21b)} \quad a = \tilde{\omega}^2\left[1 + \left[\frac{\rho}{z}\right]^2\left[1 - \left[\frac{\omega}{\tilde{\omega}}\right]^2\right]^2\right]^{3/2} z,$$

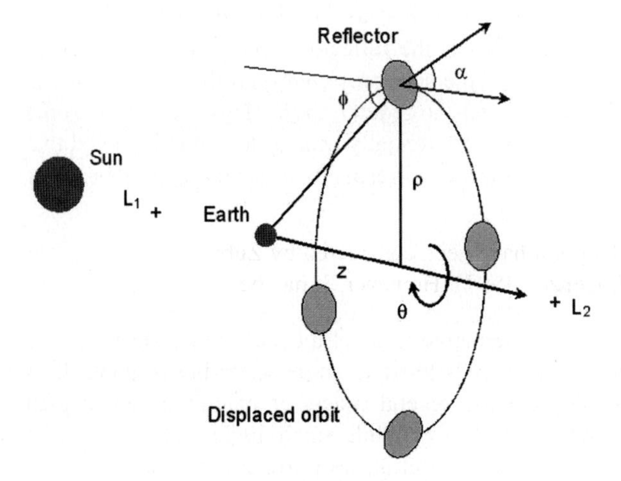

Figure 9. Orbit displaced along the Sun-Earth line

For the reflector to direct reflected light towards the Earth, it is clear from Fig. 9 that the reflector pitch angle α is related to ϕ as $\alpha = \phi/2$. Assuming a perfect specular reflector, an image of the solar disk will then be projected onto the Earth, with the image centre on the Earth-reflector line. The required reflector pitch angle is therefore defined as

$$\tan\alpha = \tan\left[\frac{1}{2}\tan^{-1}\left[\frac{\rho}{z}\right]\right]. \tag{22}$$

Since this orientation is defined a priori, the orbital angular velocity of the reflector must now be a variable of the problem in order to ensure correct illumination of the Earth by the reflector. Using Eq. (21a) it can be seen that the orbital angular velocity required for this desired reflector orientation is given by

$$\omega = \tilde{\omega}\left[1 - \left[\frac{z}{\rho}\right]\tan\alpha\right]^{1/2}. \tag{23}$$

The corresponding reflector acceleration can now be obtained from Eq. (21b). A section of the surfaces of constant reflector acceleration generated by this family of orbits is shown in Fig. 10 in the $\rho - z$ plane. Each point on a surface corresponds to a displaced orbit with some radius ρ and displacement distance z in the anti-Sun direction. It can be seen that the surfaces of constant reflector acceleration have a rotational symmetry about the Sun-line, and that for a given reflector acceleration a an orbit with a large radius and small displacement is possible ($\alpha \sim 45°$), or an orbit with a small radius and large displacement ($\alpha \sim 0°$). Clearly, displaced orbits with a small pitch angle α are desirable to maximise the projected reflector area, however, orbits with large displacements are found to be strongly perturbed by lunar gravitational perturbations. An orbit with a relatively modest displacement is still able to offset the momentum accumulated by the reflector due to solar radiation pressure by the z-component of gravitational force, while the orbit remains Sun-synchronous with the orbit plane always normal to the Sun-Earth line. While not considered further here, it is noted that if the reflector orbit period is chosen to be an integer number of sidereal days, the spot projected onto the Earth's surface by the solar reflector will follow a repeat ground track. This may have some importance if certain regions are to be preferentially heated. It should be noted that a reflector acceleration of order 10 mm/s^2 corresponds to a total areal density of order $0.9 - 1.0\,\text{g/m}^2$.

4.3 Small Angle Scattering

As discussed in Section 1.2.4, an alternative to a solid occulter or reflector is the use of a mesh micro-structure which leads to small angle scattering of light. Such meshes can have a total deployed mass several orders of magnitude lower than that required for a solid occulter or reflector. While small angle scattering can be considered for Earth climate cooling, Teller et al. also propose the use of a mesh

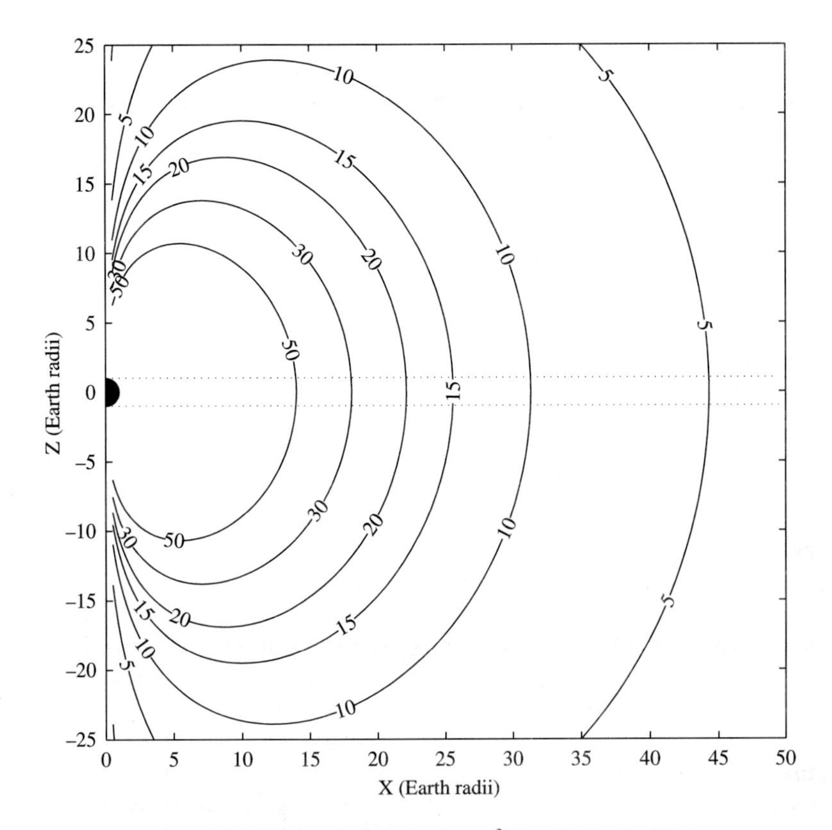

Figure 10. Surfaces of reflector acceleration a (mm/s^2) for displaced orbits

displaced off the Sun-Earth line to scatter light onto the day side of the Earth, increasing the total solar insolation (Teller et al., 2004). This is a more challenging prospect since at L_1, for example, the Earth subtends an angle of order 0.01 rad, so that extremely accurate scattering and attitude control of the mesh is required over long periods to ensure light is continually directed onto the Earth's disk.

4.4 Reflector Sizing

It was assumed in Section 4.1 that the total reflector area for climate heating is of the same order as that for climate cooling. In addition, it will be assumed that orbits with a relatively modest displacement are utilised so that $\alpha \sim 45°$. As noted in Section 4.1, the increase in solar insolation is of order 3.6%, corresponding to an increase in mean global temperature of order 4 K (in concurrence with Teller et al., 2004). If an orbit with a modest displacement is selected so that the required reflector acceleration is of order $10\,mm/s^2$, corresponding to an areal density of order $0.9-1.0\,g/m^2$, the mass of the reflector system is of order $6.6 \cdot 10^9\,kg$. As noted in Table 3, this mass is comparable with that of the great pyramid of Giza, and is an order of magnitude less than the requirements for the Three Gorges Dam.

In order to investigate the potential effect of an increase in solar insolation of order 3.6% on future glaciation episodes, it necessary to use a 1-dimensional climate model which attempts to capture the variation of temperature with latitude, as discussed in Section 2.4. The model utilized here has 9 latitude bands with heat transport between bands being proportional to the temperature difference between them. Ice albedo effects are modelled using a switch from ice covered to ice-free at a critical temperature of −10 °C. This is below the standard freezing point of water since sea ice will only form at somewhat lower temperatures (McGuffie and Henderson-Sellers, 1997).

Lastly, as discussed in Section 3.3, the reflecting disks are assumed to be fabricated from near Earth asteroid material and deployed to form a displaced ring, normal to the Sun-Earth line, as shown in Fig. 12. In order to avoid brightening the night sky, it is likely that the ring of reflectors would need to be configured such that the scattered light was incident near the rim of the Earth's disk. Such a configuration would lead to a brightening of the sky at local dawn and dusk. A swarm of reflectors could in principle use solar radiation pressure to manoeuvre between an occulting location sunward of L_1 and a heating location displaced behind the Earth to provide complete control over the solar insolation incident on the Earth.

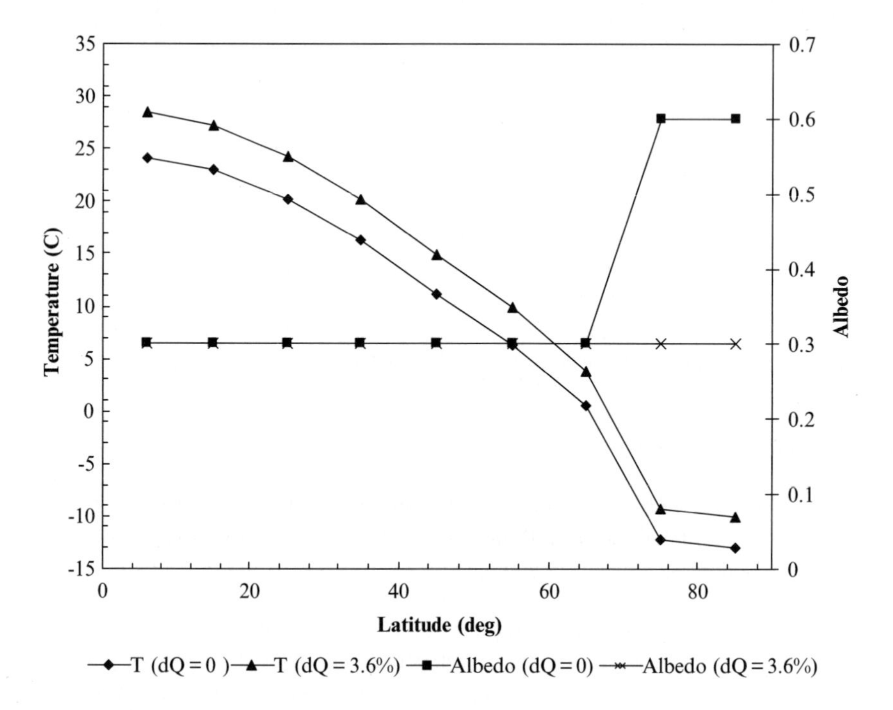

Figure 11. Effect of 3.6% increase in solar insolation using a 1-dimensional climate model

Figure 12. Solar reflectors in Earth orbit (credit: Aerospace Systems Ltd). [Please see this figure in the color section at p. 316]

Using this 1-dimensional climate model the effect of increasing the solar insolation by 3.6% can now be investigated. Figure 11 shows a comparison of the variation of temperature with latitude for $\delta Q/Q = 0$ (model mean global temperature 14.9°C) and $\delta Q/Q = 3.6\%$ (model mean global temperature 18.8°C). The key result from the model is that for $\delta Q/Q = 0$ the final two latitude bands are ice covered (representing the present day polar caps), while for $\delta Q/Q = 3.6\%$ the ice caps vanish. Therefore, for the reflector area assumed here, the present day ice caps can be removed and in principle future advance of polar ice sheets controlled.

5. ENGINEERING CHALLENGES

The first steps towards fielding large orbiting reflectors are currently taking place through the accelerating development of solar sail propulsion (McInnes, 1999). Square solar sails with a side of order 100 m and an areal density of order 5 g/m^2 are foreseen in the near future, although concepts exist for both extremely large disk sails (~1000 m) and low areal densities (~0.1 g/m^2) using thin metallic film. These terrestrially fabricated reflectors can provide a route towards the fabrication of reflectors from in-situ resources such as near Earth asteroids.

The use of reflectors for geo-engineering has been discussed in some detail in Sections 3 and 4, with a mass of order $5 \cdot 10^{10}$ kg required for useful climate cooling and a mass of order $6 \cdot 10^{9}$ kg for useful climate heating. Again, it is noted that use of small angle scattering meshes could reduce these mass requirements considerably. Clearly, the fabrication of a swarm of reflectors with a mass of order 10^{10} kg would require a capability to exploit in-situ resources such as near Earth asteroids. For example, the mass requirements for the minimum mass solar occulter can be satisfied by small M-type asteroids, which are abundant in Nickel-Iron materials (Gehrels, 1979).

It will be assumed that the occulter is fabricated from thin metallic film processed from such a body, and that it has a bulk density half that of Iron ($7860\,kg/m^3$) to account for the non-metal content. It is then found that a small M-type asteroid with a radius of order 145 m will provide the required mass for fabrication of the occulting disks. The asteroid would require to be processed in-situ, probably using solar heating, and the metallic products extruded into thin film for fabrication of the disks. If it is assumed that a terrestrially fabricated solar reflector is initially deployed, then the time required to process the asteroid can be estimated. For example, a 500 m disk reflector will intercept a solar flux equivalent to approximately 1 GW at 1 astronomical unit. If the disk has an areal density of order $1\,g\,m^{-2}$, its total mass is of order 800 kg. Assuming the asteroid material is liberated by focusing this energy to raise the local temperature to above the melting point of the metallic component (heat of fusion of Iron $\sim 2.8 \cdot 10^5\,J/kg$), the rate of production of mass is potentially of order $3.6 \cdot 10^3\,kg/s$, resulting in complete processing of the asteroid on a timescale of order 150 days. Clearly, there are significant engineering challenges associated with quickly extruding the liberated asteroid material into thin film. However, it can be shown that the processing time can in fact be greatly reduced by using the fabricated film to further increase the heat flux onto the asteroid to accelerate the processing.

6. ASTRONOMICAL ENGINEERING

6.1 Engineering the Deep Future

The previous sections have considered the use of solar reflectors to actively geo-engineer the Earth's climate. While such possibilities are daunting, even larger scale macro-engineering schemes can be contemplated. A scheme has been proposed by Korycansky et al., (2001) to gradually modify the Earth's orbit over a timescale of order 10^9 years to compensate for the slowly increasing solar luminosity (also Birch, 1993; Shkadov, 1987). Such a feat of truly astronomical engineering could in principle preserve the viability of the Earth's biosphere into the distant future. The scheme requires repeated $\sim 10^6$ fly-pasts of the Earth by a small solar system body (Kuiper belt object or main belt asteroid $\sim 10^{19}$ kg) to transfer energy from the body to the Earth, thus raising the Earth's orbit radius. Intermediate fly-pasts of Jupiter are used to re-target the small body, thus effectively transferring orbital energy from Jupiter to the Earth. Although such a scheme can leverage significant capability from a single small body, trajectory corrections require the processing of a total of $\sim 10^{19}$ kg of material to provide propulsive reaction mass to accurately target each of the $\sim 10^6$ fly-pasts. In addition, repeated grazing fly-pasts of the Earth by a body $\sim 10^{19}$ kg poses significant risks in the event of targeting errors.

In this Section an alternative strategy is proposed which provides a slow, but continuous, modification of the Earth's orbit radius and reduces the mass processing requirements of the venture by four orders of magnitude (McInnes, 2002c). A mass of order 10^{15} kg is used to fabricate a large thin metallic film to form a solar reflector to generate a propulsive force from solar radiation pressure (solar sail). Although the area of the sail is large, mass processing will be the key driver of any future large scale space engineering endeavour. The sail is located at a static equilibrium position near the Earth where the solar radiation pressure induced force exerted on the sail is essentially in balance with gravitational force of the Earth. Analysis then shows that the centre-of-mass of the Earth-sail system in fact accelerates, and so provides an extremely low but continuous acceleration which will modify the Earth's orbit radius as required. The acceleration of the centre-of-mass comes from the redirection of the flux of momentum transported by solar photons which are reflected by the sail. While the sail experiences a propulsive thrust, it is gravitationally anchored to the Earth and thus provides an acceleration of the centre-of-mass of the Earth-sail system. Much nearer term applications of solar sails have also been considered to modify the orbits of near Earth asteroid (McInnes, 2004).

6.2 Gravitational Coupling

In order to investigate the use of solar radiation pressure to modify the Earth's orbit radius, the dynamics of the centre-of-mass of the Earth-sail system C will be considered, as shown in Fig. 13. An inertial frame of reference $[X,Y,Z]$ with origin O is defined, with the Sun assumed fixed at O. The Earth and sail are located at position $\mathbf{r}_S$ and $\mathbf{r}_E$ with the centre-of-mass of the Earth-sail system C located at $\mathbf{r}_C$. The sail is assumed to be in static equilibrium relative to the Earth such that $|\mathbf{r}| = 0$, where $\mathbf{r}$ denotes the relative position of the Earth and sail.

The solar sail exerts a force $\boldsymbol{f}$ due to solar radiation pressure, where the force is directed normal to the sail surface for a specular reflector. The magnitude of this force scales as the inverse square of solar distance, where some reference force f_O will be defined at solar distance r_O such that

$$f = f_O \left(r_O / r_S\right)^2 (\hat{\mathbf{r}}_S \cdot \mathbf{n})^2 \mathbf{n}. \tag{24}$$

where again the magnitude of the solar radiation pressure induced force scales as $\cos^2 \alpha$, where α is the angle between the sail normal and the Sun-sail line, as shown in Fig. 13. Furthermore, the reference force f_O for a perfectly reflecting sail can be obtained from the solar luminosity L_S as

$$f_O = \frac{2A_S}{C} \frac{L_S}{4\pi r_O{}^2}, \tag{25}$$

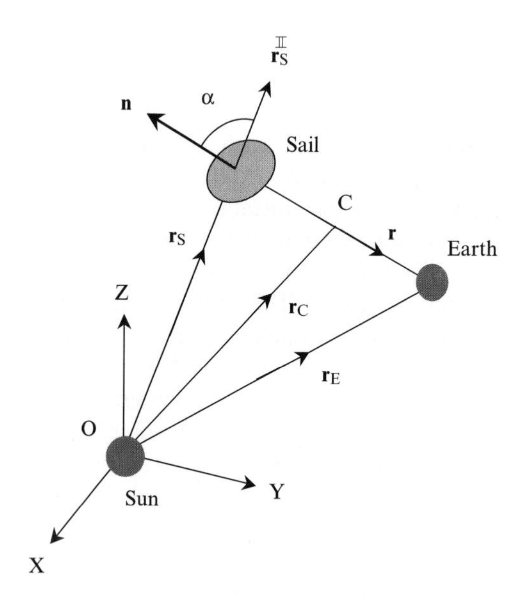

Figure 13. Sun-Earth-sail three-body system

where C is the speed of light and A_S is the sail area (McInnes, 1999). It will be assumed that the sail is an ideal specular reflector with no absorption.

Now that the force $\boldsymbol{f}$ exerted on the sail due to solar radiation pressure has been defined, the equation of motion of the sail may be written as

$$(26) \qquad m_S\ddot{\mathbf{r}}_S = -\frac{Gm_O m_S}{{r_S}^3}\mathbf{r}_S + \frac{Gm_S m_E}{r^3}\mathbf{r} + \boldsymbol{f},$$

where the Sun is assumed to be fixed at the origin O of the inertial frame of reference and $\mathbf{r} = \mathbf{r}_E - \mathbf{r}_S$. Similarly, the equation of motion for the Earth may be written as

$$(27) \qquad m_E\ddot{\mathbf{r}}_E = -\frac{Gm_O m_E}{{r_E}^3}\mathbf{r}_E - \frac{Gm_S m_E}{r^3}\mathbf{r},$$

where the last term of Eq. (27) represents the gravitational force of the sail exerted on the Earth. In order to investigate the dynamics of the centre-of-mass of the Earth-sail system, its location is defined as

$$(28) \qquad \mathbf{r}_C = \frac{m_S\mathbf{r}_S + m_E\mathbf{r}_E}{m_S + m_E}.$$

Therefore, adding Eq. (26) and Eq. (27) and using the definition of the centre-of-mass of the Earth-sail system yields

$$(29) \qquad (m_S + m_E)\ddot{\mathbf{r}}_C = -\frac{Gm_O m_S}{{r_S}^3}\mathbf{r}_S - \frac{Gm_O m_E}{{r_E}^3}\mathbf{r}_E + \boldsymbol{f}.$$

Since the sail will be located in close proximity to the Earth it is clear that $r_S \approx r_E \approx r_C$ so that

$$(30) \qquad (m_S + m_E)\ddot{\mathbf{r}}_C \approx -Gm_O(m_S + m_E)\frac{\mathbf{r}_C}{r_C{}^3} + \boldsymbol{f}.$$

Furthermore, since $s_a me$ an equation of motion for the centre-of-mass of the Earth-sail system is then obtained as

$$(31) \qquad \ddot{\mathbf{r}}_C + Gm_O\frac{\mathbf{r}_C}{r_C{}^3} \approx \frac{1}{m_E}\boldsymbol{f}.$$

It can be seen that Eq. (31) has the form of the usual two-body equation of motion with the addition of a forcing term from the solar sail. Since the forcing term is scaled by $m_E{}^{-1}$, the centre-of-mass of the Earth-sail system will experience an extremely small, but secular perturbation.

6.3 Earth Orbit Modification

Now that its has been demonstrated that the perturbing force $\boldsymbol{f}$ due to solar radiation pressure will accelerate the centre-of-mass of the Earth-sail system, an optimum strategy is required for orbit modification. Firstly, taking the scalar product of Eq. (31) with the velocity vector of the centre-of-mass yields

$$(32) \qquad \frac{d}{dt}\left[\frac{1}{2}\mathbf{v}_C \cdot \mathbf{v}_C - \frac{Gm_O}{r_C}\right] \approx \frac{1}{m_E}\boldsymbol{f} \cdot \mathbf{v}_C,$$

where $\mathbf{v}_C$ is the velocity of the centre-of-mass of the Earth-sail system. It can be seen that the term on the left is identified as the rate of change of total energy E due to the perturbing force $\boldsymbol{f}$. Assuming that $\boldsymbol{f}$ is directed within the ecliptic plane and that $\mathbf{v}_C \cdot \mathbf{r}_C \approx 0$, its direction can be optimised to maximise the rate of change of energy of the centre-of-mass of the Earth-sail system. From Eq. (32) it can then be shown that

$$(33) \qquad \dot{E} \approx \frac{1}{m_E} v_C f_o (r_o/r_S)^2 \cos^2\alpha \sin\alpha, \;\; E = 1/2\,\mathbf{v}_C \cdot \mathbf{v}_C - Gm_O/r_C.$$

The rate of change of total energy E is now maximized by setting $d\dot{E}/d\alpha = 0$ which yields an optimum pitch angle of $\tan\alpha = 1/\sqrt{2}$. With this optimum pitch angle the orbit radius of the centre-of-mass of the Earth-sail system will exhibit a secular increase, corresponding to a slow outward quasi-circular spiral. Since the resulting radial and transverse perturbing force components are essentially constant, it can be demonstrated that there is no secular increase in orbit eccentricity.

Assuming the orbit of the centre-of-mass of the Earth-sail system remains quasi-circular, the total orbit energy may be written as $E = -Gm_O/2r_C$ so that $\dot{E} \approx Gm_O\dot{r}_C/2r_C{}^2$ and $v_C \approx \sqrt{Gm_O/r_C}$. Then, using Eq. (33) and assuming again that

$r_S \approx r_C$, the evolution of the centre-of-mass of the Earth-sail system can be obtained. Although the magnitude of the solar radiation pressure induced force diminishes as an inverse square law, the solar luminosity will be increasing (indeed the objective of maneuvering the Earth's orbit is to maintain a constant flux). Therefore, it will be assumed that the magnitude of the solar radiation pressure induced force exerted on the sail is constant so that

$$\dot{r}_C \approx \frac{2 r_C^{3/2}}{\sqrt{G m_O}} \frac{f_O}{m_E} \cos^2 \alpha \sin \alpha, \tag{34}$$

where again $\tan \alpha = 1/\sqrt{2}$. This equation can now be integrated to obtain the evolution of the centre-of-mass of the Earth-sail system as

$$\sqrt{\frac{G m_O}{r_C^{\,O}}} - \sqrt{\frac{G m_O}{r_C^{\,f}}} \approx \frac{f_O}{m_E} \cos^2 \alpha \sin \alpha \left(t_f - t_O\right), \tag{35}$$

where $r_C^{\,O}$ is the starting orbit at 1 astronomical unit and $r_C^{\,f}$ is the end-point, taken as 1.5 astronomical units. The required solar radiation pressure induced force to perform this manoeuvre can therefore be obtained from Eq. (35) as

$$f_O \approx \frac{m_E}{\cos^2 \alpha \sin \alpha} \left[\sqrt{\frac{G m_O}{r_C^{\,O}}} - \sqrt{\frac{G m_O}{r_C^{\,f}}} \right] \left(t_f - t_O\right)^{-1}. \tag{36}$$

If $\left(t_f - t_O\right)$ is taken to be $6.3 \cdot 10^9$ years, to ensure a constant solar flux, as considered by Korycansky et. al., and $\tan \alpha = 1/\sqrt{2}$, the required solar radiation pressure induced force is found to be of order $4.3 \cdot 10^{11}$ N. From Eq. (25) this corresponds to a sail area of order $4.7 \cdot 10^{16}\,\mathrm{m}^2$, which is equivalent to a disc sail with a radius of $1.2 \cdot 10^8$ m (19.2 Earth radii). Although it is clear that a truly vast reflecting area is required, it is mass which is the key driver. For comparison, the longest human-made terrestrial structure is FLAG (Fibreoptic Link Around the Globe) with a length of 27,300 km (4.3 Earth radii). It should be noted that the total area need not comprise a single reflector, but could comprise a large number of much smaller reflectors. This is also likely to alleviate, to some extent, the large tidal forces exerted across a single large sail which would otherwise generate a torque aligning the sail with the local gravity gradient. The required sail mass will now be determined from the conditions for static equilibrium of the sail relative to the Earth.

6.4 Reflector Static Equilibrium

The static equilibrium of a solar sail in the Sun-Earth three-body system has been considered for near term applications of small solar sails (McInnes et al., 1994; McInnes, 1999). However, for the approximate analysis presented here, a simpler two-body analysis is sufficient. It can be assumed that the solar sail is in static equilibrium relative to the Earth such that $|\mathbf{r}| = 0$ and that $\mathrm{r}_S \approx \mathrm{r}_E$ and $m_E \sim m_S$ so that Eq. (26) and Eq. (27) yield

$$\ddot{\mathbf{r}} = \frac{Gm_E}{r^3}\mathbf{r} + \frac{1}{m_S}f. \tag{37}$$

A simple equilibrium solution to Eq. (37) can now be found if

$$f = -\frac{Gm_E m_S}{r^3}\mathbf{r}, \tag{38}$$

so that the solar radiation pressure induced force balances the local gravitational force due to the Earth. It should be noted that such equilibria are unstable, so that active control of the sail is required. While the equilibrium is unstable, it has been demonstrated elsewhere that it is strictly controllable using trims to the sail orientation (McInnes et al., 1994). A further issue is the eccentricity of the Earth's orbit which will pose an additional requirement for active control.

The sail is positioned such that the sail normal is tilted at an angle $\tan\alpha = 1/\sqrt{2}$ relative to the Sun-line. Then, $f_O \cos^2\alpha = Gm_E m_S/r^2$ where r is the Earth-sail separation. The required sail mass for static equilibrium, to gravitationally 'anchor' the sail to the Earth, can now be determined from

$$m_S = \frac{r^2}{Gm_E} f_O \cos^2\alpha. \tag{39}$$

Given the large physical extent of the sail, the Earth-sail separation will be defined as $2 \cdot 10^9$ m (of order 300 Earth radii, or 5 times the lunar distance). Then, the required sail mass is found from Eq. (39) to be $2.9 \cdot 10^{15}$ kg.

The scheme proposed by Korycansky et al. requires an object with a mass of order 10^{19} kg to be manoeuvred, with an additional 10^{19} kg of mass to be processed for reaction mass to provide propulsive thrust for trajectory correction manoeuvres.

The scheme presented here therefore requires four orders of magnitude less mass to be processed and provides a small, but continuous acceleration to the centre-of-mass of the Earth-sail system, rather than a large number of grazing fly-pasts of a 10^{19} kg object. It is suggested that these two advantages are significant, although the total area of the reflector is large. Again, it is emphasized that mass will be the driver in any future spaced-based macro-engineering venture. A comparison with natural structures is provided in Table 4.

6.5 Requirements

It has been shown that the mass processing requirements of the scheme proposed by Korycansky et al. can be reduced by some four orders of magnitude. Although this is a significant reduction in mass, the vast physical extent of the sail requires some thought.

Table 4. Mass comparison with natural structures

Scale	Mass (kg)	Natural Structure
10^{14}	$2\text{–}3 \cdot 10^{14}$	Mass ejected from Mount Tambora volcano (1815)
10^{15}	$7.2 \cdot 10^{15}$	Mass of S-type asteroid Eros ($13 \cdot 13 \cdot 33\,\text{km}^3$)
10^{18}	$5.3 \cdot 10^{18}$	Mass of Earth's atmosphere
10^{20}	$9 \cdot 10^{20}$	Mass of G-type asteroid Ceres (933 km diameter)
10^{21}	$1.4 \cdot 10^{21}$	Mass of Earth's oceans

The areal density of the sail is of order $0.06\,\text{kg/m}^2$. If the sail is assumed to be a thin metallic film, with a bulk density similar to that of Iron ($7860\,\text{kg/m}^3$), the corresponding film thickness is of order $8 \cdot 10^{-6}$ m. Thin metallic film is readily produced to a thickness of order 10^{-6} m and Nickel-Iron materials are abundant in main belt M-type asteroids, as discussed in Section 5. Again, assuming a predominantly Iron content (density half that of Iron), the fabrication of the sail would require an M-type body with a radius of order $5.6 \cdot 10^3$ m.

Forming a thin metallic film with an area of order $10^{16}\,\text{m}^2$ is clearly a challenge, although again mass is the key driver (a truly impressive sail area is associated with the scheme proposed by Shkadov, 1987, and later by Badescu and Cathcart, 2000, to modify the Sun's orbit about the galactic centre). While a single thin film can be envisaged, a more realistic proposition is a swarm of a large number of sails with a smaller area. In this way the fabrication, and indeed the control issues associated with a single vast sail can be avoided somewhat. Clearly, lunar gravitational perturbations will require compensation through active control, although the timescale of the dynamics of the problem is extremely slow.

Future studies must also consider the effect of the scheme on the long term evolution of the lunar orbit, although this is a common issue for any concept to modify the Earth's orbit. Indeed the effect of slowly displacing the Earth's orbit will likely lead to strong mutual perturbations with other planetary bodies as various resonance conditions are traversed.

7. CONCLUSIONS

The aim of this chapter has been to provide some insights into the possibilities offered by macro-engineering the Earth's climate using orbiting solar reflectors. While macro-engineering clearly requires a leap of the imagination over current large scale terrestrial engineering, the natural variability of the Earth's climate will necessitate some form of manipulation of the climate in the long term. Experience obtained from such manipulation may then show the way towards engineering the climates of other planets, in particular Mars. Again, while the scale of engineering discussed in this chapter is daunting, the availability of vast quantities of freely available solar energy in space, and the active control of such energy using orbiting solar reflectors will certainly allow the possibility of large scale manipulation of planetary climates. Whether such possibilities are indeed exploited in the future, both to mitigate against the Earth's natural climate variability and to unlock the resources of space, remains to be seen.

REFERENCES

Badescu V (2005) Regional and seasonal limitations for Mars intrinsic ecopoiesis. Acta Astronaut 56:670–680

Badescu V, Cathcart R (2000) Stellar engines for Kardashev type II civilisations. J Br Interplanet Soc 53:297–306

Benzi R, Parisi G, Sutera A, Vulpiani A (1983) A theory of stochastic resonance in climatic change. SIAM J Appl Math 43: 565–578

Berglund N, Gentz B (2001) Metastability in simple climate models: pathwise analysis of slowly driven Langevin equations. Proc. of the 2nd Workshop on Stochastic Climate Models, Chorin, Germany

Birch P (1991) Terraforming Venus quickly. J Br Interplanet Soc 44:157–167

Birch P (1992) Terraforming Mars quickly. J Br Interplanet Soc 45:331–340

Birch P (1993) How to move a planet. J Br Interplanet Soc 46:314–316

Bookless J, McInnes CR (2004) Dynamics, stability and control of displaced non-Keplerian orbits, IAC-04-A.7.09. 55th International Astronautical Congress, Vancouver

Budyko MI (1969) The effect of solar radiation variations on the climate of the Earth. Tellus 21:611–619

Cicerone RJ, Elliott S, Turco RP (1992) Global environmental engineering. Nature 356:9

Early JT (1989) Space-based solar shield to offset greenhouse effect. J Br Interplanet Soc 42:567–569

Emanuel K (2002) A simple model of multiple climate regimes. J Geophys Res (Atmospheres) 107, Issue D9, pp. ACL 4–1, CiteID 4077, DOI 10.1029/2001JD001002

Fogg M (1992) A synergestic approach to terraforming Mars. J Br Interplanet Soc 45:315–329

Free M, Robock A (1999) Global warming in the context of the little ice age. J Geophys Res 104:19057–19070

Gehrels T (ed) (1979) Asteroids. University of Arizona Press, Tucson

Gerstell MF, Francisco JS, Yung YL, Boxe C, Aaltonee ET (2001) Keeping Mars warm with new super greenhouse gases. Proc Natl Acad Sci USA 98:2154–2157

Govindasamy B, Caldeira K (2000) Geo-engineering Earth's radiation balance to mitigate CO_2-induced climate change. Geophys Res Lett 27:2141–2144

Govindasamy B, Thompson S, Duffy PB, Caldeira K, Delire C (2002) Impact of geoengineering schemes on the terrestrial biosphere. Geophys Res Lett 29. Issue 22, pp. 18–1, CiteID 2061, DOI 10.1029/2002GL015911 (GeoRL Homepage)

Hansen J, Lacis A, Ruedy R, Sato M (1992) Potential climate impact of Mount Pinatubo eruption. Geophys Res Lett 19:215–218

Higuchi K (1970) A possibility of constructing a dam to change the general oceanic circulation. 2nd International Future Research Conference, Kyoto

Hoffman PF, Schrag DP (2002) The snowball Earth hypothesis: testing the limits of global change. Terra Nova 14:129–155

Hudson H (1991) A space parasol as a countermeasure against the greenhouse effect. J Br Interplanet Soc 44:139–141

Keith DW (2000) Geoengineering the climate: history and prospect. Annu Rev Energy Environ 25:245–284

Korycansky DG, Laughlin G, Adams FC (2001) Astronomical engineering: A strategy for modifying planetary orbits. Astrophys Sp Sci 275:349–366

Mann ME, Bradley RS, Hughes MK (1998) Global-scale temperature patterns and climate forcing over the past six centuries. Nature 392:779–787

Mautner M (1991) A space-based solar screen against climatic warming. J Br Interplanet Soc 44:135–138

Mautner M, and Parks K (1990) Space-based control of the climate. Proc. Space 90. Amer Soc Civ Eng, pp 1159–1169

McGuffie K, Henderson-Sellers A (1997) A climate modeling primer. John Wiley & Sons, Chichester

McInnes CR (1999) Solar sailing: Technology, dynamics and mission applications. Springer-Verlag, London

McInnes CR (2002a) Non-Keplerian orbits for Mars solar reflectors. J Br Interplanet Soc 55:74–84

McInnes CR (2002b) Minimum mass solar shield for terrestrial climate control. J Br Interplanet Soc 55:307–311

McInnes CR (2002c) Astronomical engineering revisited: planetary orbit modification using solar radiation pressure. Astrophys Sp Sci 282:765–772

McInnes CR (2004) Deflection of near-Earth asteroids by kinetic energy impacts from retrograde orbits. Plan Sp Sci 52:587–590

McInnes CR, McDonald AJC, Simmons JFL, MacDonald EW (1994) Solar sail parking in restricted three-body systems. J Guid Dyn Cont 17:399–406

Muller RA, MacDonald GJ (1997) Glacial cycles and astronomical forcing. Science 277:215–218

Oberth H (1972) Ways to spaceflight, NASA Technical Translation TT F-622

Pearson J, Oldson J, Levin E (2002) Earth rings for planetary environmental control, IAF-02-U.1.01. 53rd International Astronautical Congress, Houston

Rahmstorf S (2003) Timing of abrupt climate change: a precise clock. Geophys Res Lett 30:1510–1514

Ruddiman WF (2003) The anthropogenic greenhouse era began thousands of years ago. Climate Change 61:261–293

Sagan C (1961) The planet Venus. Science 133:849–858

Sagan C (1973) Planetary engineering on Mars. Icarus 20:513–514

Schneider SH (2001) Earth systems engineering and management. Nature 409:417–421

Seifritz W (1989) Mirrors to halt global warming. Nature 340:603

Shkadov. LM (1987) Possibility of controlling solar system motion in the galaxy, IAA-87-613. 38th International Astronautical Federation Congress, Brighton

Teller E, Hyde R, Ishikawa M, Nuckolls J, Wood L (2004) Active climate stabilization: presently-feasible albedo-control approaches to prevention of both types of climate change, Cambridge-MIT Institute. Symposium on Macro-Engineering Options for Climate Change Management and Mitigation, Cambridge

Teller E, Wood L, Hyde R (1997) Global warming and ice ages: I. prospects for physics based modulation of global change, UCRL-231636/UCRL JC 128715. Lawrence Livermore National Laboratory

Zubrin R, McKay C (1997) Technological requirements for terraforming Mars. J Br Interplanet Soc 50:83–92

CHAPTER 12

STELLAR ENGINES AND THE CONTROLLED MOVEMENT OF THE SUN

VIOREL BADESCU[1] AND RICHARD BROOK CATHCART[2]

[1] *Candida Oancea Institute, Polytechnic University of Bucharest, Spl. Independentei 313, Bucharest 79590, Romania;*

[2] *Geographos, 1300 West Olive Avenue, Burbank, CA 91506, USA*

Abstract: A stellar engine is defined in this chapter as a device that uses the resources of a star to generate work. Stellar engines belong to class A and B when they use the impulse and the energy of star's radiation, respectively. Class C stellar engines are combinations of types A and B. Minimum and optimum radii were identified for class C stellar engines. When the Sun is considered, the optimum radius is around 450 millions km. Class A and C stellar engines provide almost the same thrust force. A simple dynamic model for solar motion in the Galaxy is developed. It takes into account the (perturbation) thrust force provided by a stellar engine, which is superposed on the usual gravitational forces. Two different Galaxy gravitational potential models were used to describe solar motion. The results obtained in both cases are in reasonably good agreement. Three simple strategies of changing the solar trajectory are considered. For a single Sun revolution the maximum deviation from the usual orbit is of the order of 35 to 40 pc. Thus, stellar engines of the kind envisaged here may be used to control to a certain extent the Sun movement in the Galaxy

Keywords: stellar engine, Kardashev type II civilization, Shkadov thruster, Dyson sphere, galaxy gravitational potential, Sun movement control strategy

1. INTRODUCTION

For various reasons, mankind may be faced in the future with the problem of changing the Sun revolution motion. Avoiding nearby supernovae or ordinary star collisions are examples. Diffuse matter clouds could also be a potential danger. Some studies suggest that during its lifetime the Sun has suffered about ten

V. Badescu et al. (eds.), Macro-Engineering: A Challenge for the Future, 251–279.

encounters with major molecular clouds (MMC) and it has had close (impact parameter less than 20 pc) encounters with more than 60 MMC of various masses (Clube and Napier, 1984; Napier, 1985). These events induce perturbations of the Oort comet cloud, known to be sensitive to the particular galactic orbit of the Sun, leading to possible comet impacts on Earth (Gonzalez, 1999).

The Sun will steadily leave the main sequence in a few billion years, as stellar evolution calculations show (see e.g. Sackmann et al., 1993). The consequences will be a "moist greenhouse" effect on Earth, which will likely spell a definite end to life on our planet well before the Sun will become a Red Giant (Kasting, 1988; Nakajima et al., 1992). A preliminary solution to preserve the present-day climate on Earth may be to change its orbit. This subject is treated in detail in literature (see, e.g. Korykansky et al., 2001; McInnes, 2002) and in Chapter 11 of this book.

Zuckerman (1985) estimates that if ancient extraterrestrial civilizations exist in the Galaxy, then between 0.01 and 0.1 of them would have been forced to vacate their native planet due to the primary star leaving the main sequence. Problems with feasibility and dynamics of mass interstellar migrations (Jones, 1981; Newman and Sagan, 1981) prompted some researchers to propose the so-called "interstellar transfer" (or "solar exchange") solution (Hills, 1984; Shkadov, 1987; Fogg, 1989). In this case the Earth (or, more generally, the home planet) is to be transformed into a planet of a different star. The interstellar transfer requires first of all a way of controlling Sun (or star) movement in the Galaxy.

In this chapter we study the amplitude of a possible human intervention on Sun revolution motion. In section 2 we give a brief overview of different proposals in the literature. Also, we define the concept of stellar engine and we give details about various stellar engine classes. In section 3 we give the background physics associated to these devices. In section 4 we develop a model for the motion of the Sun in the Galaxy, based on usual Newtonian dynamics. The details of Sun movement are complex but an "average" motion can be defined by using appropriate global Galaxy gravitational potentials. The movement is then studied in both the normal (unperturbed) case and in the perturbed case, when an additional (stellar engine) thrust force is acting on the Sun. To increase the confidence in results, two different global gravitational potentials are used. Finally, in the Conclusion section we summarize the main findings of our work.

2. PROPOSALS TO CHANGE SUN MOTION

In his 12 May 1948 Halley Lecture at Oxford University in the UK, Fritz Zwicky (1889–1974) (see Zwicky, 1957) announced the possibility of

"... accelerating ... (the Sun) to higher speeds, for instance 1000 km/s directed toward Alpha Centauri A in whose neighborhood our descendents then might arrive a thousand years hence. [Such a one-way trip] ... could be realized through the action of nuclear fusion jets, using the matter constituting the Sun and the planets as nuclear propellants".

Zwicky's Halley Lecture, which may be seen as a response to the 16 July 1945 first nuclear fission explosion in the USA, was published in *The Observatory* (68:121–143, 1948) where the author merely hinted at the technical possibilities. At that time lasers were yet to be invented—*circa* 1960—and the controlled movement of asteroids and planets was still to be scientifically theorized (Korykansky, 2004). However, during 1971 when SCIART, a blend of "Science" and "Art", was organized by Bern Porter (1911–2004) even artists started to advocate use of nuclear particle beams for peaceful projects. By 1992, the artist Francisco Infante voiced his desire that humans redesign the firmament by intentionally shifting the positions of the stars other than the Sun (Infante, 1992).

The first scientifically recorded evidence of a natural celestial body in space colliding with the Sun came on 30-31 August 1979 when a cometary nucleus (1979 XI: Howard-Koomen-Michels) was observed as it vaporized in the Sun's corona.

At the Conference on Interstellar Migration (held at Los Alamos, New Mexico, in May 1983), David Russell Criswell extrapolated from available astronomical facts that the Sun might never enter a Red Giant-stage because it will be transformed into a stable White Dwarf-stage star via anthropogenic "star lifting". Criswell speculated about, perhaps proposed, a nameless macro-project the goal of which was to annually remove $6.5 \cdot 10^{18}$ tons of solar plasma from the Sun for a period of $\sim$300 million years—about 2% of the Milky Way Galaxy's estimated age—setting aside the evicted plasma to cool by storing it near the Sun's poles in a stable form. He foresaw this macro-engineering activity commencing *circa* AD 2170–5650. Criswell's polar solar plasma lifts would be controlled and sustained versions of the Sun's natural coronal mass ejections, which occur most everywhere on that glowing celestial body's turbulent surface. Criswell's technique could be adapted to spin-up the Sun, thus causing a mixing of its materials artificially. However, a too rapid equatorial rotation could force the Sun to become dangerously unstable. Criswell did not mention moving stars in his work. His stellar husbandry and star lifting concepts essentially involved mining stars in order to divide their mass into smaller units so as to greatly extend their main sequence lifetime and the efficiency with which their radiant energy could be utilised. It was Fogg (1989) who adapted star lifting to moving stars by accelerating mass from just one stellar pole rather than both.

Oliver Knill, in 1997, suggested deliberate triggering of asymmetric fusion and fission in the Sun might be utilized to move the Sun and its cortege of planets (Knill, 2003). He referred to solar flares, both natural and man-made, as "rockets on the Sun". He alleged that if all the Sun's wind were focused in only one direction instead of being emitted globally, then the Sun might, in principle, be accelerated to a speed of 100 m/s in a year's time. Since such total harnessing of the Sun is unlikely, Knill offered that giant solar flares might be induced which would have the effect of propelling the Sun in a selected direction through space. His technical preference was to trigger huge artificial solar flares at one of the Sun's poles that perform as rocket motors, lest the induced anthropogenic solar wind cause Earth serious problems of human health or civilization's infrastructure breakdowns.

Of course, this limits the trajectory of the Sun to flight courses that may not be what human civilization most wants or needs. Like Fritz Zwicky, Knill opted for the use of nuclear particle beams as a tool of rocket motor ignition.

Zwicky, Knill and Criswell, therefore, have proposed very advanced tele-mining macro-projects that can have the planned effect of moving the Sun in some desired direction (Fogg, 1989).

Another way of controlling the Sun's movement is based on the concept of *stellar engine*. A stellar engine was defined in Badescu and Cathcart (2000) as a device that uses a significant part of a star's resources to generate work. Three types of stellar engines were identified and denoted as class A, B and C, respectively.

A class A stellar engine uses the impulse of the radiation emitted by a star to produce a thrust force. When acting through a finite distance the thrust force generates work. As example of class A stellar engine we refer to the Sun thruster proposed in Shkadov (1987), which consists of a mirror placed at some distance from the Sun (Fig. 1a). The mirror is situated such that the central symmetry of the solar radiation in the combined mirror-Sun system is violated and, as a consequence, a certain thrust force will arise. For a mirror of given surface mass density a balance exists between the gravitational force and the force due to solar radiation pressure at a certain mirror-Sun distance which remains constant. It may be shown that the equilibrium does not depend on the distance between mirror and the Sun, since both the gravitational force and the force of the solar light pressure per unit mirror surface are inversely proportional to the square of the radius. A mirror with given geometry located at 150 million km from the Sun requires a surface mass density of about $1.55 \cdot 10^{-3}$ kg/m^2 while its total mass amounts $10^{19} - 10^{20}$ kg (which may be compared with the mass of the Earth, which is $5.977 \cdot 10^{24}$ kg). Detailed calculations may be found in Shkadov (1987).

A class B stellar engine uses the energy flux of the radiation emitted by a star to generate mechanical power. An example of class B stellar engine was proposed in Badescu (1995). It consists of two concentric spherical "shells" centered on the star. The "shells" have not necessarily continuous boundaries but they could be as well as imaginary envelopes of a very large number of smaller 3D bodies englobing the star. The inner surface acts as a solar energy collector. The outer surface is a thermal radiator. The two surfaces have different but rather uniformly distributed temperatures, T_p and T_r, respectively. The existing difference of temperature $T_p - T_r$ determines a heat flux from the inner towards the outer surface. This flux entering the thermal engine is used for power generation.

A class C stellar engine was defined in Badescu and Cathcart (2000) as a combination of a class A and class B stellar engine (Fig. 1b). It uses the impulse and the energy of the star radiation to provide both a thrust force and mechanical power for its owning civilization. Note that class B and C stellar engines are normally built by using the material of the inner planets (see Section 2.2). Of course, in this case the entire human population has to leave the Earth and move on the stellar engine.

For completeness here we define a new stellar engine as follows. A class D stellar engine uses a star's mass to propel the star. A particular class D stellar engine is

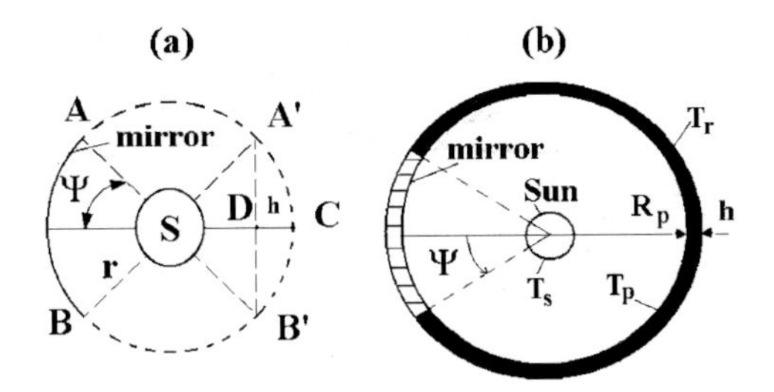

Figure 1. (a). A class A stellar engine (Shkadov thruster). r – distance between star S and the mirror. Ψ – mirror rim angle. (b). The class C stellar engine proposed in Badescu and Cathcart (2000). R_p – distance between star and inner surface, h – distance between inner and outer surfaces, T_S – star temperature; T_p, T_r – temperatures of the inner and outer surfaces, respectively

the stellar rocket described in Fogg (1989), based on a modification of the concept of "star lifting" proposed in Criswell (1985).

3. THERMODYNAMICS OF STELLAR ENGINES

3.1 Class A Stellar Engines

The energy radiated by a star is due to the nuclear reactions taking place in the nucleus. A steady-state star is characterized by a permanent balance between the energy flux generated during the nuclear reactions and the energy flux emitted at star's surface in all directions.

The bolometric luminosity $\tilde{L}_S$ of the Sun (i.e. its energy radiated on all wavelengths per unit time) is in present times (Ureche, 1987, p. 102):

$$\tilde{L}_S = 3.826 \cdot 10^{26} \text{ W}. \tag{1}$$

Let us consider the class A stellar engine of Fig. 1a. The star is prevented from losing energy on the solid angle covered by the mirror, as the energy emitted on that direction is returned to the star together with the reflected radiation. As the nuclear reaction rate doesn't change, the same energy flux $\tilde{L}_S$ has to be dissipated in space but this time from the effective (not covered by the mirror) star surface only. Consequently, the photosphere temperature will increase and it is expected that the star will change gradually to a different steady state. This effect was neglected in Shkadov (1987).

One denotes by R_S and $\tilde{T}_S$ the Sun's ray and its present-day temperature, respectively. The area of the Sun surface (S_S) and the surface of the Sun covered by the mirror ($S_{S,covered}$) are, respectively (Fig. 1a)

$$S_S = 4\pi R_S^2 \tag{2}$$

$$S_{S,covered} = 2\pi R_S h \tag{3}$$

Here h can be easily computed as a function of the mirror rim angle Ψ

(4) $$h = R_S(1 - \cos\Psi).$$

The effective (not covered by the mirror) Sun surface area, $S_{S,eff}$, is:

(5) $$S_{S,eff} = S_S - S_{S,covered}.$$

One supposes the Sun is a blackbody, both before and after mirror installation. Then the steady-state Sun temperature after mirror installation (T_S) has to obey the following energy balance equation:

(6) $$\tilde{L}_S = S_S \sigma \tilde{T}_S^4 = S_{S,eff} \sigma T_S^4.$$

By using Eqs. (5) and (6) one obtains

(7) $$T_S = \frac{\tilde{T}_S}{\left(1 - S_{S,covered}/S_S\right)^{1/4}}.$$

Using Eqs. (2)–(4) and Eq. (7) allows us to obtain the dependence of the Sun's temperature T_S on the mirror rim angle Ψ. Results are shown in Fig. 2. By increasing the mirror's rim angle the spectral class of the Sun gradually changes from G2 towards F2 (Harvard classification).

The increase in the Sun's photosphere temperature is accompanied by a change in its present absolute bolometric magnitude $\tilde{M}_b$. This change is governed by the equation (see Eq. (5.23) in Ureche, 1987, p. 109):

(8) $$M_b = \tilde{M}_b - 10\lg\left(\frac{T_S}{\tilde{T}_S}\right).$$

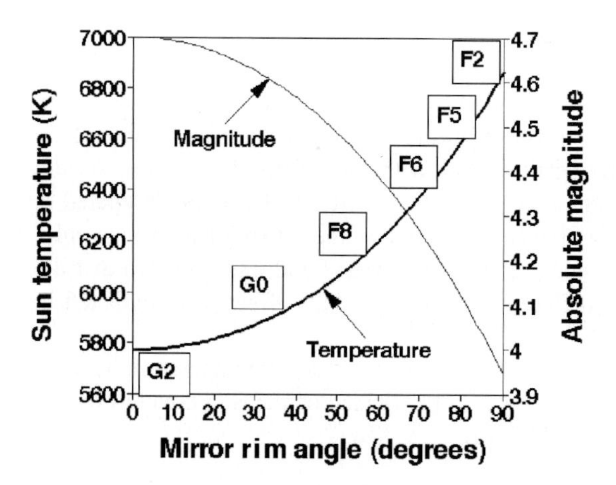

Figure 2. Dependence of Sun's photosphere temperature T_s and absolute magnitude M_b on the mirror rim angle Ψ (see Fig. 1a). The relation between temperature and star spectral classes (Harvard classification) is also shown

Figure 2 shows the dependence of the absolute bolometric magnitude of the Sun, M_b, as a function of mirror rim angle Ψ. We have taken into account that the Sun's present absolute bolometric magnitude is $\tilde{M}_b = 4.7$.

One can see that for the rim angle considered by Shkadov (1987) in his calculations (i.e. $\Psi = 30°$) both the photosphere temperature T_S (and its associated spectral class) and the absolute bolometric magnitude M_b remains quite close to the present-day values.

The mass of the mirror is distributed over a very large surface and, as a consequence, its influence on the orbit of the Earth is expected to be small. However, the Earth temperature may be affected in case of mirrors with large rim angle. Therefore, the mirror should be placed and kept in such a position that the orbit and temperature of the Earth are not affected significantly (for example, the mirror-Sun direction may be kept perpendicular on Earth orbit).

3.2 The Dyson Sphere Revisited

In this section we shall consider a 'usual' thin Dyson sphere (DS) englobing the Sun (Dyson, 1966). The inner DS surface constitutes the habitat of mankind. Due to its symmetry, the Dyson sphere will have a rather uniform surface temperature. The DS material is assumed to have a good thermal conductivity. Consequently, one could neglect the thermal gradients on material's thickness.

The steady-state energy balance per unit DS area is:

$$(9) \qquad a\frac{B_S}{\pi}\sigma T_S^4 + a\left(1-\frac{B_S}{\pi}\right)e_{\text{int}}\sigma T_p^4 = (e_{\text{int}} + e_{ext})\sigma T_p^4.$$

Here a is the absorptance of DS inner surface while e_{int} and e_{ext} is the emittance of DS inner and outer surfaces, respectively. Also, T_S and T_p is Sun and DS temperature, respectively. The first term in the l.h.s. of Eq. (9) is the energy flux density absorbed from the Sun while the second term is the energy flux density absorbed from the whole Dyson sphere. The r.h.s. of Eq. (9) contains the energy flux densities emitted by the DS inner and outer surfaces, respectively.

The geometric factor B_S in Eq. (9) may be computed as in Landsberg and Badescu (1998):

$$(10) \qquad B_S = \int_0^{\delta} \cos\theta \sin\theta d\theta \int_0^{2\pi} d\alpha = \pi \sin^2\delta,$$

where δ is the half-angle of the cone subtended by the Sun when viewed from an arbitrary point placed on DS inner surface. One can simply prove that

$$(11) \qquad \sin^2\delta = \left(\frac{R_S}{R_p}\right)^2,$$

where R_S and R_p are Sun and DS radii, respectively. One denotes:

$$(12) \qquad x \equiv \frac{B_S}{\pi} = \left(\frac{R_S}{R_p}\right)^2.$$

The steady-state energy balance for the Sun's surface is:

$$(13) \qquad 4\pi R_S^2 \sigma T_S^4 - 4\pi R_p^2 e_{\mathrm{int}} \sigma T_p^4 = \tilde{L}_S.$$

The first term in the r.h.s. of Eq. (13) is the energy flux emitted by the whole surface of the Sun while the second term is the energy flux received by the whole surface of the Sun from the Dyson's sphere. If one takes into account, on one hand, the multiple reflections of solar radiation on DS inner surface, and, on the other hand, the DS symmetry, one concludes that the absorptance $a \approx 1$. This is only true if one neglects that part of the radiation reflected by DS inner surface which is incident on the Sun's surface. By solving the Eqs. (9) and (13) and taking into account Eq. (12) one obtains:

$$(14) \qquad T_p = \left(\frac{\tilde{L}_S}{4\pi R_p^2 \sigma e_{ext}}\right)^{1/4}$$

$$(15) \qquad T_S = \left[\left(1 + x\frac{e_{\mathrm{int}}}{e_{ext}}\right)\frac{\tilde{L}_S}{4\pi R_S^2 \sigma}\right]^{1/4}.$$

These relations are valid under the condition $T_S > T_p$, which may be re-written (by using Eqs. 12, 14 and 15) as:

$$(16) \qquad R_p \geq \left(\frac{1 - e_{\mathrm{int}}}{e_{ext}}\right)^{1/2} R_S.$$

Figure 3 of Badescu and Cathcart (2000) shows the dependence of DS temperature T_p on the radius for various values of DS surface emittance $e = e_{\mathrm{int}} = e_{ext}$. A surface temperature comparable with present-day average ground surface temperature (~300 K) corresponds to high values of surface emittance. A number of conclusions may be drawn. First, small radii increase the feasibility of a DS project as the amount of material required is proportional to R_p^2. Second, the inner planets seem to be the best source of material in this case due to the shorter distance between their orbit and the place of the future Dyson sphere. The material of the inner planets has a relatively low albedo (between 0.07 in case of Mercury and 0.39 in Earth case (Moore, 1970); Venus' high albedo is due to its cloudy atmosphere). Normally, low albedo values are associated to surfaces with high absorptance (or, which equivalent due to Kirchoff's law, to surfaces with high emittance). Therefore, the inner planets are appropriate for DS building also from the point of view of their optical properties.

3.3 Class B and Class C Stellar Engines

Due to mirror's imperfect reflection and to the finite size of the Sun, a spot of concentrated light is expected to appear on the inner surface of class B and class C stellar engines in its part opposite to the mirror. This spot is associated with a temperature peak and can be used to increase locally the work rate provided by the thermal engine. However, for convenience we shall assume: (i) the Sun has a negligible size as compared to the radius of the stellar engine and (ii) the mirror is perfect (i.e. the mirror has a unity reflectance and it reflects all the incident rays on Sun's direction). As a consequence, the mirror temperature is very low and is not considered in this work.

Now, we shall analyze a region on the inner surface of the class B stellar engine (or on that part of the class C stellar engine that is used for power generation). The steady-state energy balance per unit area of the inner surface is:

$$q_H = \frac{B_S}{\pi}\sigma T_S^4 + \left(1 - \frac{B_S}{\pi}\right) e_{\text{int}}\sigma T_p^4 - e_{\text{int}}\sigma T_p^4. \tag{17}$$

Here, q_H is the energy density flux entering the thermal engine (Fig. 3). The first and the second terms in the r.h.s of Eq. (17) is the energy flux density absorbed from the Sun and from the whole stellar engine inner surface, respectively. Here, the conservation of the etendue on the mirror surface is taken into account (see e.g. Badescu, 1993, and references therein). The third term in the r.h.s of Eq. (17) is the energy flux density emitted by the stellar engine's inner surface.

The energy balance per unit area of the outer surface is:

$$q_L = e_{ext}\sigma T_r^4. \tag{18}$$

In Eq. (18) q_L is the energy flux density leaving the thermal engine per unit surface area while T_r is the temperature of the outer surface of the stellar engine.

Here a particular case of endoreversible thermal engine is considered, namely the Chambadal-Novikov-Curzon-Ahlborn engine (CNCA engine for short). It consists of three parts (Fig. 3):

(a) a reversible part working between two heat reservoirs (one at the high temperature), say t_1, and one at the low temperature, say t_2; (usually, t_1 and t_2 are the temperatures of the working fluid during its isothermal expansion and compression, respectively).

(b) two irreversible parts containing temperature drops (i.e. the temperature fall $T_p - t_1$ accompanying q_H and the temperature fall $t_2 - T_r$ accompanying q_L). A linear relationship exists between the heat flows and the temperature gradients.

Details on endoreversible and CNCA engines may be found in the reviews by Bejan (1996) and Hoffmann et al., (1997).

The entropy balance for the CNCA engine is (De Vos, 1985):

$$\frac{q_H}{T_p^{1/2}} + \frac{q_L}{T_r^{1/2}} = 0. \tag{19}$$

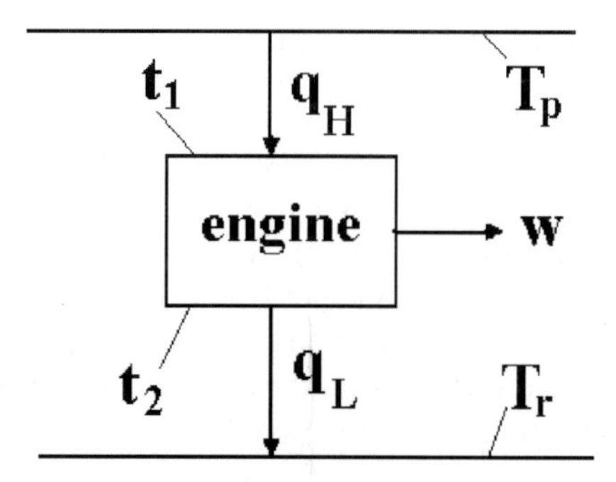

Figure 3. Power generation by using a CNCA thermal engine. q_H, q_L – heat fluxes entering and leaving the thermal engine, respectively; t_1 and t_2 – the absolute temperatures of the working fluid in contact with the two heat reservoirs. T_p, T_r – temperatures of the inner and outer surfaces, respectively; w – work rate (power)

The energy balance for the whole surface of the Sun is:

$$S_{S,eff}\sigma T_S^4 - S_S e_{\text{int}}\sigma T_p^4 = \tilde{L}_S. \tag{20}$$

The first term in the l.h.s. of Eq. (20) is the energy flux lost by the Sun; it takes into consideration that all the energy emitted by the Sun on mirror's direction is reflected back. The second term in the l.h.s. of Eq. (20) is the energy flux received by the Sun from the inner surface of the stellar engine. It takes into account that, due to the perfect mirror, each unit surface area of the Sun receives the energy flux density $e_{\text{int}}\sigma T_p^4$.

Simple computation shows that:

$$S_{S,eff} = S_S \frac{1+\cos\Psi}{2}. \tag{21}$$

One uses the following notation:

$$\theta_S \equiv \frac{T_S}{\tilde{T}_S} \quad \theta_p \equiv \frac{T_p}{\tilde{T}_S} \quad \theta_r \equiv \frac{T_r}{\tilde{T}_S}. \tag{22}$$

In the following θ_p is supposed to be known. This is a reasonable assumption as normally T_p should allow living conditions and consequently has a small variation range. By using Eqs. (12) and (17)–(22) one derives:

$$\theta_S = \left(e_{\text{int}}\theta_p^4 + \frac{2}{1+\cos\Psi}\right)^{1/4} \tag{23}$$

$$\theta_r = \left(\frac{2}{1+\cos\Psi}\frac{x}{e_{ext}}\frac{1}{\theta_p^{1/2}}\right)^{1/4}. \tag{24}$$

The two following conditions have to be fulfilled: $\theta_S > \theta_p$ and $\theta_p > \theta_r$, in order for the thermal engine to operate (i.e. to generate a positive power). By using Eqs. (22)–(24) these conditions turn out to be:

$$e_{\text{int}} \geq 1 - \frac{2}{\theta_p^4 (1 + \cos \Psi)} \tag{25}$$

$$R_p \geq R_S \left[\frac{2}{e_{ext} \theta_p^4 (1 + \cos \Psi)} \right]^{1/2} . \tag{26}$$

The constraint Eq. (25) is always fulfilled as its r.h.s. member is non-negative, because $\theta_p < 1$ (see Eq. 22) and $\cos \Psi \leq 1$. On the other hand, Eq. (26) gives a minimum limit for the radius of the stellar engine.

Let us have a look to the class B stellar engine proposed in Badescu (1995). It may be seen as a particular case of class C stellar engine (it corresponds to a missing mirror or, in other words, to $\Psi = 0°$). For an outer surface emittance $e_{ext} = 1$ one finds the minimum radius $R_{p,\min} = R_S/\theta_p^2$. This is very close to the result $R_{p,\min} = R_S(1 - \theta_p^4)^{1/2}/\theta_p^2$ derived in Badescu (1995) without taking into account that the presence of the stellar engine increases the Sun's temperature.

Figure 5 of Badescu and Cathcart (2000) shows the dependence of the Sun temperature T_S on the mirror rim angle Ψ and the radius R_p of the stellar engine. Generally, T_S increases with increasing Ψ and the radius R_p. However, this applies mainly for $R_s < 400 \cdot 10^6$ km.

Figure 4a shows the dependence of the outer surface temperature T_r on Ψ and R_p. Generally, T_r decreases by increasing R_p and decreasing the mirror rim angle Ψ.

Knowledge of the temperature T_r is important in case of searching for extraterrestrial intelligence (SETI). Indeed, it is (practically) the only information that outside world receives from a Kardashev type II civilisation. One reminds that according to the classification proposed by Kardashev a technological civilisation is of type I, II or III if it has under its control the materials and energy resources of a planet, star, or galaxy, respectively (see Kardashev, 1964; Badescu and Cathcart, 2000). From Fig. 4a one learns that galactic IR sources corresponding to temperatures lower than 300 K should not be overlooked during SETI activities. For more information about the thermal signature of possible extraterrestrial civilizations in the Galactic context see Chapter 13 in this book.

The heat flux density q_H is obtained by using Eqs. (12), (17), (22)–(24):

$$q_H = x\sigma \tilde{T}_S^4 \frac{2}{1 + \cos \Psi} . \tag{27}$$

The well-known CNCA efficiency is (De Vos, 1985)

$$\eta_{CNCA} = 1 - \left(\frac{T_r}{T_p} \right)^{1/2} = 1 - \left(\frac{\theta_r}{\theta_p} \right)^{1/2} . \tag{28}$$

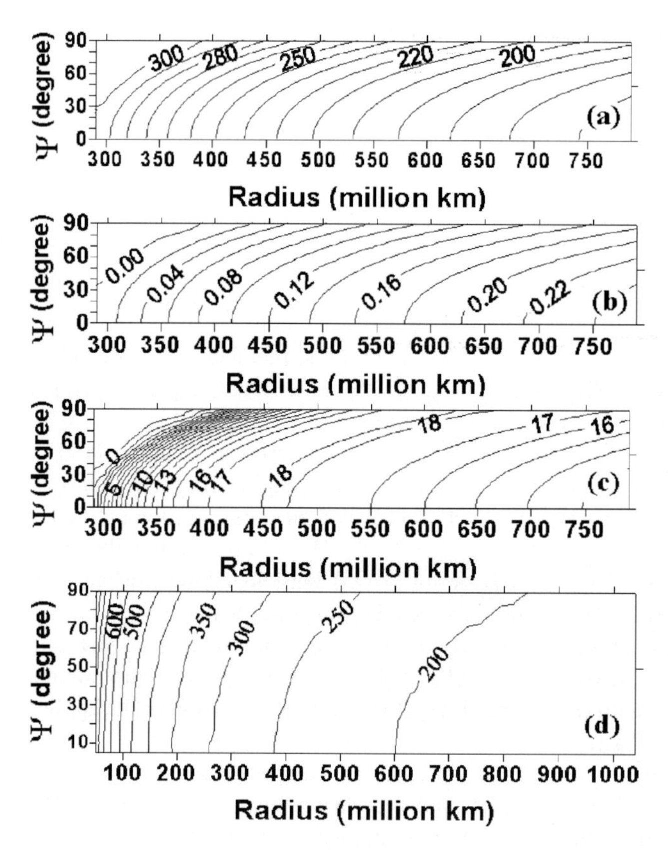

Figure 4. Dependence of various quantities on mirror rim angle Ψ and radius R_p in case of a class C stellar engine. (a) Outer surface temperature T_r (K); (b) Thermal engine efficiency η_{CNCA}; (c). Power density w_{CNCA} (W/m^2); (d) Optimum inner surface temperature T_p (K). In cases (a), (b), (c) the temperature of the inner surface is $T_p = 300$ K and the emittance of both inner and outer surfaces is $e_{int} = e_{ext} = 0.8$. In case (d) $e_{int} = e_{ext} = 1$

Figure 4b show the dependence of η_{CNCA} on Ψ and R_p. This performance indicator increases by increasing the radius R_p and decreasing the mirror rim angle Ψ. One can see that the efficiency vanishes and tends to become negative for R_p values smaller than the limit predicted by Eq. (26) (see the top left corner of Fig. 4b, where the associated "critical" rim angle may be easily evaluated). Generally, η_{CNCA} is smaller than the efficiency (which may exceed 0.5) of common terrestrial power plants working at large temperature differences but it is comparable with the efficiency of Stirling engines working at small differences of temperature (tens of Kelvin)(see Badescu, 2004).

Figure 8 of Badescu and Cathcart (2000) shows the dependence of η_{CNCA} on the outer surface emittance e_{ext}. The efficiency increases by increasing e_{ext}. This can be explained as follows. Increasing the emittance e_{ext} makes the temperature T_r decrease (see Fig. 5) and this finally leads to an increase in the efficiency. This has

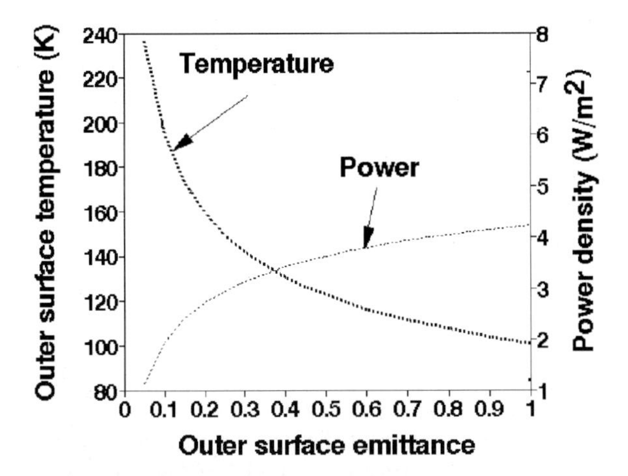

Figure 5. Dependence of the power density w_{CNCA} and of the outer surface temperature T_r on the outer surface emittance e_{ext} in case of a class C stellar engine. $R_p = 300$ millions km and $\Psi = 30$ degrees. Other inputs as in Fig. 4a

again consequences for SETI activities. Indeed, the thermal signature of possible extraterrestrial civilizations may be at a lower level than commonly expected.

The work rate (power) density w_{CNCA} is given by:

$$w_{CNCA} = q_H \eta_{CNCA}. \tag{29}$$

Figure 4c shows that for small R_p values the power density w_{CNCA} decreases by increasing the mirror rim angle Ψ. However, at high R_p values the reverse happens. There is a maximum maximorum power density which corresponds in both cases to a R_p radius of about $450 \cdot 10^6$ km. This means that there is an optimum stellar engine radius. That optimum radius is obviously larger than the radius of commonly proposed Dyson spheres, which is of the order or Earth orbit radius. One can notice that for some values of R_p and Ψ the power density w_{CNCA} becomes negative (left top corner of Fig. 4c).

Figure 5 shows the dependence of the power density w_{CNCA} and of the temperature T_r on the emittance of the outer surface e_{ext}. As expected, the temperature T_r decreases by increasing e_{ext}. But decreasing T_r leads to an increase in the efficiency (see Fig. 8 of Badescu and Cathcart, 2000) and finally this is associated with an increase in the power density w_{CNCA}.

Practically, w_{CNCA} and T_r do not depend on the emittance of the inner surface e_{int}.

3.4 The Thrust Force Acting on the Sun

We showed in Section 3.1 that the presence of the mirror makes the Sun's temperature increase. This has consequences on the radiation impulse and finally on the thrust force acting on the Sun. In this section one evaluates the thrust force in case of both class A (Shkadov thruster) and class C stellar engines.

3.4.1 Class A stellar engine (Shkadov thruster)

The impulse of the radiation per unit time leaving the Sun is proportional to the energy emitted (see Shkadov, 1987):

$$p = \frac{S_S \sigma T_S^4}{c}. \tag{30}$$

By taking into account the Eqs. (5)–(7), (21) and (30) one obtains:

$$p = \frac{\tilde{L}_S}{c} \frac{2}{1+\cos\Psi}. \tag{31}$$

Note that in Shkadov (1987) the increase in Sun temperature due to mirror's existence is not considered and the following approximate relation is used: $p = \tilde{L}_S / c$.

The thrust force $\bar{f}$ per unit area in the direction of the normal $\bar{n}$ to an arbitrary unit area placed on the base surface of the spherical cone A'SB' of Fig. 1a is (see also Fig. 1 of Shkadov, 1987):

$$f = \frac{\tilde{L}_S}{c} \frac{2}{1+\cos\Psi} \frac{1}{4\pi R_p^2}. \tag{32}$$

When $\Psi = 0$, Eq. (32) reduces to Eq. (1) of Shkadov (1987). The correction factor $2/(1+\cos\Psi)$ takes into account the increase in Sun's temperature.

The thrust F being produced by the Sun-mirror system due to the non-symmetric radiation field is given by Shkadov (1987)

$$F = \int \bar{f} \cdot \bar{n}\tau \, dS, \tag{33}$$

where S is the base area of the spherical cone A'SB' in Fig. 1a while τ is the unit vector along the axis of that cone. After integration one obtains:

$$F = \frac{\tilde{L}_S}{2c}(1-\cos\Psi). \tag{34}$$

When $\Psi = 0$, Eq. (34) reduces to Eq. (2) of Shkadov (1987). The thrust force F increases by increasing the mirror rim angle Ψ, as expected. The original result is $4cF = \tilde{L}_S \sin^2\Psi$ (see Eq. 2 of Shkadov, 1987). One can see that our Eq. (34) generally estimates a higher thrust force, which, in the particular case $\Psi = 90°$, doubles the result obtained by using Eq. (2) of Shkadov (1987). The main consequence is the fact that the lateral deviation during one orbital period of the Sun, evaluated by Shkadov (1987) to about 4.4 parsec, is underestimated. The value estimated by Shkadov (1987) for the acceleration induced by the thrust force F on the solar system motion is $6.5 \cdot 10^{-13}$ m/s^2. This is half of the result obtained by using the improved model from this chapter. Both values have to be compared with the gravitational acceleration of the galactic field, which is about $1.85 \cdot 10^{-10}$ m/s^2 (Shkadov, 1987). One concludes that the magnitude of the disturbing force created by the sun-mirror system is small, as expected.

3.4.2 Class C stellar engine

The impulse of the radiation emitted by the inner surface of a class C stellar engine at temperature T_p impinging on the Sun, in case that no mirror exists, is:

$$p_D = \frac{4\pi R_S^2 e_{\text{int}} \sigma T_p^4}{c}. \tag{35}$$

By using the notations Eqs. (22) and Eqs. (23)-(24) one obtains:

$$p_D = \frac{\tilde{L}_S}{c} e_{\text{int}} \frac{\theta_p^4}{\theta_S^4}. \tag{36}$$

Consequently, the impulse of the net flux of radiation leaving the Sun is:

$$p_{net} = p - p_D. \tag{37}$$

By taking into account the Eqs. (35)–(37) one obtains:

$$p_{net} = \frac{\tilde{L}_S}{c}\left(1 - e_{\text{int}}\frac{\theta_p^4}{\theta_S^4}\right)\frac{2}{1+\cos\Psi}. \tag{38}$$

The thrust force f per unit area in the direction of the normal $\bar{n}$ to the base surface of the circular cone A'SB' (see Fig 1a) is:

$$f = \frac{\tilde{L}_S}{c}\left(1 - e_{\text{int}}\frac{\theta_p^4}{\theta_S^4}\right)\frac{2}{1+\cos\Psi}\frac{1}{4\pi R_p^2}. \tag{39}$$

The thrust force F is obtained after computing the integral in Eq. (33):

$$F = \frac{\tilde{L}_S}{2c}\left(1 - e_{\text{int}}\frac{\theta_p^4}{\theta_S^4}\right)(1-\cos\Psi). \tag{40}$$

Generally, the thrust force F increases by increasing the mirror rim angle Ψ. One has to remind, however, that increasing Ψ leads to a decrease in the efficiency η_{CNCA} (see Fig. 4b). Note that F is dependent on the temperature T_p via the dimensionless parameter θ_p. The optimum value of T_P which maximizes F is shown in Fig. 4d for $e_{\text{int}} = e_{ext} = 1$. Let us consider an optimum temperature $T_p \sim 300\ K$ (appropriate for common living conditions on Earth). Then $F$is a maximum for a radius R_p around 300 millions km.

In the next sections one shall need the thrust force F' per unit mass of the solar system. This implies dividing the Eqs. (34) and (40), respectively, by $M_S + M_{planets}$, where $M_S = 1.989 \cdot 10^{30}$ kg and $M_{planets} = 2.7 \cdot 10^{27}$ kg are Sun mass and the mass of Solar System's planets, respectively. Then, the expressions of F' for class A and class C stellar engines are, respectively:

$$F'_A = \frac{\tilde{L}_S}{2c} \frac{1-\cos\Psi}{M_S + M_{planets}} \tag{41}$$

$$F'_C = \frac{\tilde{L}_S}{2c}\left(1 - e_{\text{int}}\frac{\theta_p^4}{\theta_S^4}\right)\frac{1-\cos\Psi}{M_S + M_{planets}}. \tag{42}$$

When used in case of the Sun, the Eqs. (41) and (42) lead to (practically) the same numerical results. This is due to the fact that $e_{\text{int}}(\theta_p/\theta_S) \propto (300/5760)^4$ is very close to zero. Consequently, the results reported below apply to both types of stellar engines and the indexes A or C will be removed for convenience.

4. CHANGE OF SUN MOVEMENT IN GALAXY

The Sun's galactic orbit is described here by ignoring the perturbations due to Galaxy spiral arms and the encounters with massive dust/molecular clouds. The gravitational forces acting on the Sun in the absence of a stellar engine are modeled as being derived from scalar potentials. Various gravitational potentials were proposed and studied in the relevant literature. It is not our aim to decide which of these potentials is more appropriate to be used in practice. Here we shall use the simple spherical potential adopted earlier by Shkadov (1987) (see section 4.1 below). However, some authors consider it to be helpful to decompose the Sun's motion into two orthogonal components: a motion in the galactic mid-plane and a motion perpendicular to the plane (Gonzalez, 1999). Therefore, a cylindrical gravitational potential will be used in Section 4.2. It is a generalization of a Plummer potential, previously used by Carlberg and Innanen (1987). We shall see that both potentials predict results of the same order of magnitude and this may act as a sort of cross-checking.

The stellar engine thrust force F is superposed on the Galaxy gravitational forces acting on the Sun. As a result, a *perturbed* Sun trajectory will result. It is the scope of the present section to evaluate the distance between the perturbed position and the Sun's usual (average) position.

4.1 Movement in Curvilinear Coordinates

A few results of vector analysis are used here to describe Sun motion. We define a cartesian system of coordinates $\{x^i\}$ $(i = 1, 2, 3)$ with the plane Ox^1x^2 in the equatorial plane of the Galaxy. The Sun movement in the Galaxy will be given by three parametric functions, say $x^i = x^i(t)$ $(i = 1, 2, 3)$. A curvilinear coordination system q^i $(i = 1, 2, 3)$ is then introduced. The transformation $\{x^i\} \to \{q^i\}$ defines a metric tensor g_{ij} $(i, j = 1, 2, 3)$. One denotes by $\dot{q}^i$ the usual first order time derivatives of the coordinates q^i $(i = 1, 2, 3)$. They are called *generalized* velocities. Note that their dimensions are not necessarily length per time. The connection between $\dot{q}^i$ and the components of Sun's velocity v^i (i.e. the projections on the coordinates q^i $(i = 1, 2, 3)$) are given by the usual relationships (Beju et al., 1976, p 173):

$$(43) \qquad v^i = H_i \dot{q}^i \qquad (i = 1, 2, 3),$$

where H_i are the Lamé coefficients, which can be obtained by (Beju et al., 1976, p 172):

$$(44) \qquad H_i = g_{ii}^{1/2} \qquad (i = 1, 2, 3).$$

The contravariant time derivative D/Dt of the generalized velocities $\dot{q}^i$ is given by (Beju et al., 1976, p 183)

$$(45) \qquad \frac{D\dot{q}^i}{Dt} = \frac{d\dot{q}^i}{dt} + \Gamma^i_{jk}\dot{q}^j\dot{q}^k,$$

where Γ^i_{jk} are Christoffel coefficients of the second kind. Here the Einstein convention for summation was used. The equations of movement of the Sun have the covariant form:

$$(46) \qquad a^i = H_i \frac{D\dot{q}^i}{Dt} = G^i + F'^i \qquad (i = 1, 2, 3),$$

where a^i is the i-th contravariant component of Sun's acceleration while G^i and F'^i are the i-th contravariant components of the gravitational force and of the stellar engine thrust, respectively, both of them per unit mass of the Solar System.

The Sun motion is described first by mean of the spherical coordinate system (R, λ, φ) used in Shkadov (1987). The change of coordinates $\left(x^1, x^2, x^3\right) \to (R, \lambda, \varphi)$ is:

$$(47) \qquad x^1 = R\cos\varphi\cos\lambda \quad x^2 = R\cos\varphi\sin\lambda \quad x^3 = R\sin\varphi,$$

where $0 \le \lambda \le 2\pi$ and $-\pi/2 \le \varphi \le \pi/2$. The equatorial plane of the Galaxy is associated to $\varphi = 0$. Details about the metric tensor g_{ij}, the contravariant tensor g^{ij} $(i, j = R, \lambda, \varphi)$, the Lamé coefficients and the Christoffel symbols of first and second kind, respectively, may be found in Badescu and Cathcart (2006).

Use of Eqs. (43) allows to obtain the components v^i $(i = R, \lambda, \varphi)$ of Sun's velocity:

$$(48) \qquad v^R = \dot{R} \quad v^\lambda = R\cos\varphi\dot{\lambda} \quad v^\varphi = R\dot{\varphi},$$

while use of Eqs. (44)–(46) and (48) allow to obtain the components $\dot{v}^i$ (R, λ, φ) of Sun's acceleration:

$$(49) \qquad \begin{aligned} \dot{v}^R &= \frac{\left(v^\lambda\right)^2 + \left(v^\varphi\right)^2}{R} + G^R + F'^R \\ \dot{v}^\lambda &= -\frac{v^R v^\lambda - v^\lambda v^\varphi \tan\varphi}{R} + G^\lambda + F'^\lambda \\ \dot{v}^\varphi &= -\frac{v^\lambda v^\varphi + \left(v^\lambda\right)^2 \tan\varphi}{R} + G^\varphi + F'^\varphi \end{aligned}.$$

The Sun motion is described now by mean of the cylindrical coordinate system (r, θ, z) used in Carlberg and Innanen (1987). The change of coordinates $\left(x^1, x^2, x^3\right) \rightarrow (r, \theta, z)$ is:

$$(50) \qquad x^1 = r\cos\theta \quad x^2 = r\sin\theta \quad x^3 = z,$$

where $0 \leq \theta \leq 2\pi$. The equatorial plane of the Galaxy is associated to $z = 0$. Again, details about the metric tensor g_{ij}, the contravariant tensor g^{ij} $(i, j = r, \theta, z)$, the Lamé coefficients and the Christoffel symbols of first and second kind, respectively, may be found in Badescu and Cathcart (2006).

Use of Eqs. (43) allows to obtain the components v^i $(i = r, \theta, z)$ of Sun's velocity:

$$(51) \qquad v^r = \dot{r} \quad v^\theta = r\dot{\theta} \quad v^z = \dot{z},$$

while use of Eqs. (44)–(46) and (51) allow to obtain the components $\dot{v}^i$ (r, θ, z) of Sun's acceleration:

$$(52) \qquad \begin{aligned} \dot{v}^r &= \frac{\left(v^\theta\right)^2}{r} + G^r + F'^r \\ \dot{v}^\theta &= -\frac{v^r v^\theta}{r} + G^\theta + F'^\theta . \\ \dot{v}^z &= G^z + F'^z \end{aligned}$$

The above theory will be used now in case of two Galaxy gravitational potentials.

4.2 First Galaxy Gravitational Potential

As a first axi-symmetrical gravitational potential per unit mass of the solar system, $\Phi(R, \lambda, \varphi)$, we shall adopt:

$$(53) \qquad \Phi(R, \lambda, \varphi) = \frac{A}{B + (B^2 + R^2)^{1/2}} - \frac{C^2 \tan^2(\varphi/2)}{R^2\left(1 + D^2\tan^2(\varphi/2)\right)^{1/2}}.$$

Here a spherical system of coordinates (R, λ, φ) was used. The constants in Eq. (53) are as follows: $A = 3.18 \cdot 10^{22}$ km^3/s^2, $B = 8.6 \cdot 10^{16}$ km, $C = 3.27 \cdot 10^{20}$ km^2/s and $D = 30.8$ (Shkadov, 1987).

The components of the gradient of Φ are projections of the gravitational acceleration vector G:

$$(54) \qquad G^R = \frac{1}{H_R}\frac{\partial\Phi}{\partial R} \qquad G^\lambda = \frac{1}{H_\lambda}\frac{\partial\Phi}{\partial\lambda} = 0 \qquad G^\varphi = \frac{1}{H_\varphi}\frac{\partial\Phi}{\partial\varphi}.$$

Here the Lamé coefficients H_i $(i = R, \lambda, \varphi)$ were used.

One denotes the components of Sun's velocity by v^i $(i = R, \lambda, \varphi)$ and one defines the following dimensionless variables:

$$(55) \qquad \tilde{t} \equiv \frac{t}{T_0} \quad \tilde{R} \equiv \frac{R}{r_0} \quad \tilde{v}^R \equiv \frac{v^R}{v_0} \quad \tilde{v}^\lambda \equiv \frac{v^\lambda}{v_0} \quad \tilde{v}^\varphi \equiv \frac{v^\varphi}{v_0},$$

where T_0 $(= 220 \cdot 10^6\ \text{yr})$, r_0 $(= 8500\ \text{pc})$ and v_0 $(= 12\ \text{km/s})$ are appropriate scaling values.

In the dimensionless notation Eq. (55), the Eqs.(48) and (49) describing the Sun movement are:

$$
\begin{aligned}
&\dot{\tilde{R}} = \tilde{v}^R \qquad \dot{\lambda} = \frac{\tilde{v}^\lambda}{\tilde{R}\cos\varphi} \qquad \dot{\lambda} = \frac{\tilde{v}^\lambda}{\tilde{R}\cos\varphi} \\
&\dot{\tilde{v}}^R = \tilde{D}_1 \frac{\left(\tilde{v}^\lambda\right)^2 + \left(\tilde{v}^\varphi\right)^2}{\tilde{R}} - \tilde{D}_{2a} \frac{\tilde{R}}{\left(\tilde{B}^2 + \tilde{R}^2\right)^{1/2} \left[\tilde{B}^2 + \left(\tilde{B}^2 + \tilde{R}^2\right)^{1/2}\right]^2} \\
&\qquad + \tilde{D}_{2b} \frac{2\tan^2(\varphi/2)}{\tilde{R}^3 \left(1 + D^2 \tan^2(\varphi/2)\right)^{1/2}} + f^R \tilde{D}_3 \\
&\tilde{v}^\lambda = -\tilde{D}_1 \frac{\tilde{v}^R \tilde{v}^\lambda - \tilde{v}^\lambda \tilde{v}^\varphi \tan\varphi}{\tilde{R}} + f^\lambda \tilde{D}_3 \\
&\dot{\tilde{v}}^\varphi = -\tilde{D}_1 \frac{\tilde{v}^\lambda \tilde{v}^\varphi + \left(\tilde{v}^\lambda\right)^2 \tan\varphi}{\tilde{R}} \\
&\qquad - \tilde{D}_{2b} \frac{1 + \frac{D^2}{2}\tan^2(\varphi/2)}{\tilde{R}^3 \left(1 + D^2 \tan^2(\varphi/2)\right)^{1/2}} \frac{\tan^2(\varphi/2)}{\cos^2(\varphi/2)} + f^\varphi \tilde{D}_3 .
\end{aligned}
\tag{56a-e}
$$

Here the Eqs. (42) and (30) and the dimensionless parameters defined below were also used:

$$
\begin{aligned}
&\tilde{B} \equiv \frac{B}{r_0} \quad \tilde{D}_1 \equiv \frac{v_0 T_0}{r_0} \quad \tilde{D}_{2a} \equiv \frac{T_0 A}{v_0 r_0^2} \quad \tilde{D}_{2b} \equiv \frac{T_0 C^2}{v_0 r_0^3} \\
&\tilde{D}_3 \equiv \frac{T_0}{v_0} \frac{L_s}{2c} \frac{1 - \cos\Psi}{M_{sun} + M_{planets}} \left(1 - e_{\text{int}} \frac{\theta_p^4}{\theta_s^4}\right)
\end{aligned}
\tag{57}
$$

In Eqs. (56) the unit vector $\left(f^R, f^\lambda, f^\varphi\right)$ gives the direction of the thrust force per unit mass F' (see Eq. 42) in the coordinate system (R, λ, φ).

A simplifying hypothesis was adopted in Shkadov (1987) to allow an analytical solution for the perturbed motion of the Sun. Thus, one considered a particular set of initial conditions that makes the usual (unperturbed) motion of the Sun to be along a circular orbit in the equatorial plane $\varphi = 0$ of the Galaxy. One proved that the ratio of the maximum acceleration generated by the Sun-mirror system to the Galaxy gravitational acceleration is less than one percent. One concluded that the magnitude of the thrust force is (relatively) small and the small parameter method can be used to solve the equations of the perturbed motion. Consequently, the perturbed motion of the Sun was described mathematically in Shkadov (1987) as a variational problem with respect to the unperturbed (circular) orbit.

In this chapter the Eqs. (56) are solved numerically by using the ODE-solver SDRIV3 from the SLATEC library (Fong et al., 1993).

A few details about the initial values used to solve the Eqs. (56) follow. The (absolute) Sun velocity is usually obtained by adding the (average) near circular velocity of the Galaxy at the Sun to the Sun's velocity in the local standard of rest (LSR). Discussions on various ways of defining the LSR can be found in (Bash, 1986, p. 42). One knows that the Sun is located near the corotation circle, where in a spiral galaxy such as ours the angular speeds of the spiral pattern and the stars are equal (Mishunov and Zenina, 1999). Consequently, one expects a rather small value for Sun's velocity in the LSR. Indeed, HIPPARCOS-based studies give the mean value 13.4 ± 0.4 km/s (Dehnen and Binney, 1998; Kovalevsky, 1999; Bienayme, 1999). The LSR velocity is higher for older than for younger stars due to accumulation of perturbations to a star's trajectory. Consequently, the present-day orbit differs from the originally nearly circular motion in plane (Gonzalez, 1999). One defines the Sun's velocity components (u, v, w) in the LSR as follow: u is the velocity positive outward away from the galactic center; v is the velocity in the galactic plane positive in the sense of the galactic rotation and w is the velocity in the direction perpendicular on the galactic plane, positive toward the north galactic pole (Bash, 1986, p. 36). In this convention $(u, v, w) = (0, 0, 0)$ characterizes a body at Sun position, moving in the galactic plane on a circular orbit. The initial components of LSR Sun velocity are denoted (u_0, v_0, w_0). In computation we used the rather popular values (in km/s):$(u_0, v_0, w_0) = (-9, 12, 7)$ (Bash, 1986, p. 36; Darling, 2004). The local components (U, V, W) of Galaxy's velocity are defined in a similar coordinate system, with the origin in the center of the Galaxy. One denotes by (U_0, V_0, W_0) the Galaxy's velocity at Sun position. In computations we used the values (in km/s): $(U_0, V_0, W_0) = (44, 235, 30)$ (Carlberg and Innanen, 1987). Therefore, the components of the initial (absolute) Sun velocity are: $v^R(t = 0) = u_0 + U_0$, $v^\lambda(t = 0) = v_0 + V_0$ and $v^\varphi(t = 0) = w_0 + W_0$.

Note that the estimated Sun orbit is rather sensitive on the initial velocity. For example, in Bash (1986) the chosen values of both u_0 and v_0 were increased by 3 km/s and the orbit was integrated again. After 100 Myr the Sun's position was found to differ by 400 pc.

The initial coordinates of the Sun are as follows. At time $t = 0$ the Sun is found in the equatorial plane of the Galaxy. We thus have $\varphi(t = 0) = 0$. This is reasonable as the Sun crosses the equatorial plane during its movement in the Galaxy (see e.g. Fig. 6a). Note that the present-day position of the Sun is estimated to about 10 to 20 pc above the equator plane (Pal and Ureche, 1983; Gonzalez, 1999), which is rather close to it. Other initial conditions are $\lambda(t = 0) = 0$ and $R(t = 0) = r_0$ (Carlberg and Innanen, 1987).

It is useful now to estimate how long one can safely integrate the Sun orbit. Indeed, the velocity dispersion of the Galaxy's disk stars increases with time, due to rather random encounters with interstellar clouds and periodical encounters with the spiral arms of stars. For example, during its lifetime the Sun has crossed the Galaxy's spiral arms about 17 times (Bash, 1986, p. 42). It may not be wise to integrate the Sun's orbit, using a global potential, past one spiral arm's passage. Therefore, the time between spiral arm passages, which is about 260 Myr, is the maximum integration time accepted here.

First, the Eqs. (33) were solved in the case $f^R = f^\lambda = f^\varphi = 0$. This corresponds to the usual (unperturbed) motion of the Sun. Using the solution of Eqs. (33) one could obtain from Eqs. (47) the cartesian coordinates $x^i(t)$ $(i = 1, 2, 3)$ of the Sun on the unperturbed orbit. Second, the Eqs. (56) were solved in case the Sun motion is perturbed by the stellar engine thrust force. This requires of course using a non-null unit vector $(f^R, f^\lambda, f^\varphi)$ in Eqs. (56). The cartesian coordinates of the Sun on the perturbed orbit are denoted $x_p^i(t)$ $(i = 1, 2, 3)$.

Since the local force law is not inverse-square, the galactic orbit of the Sun is not expected to be a close Keplerian ellipse. The best-fitting, approximate, Keplerian ellipse to the Sun's current orbit shows $a \approx 1.07r_0$ and $e \approx 0.07$, where a and e are the semi-major axis and the eccentricity of the orbit, respectively (Bash, 1986, p. 42). There is a reasonable concordance between these previous findings and the results obtained here (see Fig. 6a). The coordinate R lies between $0.903r_0$ and $1.136r_0$. A complete rotation of the Sun around the center of the Galaxy corresponds to a variation of λ between 0 and 360 degrees. It takes about 225 Myr. The Sun trajectory is placed both above and below the equatorial plane of the Galaxy (i.e. at positive and negative φ values, respectively) (Fig. 6a). The angular deviation from the equatorial plane is, however, small in absolute values (less than 5 degrees). The Sun trajectory crosses the equatorial plane two times during a complete rotation. One reminds that the simplified unperturbed Sun motion studied in Shkadov (1987) is confined to the equatorial plane of the Galaxy. A periodic oscillation motion in the lateral direction to the solar orbital plane was emphasized in the quoted paper only in case the Sun motion is perturbed by the mirror's thrust force.

Sun velocity changes on the unperturbed orbit (Fig. 6b). The tangential velocity v^λ has a monotonous time variation between apogalacticon and perigalacticon. The

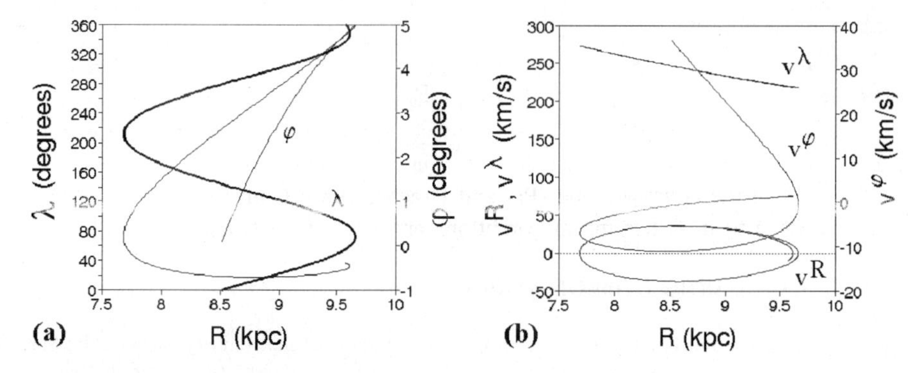

Figure 6. Solution of Eqs. (33) for one Sun revolution. (a) Dependence of angles λ and φ on R. (b) Dependence of Sun velocity components v^R, v^λ and v^φ on R

radial velocity v^R reaches its extreme values near $R = r_0$. The time variation of v^φ is slightly more complicated.

A stellar engine thrust force of constant magnitude will be considered in the following. The perturbed motion of the Sun depends, of course, on the direction of the thrust force. Three simple strategies of changing Sun movement are defined now. In the first case the thrust force is constantly acting on the (outward) direction of R and it corresponds to $f^R = 1,\ f^\lambda = f^\varphi = 0$ in Eqs. (56). The second case corresponds to $f^\varphi = 1,\ f^R = f^\lambda = 0$ and refers to a thrust force constantly acting on the direction of the generalized variable φ. A thrust force constantly acting on the direction of the generalized variable λ (i.e. $f^\lambda = 1,\ f^R = f^\varphi = 0$) is the third strategy.

In all the three above cases the time-dependent distance $\Delta R(t)$ between the perturbed and unperturbed positions of the Sun, respectively, is defined in the usual way as

$$\Delta R(t) \equiv \left[\left(x_p^1 - x^1\right)^2 + \left(x_p^2 - x^2\right)^2 + \left(x_p^3 - x^3\right)^2\right]^{1/2}. \tag{58}$$

The time dependence of ΔR is shown in Fig. 7 for the three strategies. A single rotation of the Sun around the center of the Galaxy was considered. The distance ΔR depends on the direction of the thrust force (i.e. on the strategy), as expected. None of the three strategies make the distance between the perturbed and the unperturbed Sun position increase linearly in time. An optimal control strategy for the thrust force direction is required for this purpose. The second strategy (i.e. $f^\lambda = 1$) yields the largest values of ΔR during the time interval considered here. The deviation from the unperturbed orbit could be as large as 40 pc. Note the maximum ΔR is obtained with the $f^R = 1$ strategy about 140 Myr after stellar engine implementation.

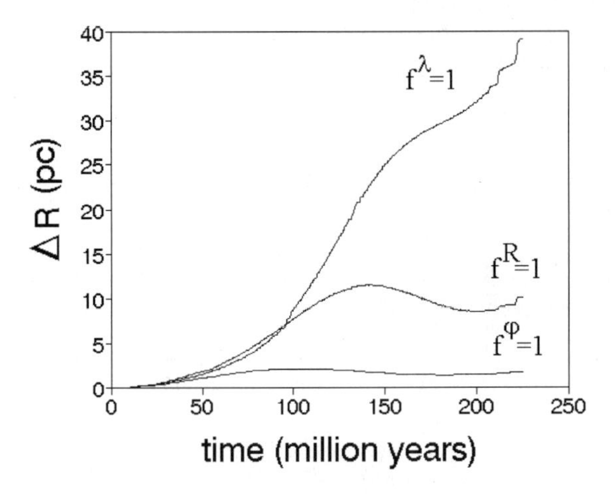

Figure 7. Time variation of distance $\Delta R(t)$ between the perturbed and unperturbed positions of the Sun, respectively, during one Sun galactic revolution. Solutions of Eqs. (56) were used. Three strategies of changing Sun movement are considered: (i) $f^R = 1$ (stellar engine thrust force is acting on the (outward) direction of R), (ii) $f^\lambda = 1$ (thrust force acting on the direction of λ), (iii) $f^\varphi = 1$ (thrust force acting on the direction of φ)

It is interesting to compare our findings with early results obtained in Shkadov (1987) by using a simplified analytical model. In the quoted paper the mirror axis forms a right angle with the radius-vector of the Sun that is assumed to move on a circular orbit in the equatorial plane of the Galaxy. This means that the thrust vector is always acting in the equatorial plane of the Galaxy and it is directed along the tangent to the solar orbit. In Shkadov (1987) the radius of the circular orbit of the Sun is estimated to 10 kpc while the period of one Sun revolution in the Galaxy is assumed to be 200 Myr. The quoted author found a Sun radial deviation from its orbit of about 12 pc. This is about three times smaller than the results obtained in this chapter by using a more accurate treatment.

4.3 Second Galaxy Gravitational Potential

Another axi-symmetric gravitational potential per unit mass of the solar system, $\Phi(r, \theta, z)$, will be used in this section. It consists of a disk-halo Plummer potential supplemented with some spherical potentials (see Carlberg and Innanen, 1987):

$$\Phi(r,\theta,z) = -\frac{\alpha_1 M_{eff} g}{\left\{\left[a+\sum_{i=1}^{3}\beta_i\left(z^2+h_i^2\right)^{1/2}\right]^2+b_1^2+r^2\right\}^{1/2}} - \sum_{j=2}^{4}\frac{\alpha_j M_{eff} g}{\left(b_j^2+r^2\right)^{1/2}}. \tag{59}$$

Here a cylindrical system of coordinates (r, θ, z) was used. Other notations in Eq. (59) are: g $\left(= 6.67 \cdot 10^{-11} \mathrm{m^3 kg^{-1} s^{-2}}\right)$ is the gravitational constant, $M_{eff} = 9.484 \cdot 10^{11} M_S$ is the effective Galaxy mass influencing Sun's movement, α_j $(j = 1, 2, 3, 4)$ are mass weighting coefficients for various potential components, a and b_1 are the scale length and the core radius of the disk-halo, respectively, β_i $(i = 1, 2, 3)$ and h_i $(i = 1, 2, 3)$ correspond to the scale heights of various disk-halo components while b_j $(j = 2, 3, 4)$ are the core radii of the additional spherical potentials (for bulge, nucleus and dark halo, respectively). Table 1 shows the data.

Table 1. Data for the Galaxy gravitational potential of (Carlberg and Innanen, 1987)

Component j	Disk-halo $(j=1)$	Bulge $(j=2)$	Nucleus $(j=3)$	Dark-halo $(j=4)$
α_j	0.1554	0.0490	0.0098	0.7859
b_j (kpc)	8.0	3.0	0.25	35.0
a (kpc)	3.0	0	0	0
β_1	0.4	0	0	0
β_2	0.5	0	0	0
β_3	0.1	0	0	0
h_1 (kpc)	0.325	0	0	0
h_2 (kpc)	0.090	0	0	0
h_3 (kpc)	0.125	0	0	0

The components of the gradient of Φ in the coordinate system (r, θ, z) are defined as projections of the gravitational acceleration vector G:

$$G^r = \frac{1}{H_r}\frac{\partial \Phi}{\partial r}, \quad G^\theta = \frac{1}{H_\theta}\frac{\partial \Phi}{\partial \theta} = 0, \quad G^z = \frac{1}{H_z}\frac{\partial \Phi}{\partial z}. \tag{60}$$

The Lamé coefficients H_i $(i = r, \theta, z)$ are used here. One denotes the components of Sun's velocity by v^i $(i = R, \lambda, \varphi)$ and one defines the new dimensionless variables:

$$\tilde{r} \equiv \frac{r}{r_0} \quad \tilde{v}^r \equiv \frac{v^r}{v_0} \quad \tilde{v}^\theta \equiv \frac{v^\theta}{v_0} \quad \tilde{v}^z \equiv \frac{v^z}{v_0}. \tag{61}$$

In the dimensionless notation Eqs. (61), the Eqs. (51)-(52) of Sun movement are:

$$\dot{\tilde{r}} = \tilde{D}_1 \tilde{v}^r \qquad \dot{\theta} = \tilde{D}_1 \frac{\tilde{v}^\theta}{\tilde{r}} \qquad \dot{\tilde{z}} = \tilde{D}_1 \tilde{v}^z \tag{62a-c}$$

$$\dot{\tilde{v}}^r = \tilde{D}_1 \frac{\left(\tilde{v}^\theta\right)^2}{\tilde{r}} + \tilde{D}_4 \left\{ \frac{\alpha_1 \tilde{r}}{\left\{\left[\tilde{a} + \sum_{i=1}^{3} \beta_i \left(\tilde{z}^2 + \tilde{h}_i^2\right)^{1/2}\right]^2 + \tilde{b}_1^2 + \tilde{r}^2\right\}^{3/2}} + \sum_{j=2}^{4} \frac{\alpha_j \tilde{r}}{\left(\tilde{b}_j^2 + \tilde{r}^2\right)^{3/2}} \right\} + f^r \tilde{D}_3 \tag{62d}$$

$$\dot{\tilde{v}}^\theta = -\tilde{D}_1 \frac{\tilde{v}^r \tilde{v}^\theta}{\tilde{r}} + f^\theta \tilde{D}_3 \tag{62e}$$

$$\dot{\tilde{v}}^z = \tilde{D}_4 \frac{\alpha_1 \tilde{z} \left[\tilde{a} + \sum_{i=1}^{3} \beta_i \left(\tilde{z}^2 + \tilde{h}_i^2\right)^{1/2}\right]}{\left\{\left[\tilde{a} + \sum_{i=1}^{3} \beta_i \left(\tilde{z}^2 + \tilde{h}_i^2\right)^{1/2}\right]^2 + \tilde{b}_1^2 + \tilde{r}^2\right\}^{3/2}} \left[\sum_{i=1}^{3} \frac{\beta_i}{\left(\tilde{z}^2 + \tilde{h}_i^2\right)^{1/2}}\right] + f^z \tilde{D}_3 \tag{62f}$$

Here the Eqs. (42) and (59) and following dimensionless constants were also used:

$$\tilde{a} \equiv \frac{a}{r_0}, \quad \tilde{b}_1 \equiv \frac{b_1}{r_0}, \quad \tilde{h}_i \equiv \frac{h_i}{r_0} \; (i = 1, 2, 3), \quad \tilde{D}_4 \equiv \frac{T_0 M_{eff} g}{v_0 r_0^2}. \tag{63}$$

In Eqs. (62) the unit vector $\left(f^r, f^\theta, f^z\right)$ gives the direction of the thrust force F' (see Eq. 42) in the coordinate system (r, θ, z).

The Eqs. (62) are solved numerically by using the ODE-solver SDRIV3 (Fong et al., 1993). One assumes again that initially (i.e. at time $t = 0$) the Sun is found in the equatorial plane of the Galaxy. This makes possible to use the same initial conditions we used in Section 4.1. In cylindrical coordinates, this means $r\,(t = 0) = r_0$, $\theta\,(t = 0) = 0$ and $z\,(t = 0) = 0$. The components of the initial Sun velocity are: $v^r\,(t = 0) = u_0 + U_0$, $v^\theta\,(t = 0) = v_0 + V_0$ and $v^z\,(t = 0) = w_0 + W_0$.

To solve the unperturbed motion of the Sun requires using $f^r = f^\theta = f^z = 0$ in Eqs. (62). Results are shown in Fig. 8. The coordinate r lies between $0.942r_0$ and $1.193r_0$. This is in reasonably good agreement with results given in Bash (1986), where an approximate Sun motion confined to the galactic mid-plane was studied. A perturbation potential due to standard spirals pattern was however included in that model. The perturbation potential was assumed to be 5% of the global axi-symmetric potential. The initial values were slightly different from those we used here. One found that the Sun reaches perigalacticon at $r = 0.995\hat{r}_0$ and apogalacticon at $r = 1.145\hat{r}_0$, where $\hat{r}_0$ is the initial value of r used in Bash (1986).

A complete rotation of the Sun around the center of the Galaxy (that corresponds to a variation of θ between 0 and 360 degrees) takes about 248.5 Myr. This is about 10 % longer than in case of the gravitational potential used in section 4.1.

The rotation motion of the Sun is qualitatively similar for both Galaxy gravitational potentials we considered in this paper (compare Fig. 8 and Fig. 6, respectively). Differences exist however in the predictions about the vertical motion.

A usual simplified orbit integration procedure is to separate the motion in the mid-galactic plane from the Sun's vertical motion. Sometimes the last motion is modeled as a simple harmonic oscillation. A vertical oscillation period of 66 Myr was accepted, for example, in Bash (1986). In this case the Sun would cross the mid-plane every 33 Myr, i.e. between seven and eight times during a complete revolution.

In the present work there is no decomposition of Sun motion and, therefore, more accurate results are expected. Figure 8a shows that the Sun deviation from the equatorial plane of the Galaxy lies between -80 pc and $+80$ pc and the Sun trajectory crosses the equatorial plane four times during a complete revolution.

Note that in Gonzalez (1999) one estimates that the Sun spends most of its time at least 40 pc from the Galactic mid-plane. Also, some studies reported for the maximum distance z_{max} between the Sun and Galaxy's equatorial plane values ranging from 76.8 to 81.8 pc (Bash, 1986), which is in good concordance with our results.

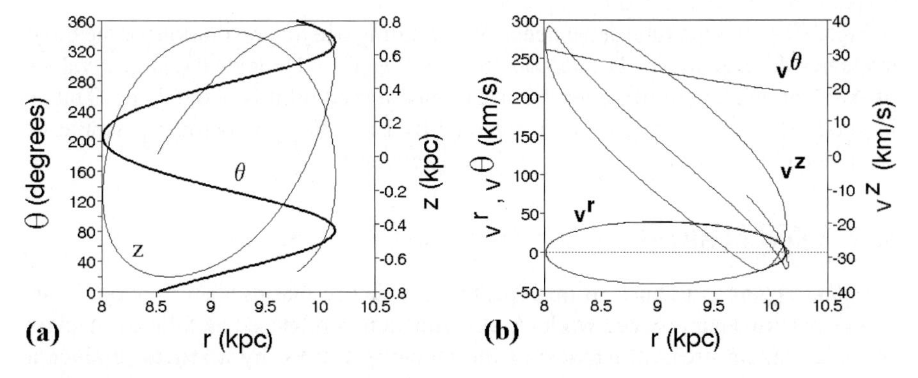

Figure 8. Solution of Eqs. (62) for one Sun revolution. (a) Dependence of variables θ and z on r. (b) Dependence of Sun velocity components v^r, v^θ and v^z on r

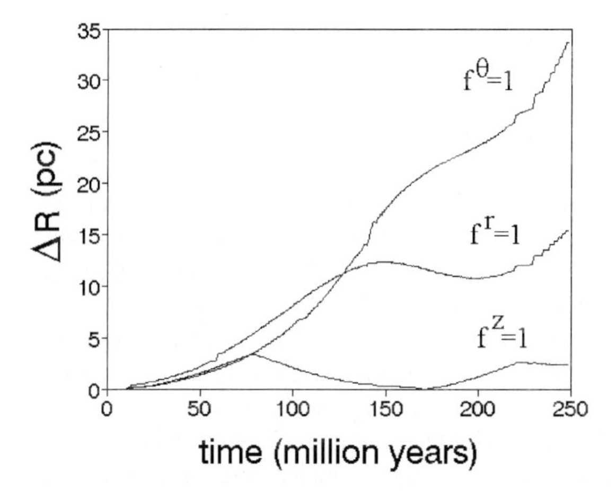

Figure 9. Time variation of the distance $\Delta R(t)$ between the perturbed and unperturbed positions of the Sun, respectively, during one Sun galactic revolution. Solutions of the equations system (62) were used. Three strategies of changing Sun movement are considered: (i) $f^r = 1$ (stellar engine thrust force is acting on the (outward) direction of r), (ii) $f^\theta = 1$ (thrust force acting on the direction of the generalized variable θ), (iii) $f^z = 1$ (thrust force acting on z direction)

The tangential velocity v^θ and the radial velocity v^r have a monotonous time variation between their minimum and maximum values but the time variation of v^z is slightly more complicated (Fig. 8b).

Three simple strategies of changing Sun movement are again considered here. In the first case the stellar engine thrust force is constantly acting on the (outward) direction of r and it corresponds to $f^r = 1$, $f^\theta = f^z = 0$ in Eqs. (62). The second case corresponds to $f^\theta = 1$, $f^r = f^z = 0$ and refers to a thrust force constantly acting on the direction of the generalized variable θ. A thrust force acting on z direction (i.e. $f^z = 1$, $f^r = f^\theta = 0$) is the third strategy.

The time-dependent distance $\Delta R(t)$ between the perturbed and unperturbed positions of the Sun is defined by Eq. (58).

Figure 9 shows the time dependence of ΔR during one Sun revolution for the three strategies defined above. The second strategy (i.e. $f^\theta = 1$) yields the largest values of ΔR. The maximum deviation from the unperturbed orbit is about 35 pc. This is in good agreement with the result obtained by using the gravitational potential of section 4.1 (compare Fig. 9 and Fig. 7, respectively).

5. CONCLUSIONS

A stellar engine is defined in this chapter as a device that uses an important part of star resources to produce work. A classification is proposed as follows. A class A stellar engine uses *the impulse* of the radiation emitted by a star to produce a thrust force. When acting on a finite distance the thrust force generates work. Class A stellar engines can be used for interstellar travel. As example we quote the Sun thruster proposed by Shkadov (1987). A class B stellar engine uses *the energy* of the radiation emitted by a star to generate mechanical power. It is based on the concept of the Dyson sphere (DS). As example we cite the stellar engine proposed in Badescu (1995). A class C stellar engine is a combination between class A and class B stellar engines and provides a Kardashev type II civilisation with both power and the possibility of interstellar travel.

A class A stellar engine is associated with an increase in the affected star's photosphere temperature. For instance, by increasing the mirror's rim angle from 0 to 90 degrees the spectral class of the Sun changes from G2 towards F2. However, a reasonable rim angle value ($\Psi = 30°$) keeps the photosphere temperature (and its associated spectral class) quite close to the present-day values.

A number of conclusions may be drawn regarding the manufacturing of a class B stellar engine in our solar system. First, small radii increase the feasibility of the project as the amount of material required is proportional to the square of the radius. Second, the inner planets seem to be the best source of material because of the shorter distance between their orbit and the place of the future construction.

The efficiency of a class C stellar engine increases by increasing its radius and decreasing the mirror rim angle Ψ. There is a minimum radius for such engine to provide useful power. The important fact is that there is an optimum stellar engine radius as far as the provided power density is concerned. For values adopted here this optimum radius is around 450 millions km (see Fig. 4c).

The mirror of class A and class C stellar engines makes the Sun's temperature increase and this has consequences on the thrust force acting on the Sun. In both cases the thrust force F increases by increasing the mirror rim angle Ψ, as expected. The thrust force of the class A stellar engine is, however, larger than first estimated by Shkadov (1987) without taking into account the increase in the Sun's temperature.

Changing into a controllable way the trajectory of the Sun in the Galaxy is of great potential interest for humanity. In this chapter we have studied in some detail how class A or class C stellar engines could be used for this purpose. One has proved in Section 3.4 that both types of stellar engines provide practically the same thrust force when used to change Sun orbit.

A simple dynamic model for Sun motion in the Galaxy was developed in Section 4. It takes into account the (perturbation) thrust force provided by the stellar engine, which is superposed on the usual gravitational forces. The model allowed us to evaluate the distance between the perturbed position of the Sun and the usual Sun position. To increase confidence in results two different Galaxy gravitational potential models were used in calculations. In both cases, the results obtained show similar qualitative features for Sun's unperturbed motion.

Three simple strategies of changing Sun orbit were considered. A constant module thrust force was always assumed and the difference consisted in the force direction. None of these strategies make the distance between the perturbed and the unperturbed Sun position increase linearly in time. For this purpose an optimal control strategy is to be used.

For a single Sun revolution the maximum estimated deviation is between 35 and 40 pc (depending on the gravitational potential involved). Both Fig. 7 and Fig. 9 show that the stellar engine gives a 10 pc deviation of the Sun in less than 150 Myr. The number density of stars in the solar neighborhood is about 0.104 per cubic pc and so within a 10 pc radius sphere we would find around 400 stars, about 10 of which would be single solar-type stars (see e.g. (Fogg, 1995, p. 461)). This gives some perspective to future interstellar transfer macro-projects. The duration involved is, however, large and other kinds of stellar engines than those we studied must also be considered. One concludes that class A or class C stellar engines may be used to control in a certain extent the Sun's movement in the Galaxy.

ACKNOWLEDGMENTS

This work is dedicated to our friend Henrik Farkas (Technical University of Budapest) who passed away on 21 July 2005.

REFERENCES

Badescu V (1993) Maximum concentration ratio of direct solar radiation. Appl Optics 32(12):2187–2189

Badescu V (1995) On the radius of the Dyson sphere. Acta Astronautica 36:135–138

Badescu V (2004) Simulation of a solar Stirling engine operation under various weather conditions on Mars. J Solar Energy Eng 126:812–818

Badescu V, Cathcart RB (2000) Stellar engines for Kardashev's type II civilisations. J Br Interplanet Soc 53:297–306

Badescu V, Cathcart RB (2006) Use of class A and class C stellar engines to control sun movement in the galaxy. Acta Astronautica 58:119–129

Bash F (1986) The present, past and future velocity of nearby stars: the path of the Sun in 10^8 years. In: Smoluchowski R, Bahcall JN, Matthews MS (eds) The Galaxy and the Solar System. The University of Arizona Press, Tucson

Bejan A (1996) Entropy generation minimization: the new thermodynamics of finite-size devices and finite-time processes. J Appl Phys 79:1191–1218

Beju I, Soos E, Teodorescu PP (1976) Tehnici de Calcul Vectorial Cu Aplicatii, Editura Tehnica, Bucuresti

Bienayme O (1999) The local stellar velocity of the galaxy: galactic structure and potential. Astronomy Astrophysics 341:86–97

Carlberg RG, Innanen KA (1987) Galactic chaos and the circular velocity at the sun. Astron J 94:666–670

Clube SVM, Napier WM (1984) Terrestrial catastrophism: Nemesis or galaxy? Nature 311:635–636

Criswell DR (1985) Solar system industrialization: implications for interstellar migrations. In: Finney R, Jones EM (eds) Interstellar Migration and the Human Experience. University of California Press, Berkeley, pp 50–87

Darling D (2004) The Universal Book of Astronomy. Wiley, New York, pp 456

De Vos A (1985) Efficiency of some heat engines at maximum-power conditions. Am J Phys 53(5):570–573

Dehnen MW, Binney JJ (1998) Local stellar kinematics from HIPPARCOS data. Monthly Notices Royal Astronom Society 298:387–394

Dyson FJ (1966) The search for extraterrestrial technology. In: Marshak RE (ed) Perspectives in modern physics. Interscience Publishers, New York, pp 641–655

Fogg MJ (1989) Solar exchange as a means of ensuring the long term habitability of the Earth. Spec Sci Technol 12:153–157

Fogg MJ (1995) Terraforming: Engineering Planetary Environments. SAE, Warrendale
Fong KW, Jefferson TH, Suyehiro T, Walton L (1993) Guide to the SLATEC common mathematical library. Energy Science and Technology Software Center, PO Box 1020, Oak Ridge, TN 37831, USA
Gonzalez G (1999) Is the Sun anomalous? Astronomy Geophys 40:5.25–5.25.29
Hills JG (1984) Close encounters between a star-planet system and a stellar intruder. Astron J 89:1559–1564
Hoffmann KH, Burzler JM, Schubert E (1997) Endoreversible Thermodynamics. J Non-Equilib Thermodyn 22:311–355
Infante F (1992) Projects for the reconstruction of the firmament. Leonardo 25:11
Jones EM (1981) Discrete calculations of interstellar migration and settlement. Icarus 46:328–336
Kardashev NS (1964) Transmission of information by extraterrestrial civilisations, Astron Zh 8:217
Kasting JF (1988) Runaway and moist greenhouse atmospheres and the evolution of Earth and Venus. Icarus 74:472–494
Knill O (2003) Moving the solar system. http://www.dynamical.systems.org/zwicky/Essay.html
Korykansky DG (2004) Astroengineering, or how to save the Earth in only one billion years. Rev Mex A A (Serie Conferencias) 22:117–120
Korykansky DG, Laughlin G, Adams FC (2001) Astronomical engineering: a strategy for modifying planetary orbits. Astrophys Sp Sci 275:349–366
Kovalevsky J (1999) First results from HIPPARCOS. Annu Rev Astronomy Astrophys 36:99–130
Landsberg PT, Badescu V (1998) Solar energy conversion: list of efficiencies and some theoretical considerations. Part I – Theoretical considerations. Prog Quantum Electronics 22:211–230
McInnes CR (2002) Astronomical engineering revisited: planetary orbit modification using solar radiation pressure. Astrophys Sp Sci 282:765–772
Mishunov Yu N, Zenina IA (1999) Yes, the Sun is located near the corotation circle. Astronomy Astrophys 341:81–85
Moore P (1970) Atlas of the Universe. Mitchell Beazley Ltd, London, pp 35,155,159,169
Nakajima S, Hayashi YY, Abe Y (1992) A study of the 'runaway greenhouse effect' with a one-dimensional radiative convective equilibrium model. J Atmos Sci 49:2256
Napier WN (1985) Comet formation in molecular clouds. Icarus 62:384
Newman WI, Sagan C (1981) Galactic civilizations: population dynamics and interstellar diffusion. Icarus 46:293–327
Pal A, Ureche V (1983) Astronomie. Ed. Didactica si Pedagogica, Bucharest, pp. 161
Sackmann I-J, Boothroyd AI, Kraemer KE (1993) Our Sun III: Present and future. Astrophys J 418:457
Shkadov LM (1987) Possibility of controlling solar system motion in the galaxy. 38th Congress of IAF, paper IAA-87-613. Brighton, UK, October 10–17
Ureche V (1987) Universul. Astrofizica, vol 2. Dacia, Cluj-Napoca
Zuckerman B (1985) Stellar evolution: motivation for mass interstellar migrations. Q J R Astr Soc 26:56–59
Zwicky F (1957) Morphological astronomy. Springer-Verlag, Berlin, pp. 260

CHAPTER 13

MACRO-ENGINEERING IN THE GALACTIC CONTEXT

A new agenda for astrobiology

MILAN M. ĆIRKOVIĆ

Astronomical Observatory of Belgrade, Volgina 7, 11160 Belgrade, Serbia

Abstract: We consider the problem of detectability of macro-engineering projects over interstellar distances, in the context of Search for ExtraTerrestrial Intelligence (SETI). Freeman J. Dyson and his imaginative precursors, like Konstantin Tsiolkovsky, Olaf Stapledon or John B. S. Haldane, suggested macro-engineering projects as focal points in the context of extrapolations about the future of humanity and, by analogy, other intelligent species in the Milky Way. We emphasize that the search for signposts of extraterrestrial macro-engineering projects is not an optional pursuit within the family of ongoing and planned SETI projects; *inter alia*, the failure of the orthodox SETI thus far clearly indicates this. Instead, this approach (for which we suggest a name of "Dysonian") should be the front-line and mainstay of any cogent SETI strategy in future, being significantly more promising than searches for directed, intentional radio or microwave emissions. This is in accord with our improved astrophysical understanding of the structure and evolution of the Galactic Habitable Zone, as well as with the recent wake-up call of Steven J. Dick to investigate consequences of postbiological evolution for astrobiology in general and SETI programs in particular. The benefits this multidisciplinary approach may bear for macro-engineers are also briefly highlighted

Keywords: astrobiology; macro-engineering; extraterrestrial intelligence; history and philosophy of science

1. INTRODUCTION: MACRO-ENGINEERING AS ASTRO-ENGINEERING

Previous chapters have clearly demonstrated a wide spectrum of views and approaches to the new multidisciplinary topic of macro-engineering. All these dynamic approaches are worthy successors to the ultimate goal set in Francis Bacon's posthumous *New Atlantis* (1626): "The end of our foundation is the knowledge of causes, and secret motions of things; and the enlarging of the bounds

V. Badescu et al. (eds.), Macro-Engineering: A Challenge for the Future, 281–300.

of human empire, to the effecting of all things possible." Here we shall attempt to offer a new twist on the same old theme, a twist that is likely to open multiple new and unexpected vistas of research. In order to expose this novel perspective, we need to take a brief detour through a seemingly unrelated, but in fact quite relevant and vigorously expanding discipline.

We are lucky enough to live in an epoch of great progress in the nascent field of *astrobiology*, which deals with three canonical questions: How does life begin and develop? Does life exist elsewhere in the universe? What is the future of life and intelligence on Earth and in space? A host of discoveries have been made during the last decade or so, the most important certainly being a large number of extrasolar planets; the existence of many extremophile organisms at the deep ocean hydrothermal vents, possibly vindicating the "deep hot biosphere" of Thomas Gold (1998); the discovery of subsurface water on Mars and the huge ocean on Europa, and possibly also Ganymede and Callisto; the unequivocal discovery of amino-acids and other complex organic compounds in meteorites; modelling organic chemistry in Martian and Titan's atmosphere; the quantitative treatment of the Galactic Habitable Zone; the development of a new generation of panspermia theories, spurred by experimental verification that even terrestrial microorganisms survive conditions of an asteroidal or a cometary impact; progress in philosophy and methodology, etc. (for recent beautiful reviews see Des Marais and Walter, 1999; Darling, 2001; Grinspoon, 2003). All this deserves the label of a true astrobiological revolution.

Perhaps the most fascinating field in the multidisciplinary astrobiological spectrum is the Search for ExtraTerrestrial Intelligence (henceforth SETI). At the beginning of XXI century it remains the oldest and perhaps the most intriguing scientific problem. Starting with the pioneering work of Frank Drake, Carl Sagan, Iosif Shklovsky, and others (e.g., Drake, 1965), as well as the historical OZMA project, SETI studies have had their ebb and flow of tides over the last four decades (Dick, 1996). During that time a set of ideas which can be characterized as "orthodox" SETI has emerged. In a simplified form, it can be summarized as follows. Life is common in the universe. Emergence of intelligence and technology is, if not necessary, then at least typical outcome of biological evolution throughout the Milky Way. A sizeable fraction of technological species are interested in communication with other intelligent creatures. It makes sense to listen for intentional radio or optical messages from out there and to transmit messages in return. It makes no sense to travel across interstellar distances or to expect such interstellar visitors. What we can hope to achieve is slow and benign exchange of messages, the greatest beneficiaries in such exchange being the youngest newcomers to the "Galactic Club" (Bracewell, 1975), such as humans. Basic tenets of this view have crystallized by mid-1970s, decades before the astrobiological revolution.

Times are changing. *In addition to* the astrobiological revolution itself, some of the important recent developments of relevance to the SETI endeavour (in the widest sense) are:

1. The rise of *digital perspective* in various fields, starting with fundamental physics and computer science (e.g., Chaitin, 1987; Toffoli, 1998; Chaisson, 2001;

Fredkin, 2003), to biological and social sciences (e.g., Kauffman, 1995; Maynard Smith and Szathmary, 1997; Adami et al., 2000; Carroll, 2001; Adams, 2003). In particular, this includes understanding of the crucial importance of information dynamics, open systems, complexity, substrate-independent dynamical laws, and interrelated evolutionary pathways.

2. Closely related, tremendous efforts in the fields of computer science and neurosciences, invested toward achieving of artificial intelligence (AI), which would offer completely new perspectives on the nature of intelligence itself (Henry, 2005), as well as the possible future evolutionary trajectory of humanity and, by analogy, other intelligent communities in the Galaxy. Coupled with the digital perspective in biological science, this raises the all-important (from the SETI point of view) issues of whether we should search for biological or *postbiological* intelligence; this is particularly forcefully put forward in a recent important paper by the distinguished historian and philosopher of science Steven J. Dick (2003). Information theory also recently highlighted all the difficulties (Lachmann et al., 2004) and inefficiencies (Rose and Wright, 2004) *inherent* in the attempts to communicate by radio signals over interstellar distances.
3. The advent of *physical eschatology*, a nascent astrophysical discipline dealing with the future of astronomical objects, including the universe itself. The groundworks were laid by Freeman J. Dyson a quarter century ago (Dyson, 1979), but the explosion of interest occured only in the last decade or so; for reviews see Adams and Laughlin (1997), Ćirković (2003).
4. Related to research in physical eschatology, though on more modest spatio-temporal scales are those aspects of future studies dealing with the future of humanity, and, in particular, the rise of new vigorous intellectual movements, which can be (in a revived term of Sir Julian Huxley; see Huxley, 1957) unified under the banner of *transhumanism* (Moravec, 1988; Kurzweil, 2000; Wright, 2000; Bostrom, 2005). These may signify an open endorsement of the postbiological paradigm, to which we shall return later.
5. Improved epistemological and methodological understanding of our properties as intelligent observers, as well as physical, chemical and other pre-conditions necessary for the existence of such observers (Carter, 1983; Livio, 1999). The latter are topics of so-called *anthropic reasoning*, the subject of much debate and controversy in cosmology, fundamental physics, and philosophy of science (Barrow and Tipler, 1986). In recent years it became clear that anthropic principle(s) can be most fruitfully construed as *observation selection effects* (Bostrom, 2002). This is a straightforward continuation of the Copernican Revolution, which emphasizes a non-special character of our cosmic habitat, and which has so immensely contributed to our scientific understanding.
6. The rise of *neocatastrophism* in Earth and planetary sciences, offering not only wealth of explanatory models and hypotheses for puzzling empirical facts (e.g., Jablonski, 1986; Raup, 1991, 1994; Courtillot, 1999; Benton, 2003), but a new philosophical perspective as well (Gould, 1985; Huggett, 1997).

7. Emergence of serious macro-engineering studies on various scales of space, time, energy and sophistication, best attested by the other contributions in this volume. (Let us note in passing that macro-engineering, as by definition the intentional change of physical environment on ever larger scales, is antithetical to the main goal of physical eschatology, which tries to track future evolution of our cosmic environment with as little complication due to the intentional influences. The epistemological and methodological relationship between the two has not been sufficiently investigated in the literature so far, although there have been several interesting studies at the crossroads; e.g., Garriga et al. 2000; Ćirković, 2004.) In particular, the aspects of macro-engineering pertaining to our astronomical surroundings (e.g., Korykansky et al., 2001; McInnes, 2002) are of relevance for our search for intelligence elsewhere.

If we now wish to ask how the actual SETI research has responded to these exciting and stimulative developments of obvious relevance to its subject matter, we are bound for severe disappointment. With honorable exceptions, the mainstream SETI (as exemplified, for instance, by activities of NASA, the SETI Institute, or the Planetary Society; e.g., Shostak 1993; Pierson 1995; McDonough 1995) has mostly ignored all of the developments listed above. SETI today is mostly in the same shape and with the same set of philosophical, methodological and technological guidelines, as it was in the time of its pioneers (Frank Drake, Carl Sagan, Giuseppe Cocconi, Philip Morrison, Iosif Shklovsky, Ronald Bracewell, Bernard Oliver, Michael Papagiannis, Nikolai Kardashev) in 1960s and 1970s. In contrast which can hardly be sharper, our views of astrophysics, planetary sciences, evolutionary biology and, especially, computer science—arguably the four key scientific pillars of SETI—changed revolutionarily, to put it mildly, since that epoch. It is both sad and ironic that the field which was once correctly identified as a paragon of originality, boldness, and vigour, has not lived to these ideals during the last 30 years or so; instead, it was gradually subverted by conservative views. Only by this odd conservativeness can be explained that, for instance, a major SETI review article at the beginning of the new millenium in by far the most authoritative publication in astronomical sciences (Tarter, 2001) can be written without even a single mention (in its 38 pages) of such crucial memes as AI, anthropic reasoning, von Neumann probes, neodarwinism, or macro-engineering. Claims that (orthodox) SETI is a strictly empirical field, should be evaluated at face value; translated into the language of modern epistemology, this would serve the counterproductive role of sorting SETI with pseudo-sciences, since amassing "empirical work" while having no theoretical basis (or hiding it) is a hallmark of pseudo-science. (Compare with cases of parapsychology, dowsing, craniometry or various nutritionisms, including the ideas of undoubted geniuses, like Linus Pauling.) The present study is an attempt to break this mold and point to serious modern alternatives to the old-fashion SETI philosophy. This is not a luxury, but a necessity in the situation when epochal changes in our views of virtually everything in science and technology are not, unfortunately, accompanied by an appropriate shift in SETI attitudes.

This sort of isolationism is never justified, not only in science, but in wider arena of intellectual life; however, as history teaches us, it could be partly mitigated by inherent short-term successes of the isolated discipline. This is certainly not the case with SETI studies. In almost four decades of SETI projects there have been no results, in spite of the prevailing "contact optimism" of 1960s and 1970s, motivated largely by uncritical acceptance of the Drake equation (Drake, 1965). Conventional estimates of that period spoke about $10^6 - 10^9$ advanced societies in the Milky Way forming the "Galactic Club" or a similar anthropocentrically conceived association; for a prototype optimistic—or naive—view of that epoch, 30 years old, see Bracewell (1975). Early SETI literature abounds in such misplaced enthusiasm. Today, even "contact-optimists" have abandoned fanciful numbers, and settled on a view that advanced extraterrestrial societies are much rarer than previously thought. One of the important factors in this downsizing of SETI expectations has been demonstrations by "contact pessimists", especially Michael Hart and Frank Tipler, that the colonization—or at least visit—of all stellar systems in the Milky Way by means of self-reproducing von Neumann probes is feasible within a minuscule fraction of the Galactic age (Hart, 1975; Jones, 1976; Tipler, 1980, 1981). In this light, Fermi's legendary question: *Where are they?* becomes disturbingly pertinent (Brin, 1983; Webb, 2002). In addition, Carter (1983) suggested an independent and powerful anthropic argument for the uniqueness of intelligent life on Earth in the Galactic context, and biologists such as Simpson (1964), Mayr (1993), or Conway Morris (2003) put forward biological arguments to the effect that complex lifeforms are quite rare in the Galaxy. These views have sometimes been subsumed within the "rare Earth" hypothesis (Ward and Brownlee, 2000). It is generally recognized in both research and the popular science circles that the "contact pessimists" have a strong position.

While fully recognizing that patience is a necessary element in any search, cosmic or else, we still wish to argue that the conventional SETI (Tarter, 2001; Duric and Field, 2003, and references therein), as exemplified by the historical OZMA Project, as well its subsequent and current counterparts (META, ARGUS, Phoenix, SERENDIP/Southern SERENDIP, etc.), notably those conveyed by NASA and the SETI Institute, is fundamentally limited and unlikely to succeed. This is emphatically **not** due to the real lack of targets, us being alone in the Galaxy, as contact-pessimists in the mold of Tipler or Mayr have argued. Quite contrary, it is due to real physical, engineering, and epistemological reasons undermining the conventional SETI philosophy. In a sense the problem has nothing to do with the universe itself, and everything to do with our ignorance and prejudices.

In particular, we wish to argue that the reality is not limited to these two extremes: **(I)** naive optimism of the SETI establishment inherited from the "founding fathers", and **(II)** blatant pessimism of the detractors, often supported by extra-scientific motives. (Examples of the latter are worrying about the magnitude of the U.S. federal debt in Mayr (1993), or various (quasi)religious elements present, e.g., in Conway Morris (2003) or in Tipler's numerous writings. Darling (2001) amusingly discusses the relationship of the "rare Earth" hypothesis of Ward and Brownlee, a mainstay of SETI skepticism today, with creationism.) As usual, the reality is much more complex, and taking into account the developments listed above will give rise to a set of more complex "middle-ground" solutions to the problem of absence (or otherwise) of extraterrestrial intelligence. Such hypotheses are, typically, not open to falsification with the standard SETI procedures (i.e., listening to intentional radio- or optical messages), but are, in general, falsifiable by a different set of SETI methods and procedures. One such "middle-ground" solution has been proposed by the present author elsewhere (Ćirković and Bradbury, 2006), but it is far from being unique in this respect; a large class of "catastrophic" solutions to Fermi's paradox belong to this category (Clarke, 1981; Annis, 1999a), as well as those based on the long-term evolutionary processes (Schroeder, 2002; Ćirković, 2006). This also applies to the ingenious idea that advanced civilizations will transfer their cognition into their environment (Karl Schroeder, private communication), following recent studies on the *distributed natural cognition* (e.g., Hutchins, 1996). In these, as in other suggested lines of postbiological evolutionary development of advanced civilizations, the approaches favored by the ongoing SETI projects will be fundamentally misguided, i.e., advanced societies remain undetectable by such approaches. All of these ideas require a sort of "breaking the mold" of conservative mainstream thinking.

Such "breaking the mold" must not be understood as leading inexorably to increased SETI optimism. A very good counterexample is the work of Raup (1992) on non-conscious SETI sources, which are likely to cause confusion in the practical SETI work. Although we hereby argue that *on the balance*, changed perspectives increase chances of actual SETI success, this is rather accidental, certainly not a necessary consequence. On the contrary, many skeptical arguments can and must be incorporated in the emerging "new" SETI paradigm.

The crucial ingredient here is exactly our development #7 above, namely *the increased awareness of the potentials inherent in macro-engineering*. The central point was clearly explicated long ago by Freeman J. Dyson, to whose ideas we shall repeatedly return (Dyson, 1966, p. 642):

> "When one discusses engineering projects on the grand scale, one can either think of what we, the human species, may do here in the future, or one can think of what extraterrestrial species, if they exist, may have already done elsewhere. To think about a grandiose future for the human species ("la nostalgie du future") is to pursue idle dreams, or science fiction. But to think in a disciplined way about what we may now be able to observe astronomically, if it should happen to be the case that technologically advanced species exist in our corner of the universe, is a serious and legitimate part of science. In this way I am able to transpose the dreams of a frustrated engineer into a framework of respectable astronomy."

While we may take a *votum separatum* on some of Dyson's views on future studies (written almost 40 years ago), the major point of this passage seems almost self-evident: one of the best ways to ascertain limits of the technically possible is to discard anthropocentrism and sample a larger volume of space and time, reasonably expecting that somewhere and somewhen it has been achieved. In the rest of this Chapter, we shall consider epistemological and methodological impact of macro-engineering (sometimes called astro-engineering in this context; we shall use the terms synonymously) on our SETI projects and *policies*. As we hope to show, this contact between macro-engineering and astrobiology can be encouraging and productive on both sides.

2. DYSONIAN APPROACH TO SETI

By their fruits ye shall know them. The Biblical proverb neatly encapsulates the proposed unconventional approach to SETI, in which the focus would be a search for manifestations and macro-engineering artifacts, instead of intentional messages. Even more, the metaphor seems particularly apt, since it warns about messages (and efforts to hear them) being actively misleading in search for the truth.

It was along these lines that, in 1960, Freeman J. Dyson suggested that the very existence of what we can term *advanced technological civilizations* (henceforth ATCs) should provide us with means of detecting them (Dyson, 1960). As is well-known, starting from the Malthusian assumptions and the well-recorded increase in power consumption with the development of technological civilization here on Earth, Dyson concluded that a truly developed society will soon face the limits regarding both living space and available energy if constrained only to planetary surfaces. On the contrary, the only way to optimize resources would be a construction of a *Dyson shell*, capturing all energy from the domicile star. This particular paper, hardly longer than a page, has not only motivated manifold subsequent visions and studies in the field of astro-engineering and is likely to continue to do so for a long time to come; we argue that it set groundwork for a different sort of SETI from the one conducted since the OZMA project in mid-1960s. That is because Dyson suggested that infrared signatures of a Dyson shell will be detectable from large distances, and will present a confirmation of the existence of ATCs. This view has been subsequently elaborated in Dyson (1966), and in the often-neglected study of Sagan and Walker (1966). Some of the very best elaborations of the Dysonian ideas have been published in the science fictional context by Stanislaw Lem (1984, 1987; for a discursive form, see Lem, 1977). For further fruitful work along Dyson's lines, see Chapter 12 of the present book.

(It is neither necessary nor desirable for our further considerations to make the notion of ATCs more precise. The diversity of postbiological evolution is likely to at least match, and probably dwarf, the diversity of its biological precedent. It is one particular feature—information processing—we assume common for the "mainstream" ATCs, in accordance with the postbiological paradigm and the Intelligence Principle of Dick (2003). Thus, whether real ATCs can most adequately be described as "being computers" or "having computers" is not of key importance for our analysis; we just suppose that in either case the desire for optimization of computations will be one of important (if not *the most* important) desires of such advanced entities. It is already clear, from the obviously short and limited human experience in astronautics, that the postbiological evolution offers significant advantages in this respect (Parkinson, 2005). However, an operational definition of ATC is clear: ATC is the community capable of grand feats of astro-engineering. This is another important link between the fields of astrobiology and macro-engineering, which certainly requires further elaboration elsewhere.)

It is thus multiply justified to call this other, unconventional paradigm the *Dysonian approach* to the problem of extraterrestrial intelligence. Some of its elements were in place, to be sure, much before Dyson and his 1960 paper. Notable sources for a future historian of ideas would be far-reaching speculations of several brilliant minds in the first few decades of the XX century: British authors and philosophers William Olaf Stapledon (1886–1950) and Herbert George Wells (1866–1946), as well as the famous biologist and polymath John B.S. Haldane (1892–1964); Russian engineer and astronautics pioneer Konstantin E. Tsiolkovsky (1857–1935); Serbian-American inventor and electrical engineer Nikola Tesla (1856–1943); as well as some others. It is impossible to give justice to the magnitude and boldness of thought of these great minds in this limited space. Let us just mention in passing that Stapledon, Wells and Haldane mused presciently upon great powers which will be available to future man in shaping both himself and his natural environment; in addition, they emphasized the need for synergistic development of both natural and social sciences as well as technology to achieve such powers, without succumbing to a catastrophe on the way. Tsiolkovsky was perhaps the first macro-engineer in the modern sense, envisioning huge space colonies and habitats (the idea later developed by Gerard O'Neill and others) serving as springboards for human space expansion and colonization; he was also a precursor to the Fermi paradox, this most perplexing SETI-related puzzle. While Tesla can claim credit for being the founder of *orthodox* SETI by first emphasizing the role of radio-waves and even doing some practical attempts to detect alien radio signals (Dick, 1996), he was also the first to perceive the link between the stage of development of a planetary society and a specific physical quantity – namely expended power. In an artistic manner, the same inspiration of macro-engineering in the cosmic context motivated some of the best graphics of one of the most prominent artist of the XX century, Maurice C. Escher (see Figs. 1, 2).

But it was Dyson, "the eclectic physicist and the frustrated engineer" by his own amusing description in a later work (Dyson, 1966), who provided the key insight which, unfortunately, almost half a century later has not been sufficiently understood in all its ramifications and consequences. It is important to emphasize here that the Dysonian approach to SETI is not limited to search for Dyson shells, but to general class of artifacts, manifestations and traces of the existence of ATCs. Some other astro-engineering feats belonging to this, potentially detectable, categories are:

- Supramundane planets, shell worlds, orbital rings, and similar circumplanetary constructions (Birch 1983, 1991; Roy et al., 2004).
- Large-scale antimatter-burning vehicles or industrial plants (Harris, 1986, 2002).
- Large artificial objects (Tsiolkovsky-O'Neill habitats, for instance) in transiting orbits, detectable through extrasolar planet searches (Arnold, 2005).

Figure 1. Astro-engineering as inspiration in arts: M.C. Escher's *Concentric Rinds* (1953) can be regarded as Dyson shell-like structures (in construction?)

Even the search for extraterrestrial artifacts in the Solar System belongs to this category. This approach, however, is still strongly in the minority in SETI circles, in contrast to its many advantages, some of which we shall briefly review.

The very obvious thing is that such an approach does not prejudice properties of target societies in the way the orthodox view does. It is self-evident that willingness of both parties is the necessary (but far from sufficient) condition for a successful communication on any scale, from human everyday life up. However, in contrast to people we encounter every day in our lives, we have not the vaguest idea of whether such willingness exists on the level of interstellar communication. It is indicative that a large portion of the early SETI literature, especially writings of the "founding fathers" consists of largely emotional attempts to make the assumption of willingness (and, indirectly, benevolence) of SETI target societies plausible (e.g., Bracewell, 1975); this is read more like wishful thinking than any real argument (Gould, 1987). To cite Dyson (1966) again: "[M]y point of view is rather different, since I do not wish to presume any spirit of benevolence or community of interest among alien societies." This, of course, does not mean that the opposite assumption (of malevolence) should be applied. Simply, such prejudicating in the nebulous

Figure 2. Another Escherian vision of a macro-project: woodcut *Tetrahedral Planetoid* (1954) reminiscent of much later O'Neill's orbital habitats. M. C. Escher was an enthusiastic amateur astronomer, who meticulously observed double stars and other celestial phenomena (cf. Locher et al., 1992)

realm of alien sociology is unnecessary in the Dysonian framework; with fewer assumptions it is easier to pass Occam's razor.

The most important advantage of the Dysonian approach concerns the spatial and temporal frames within which practical SETI is conducted. Even proponents of the SETI orthodoxy admit that the "window of opportunity" for radio- or optical laser communication is quite short; in a cogent and well-written summary of the orthodox position, Duric and Field (2003) admit:

> "Under the optimistic scenario, there may be as many as 10^6 technologically advanced civilizations in the Galaxy. However, these societies are at various stages of development. The probability that two extraterrestrial societies are at the same stage of evolution, to say within a million years, is very small."

This is obviously of crucial importance if our goal is bidirectional, intentional communication between us and an extraterrestrial civilization. It is exactly to this situation that the arguments of SETI skeptics most forcefully apply; thus, it is important to realize that what is often vaguely referred to as "anti-SETI" arguments are, indeed, only arguments against SETI *orthodoxy*. However, the same "window of opportunity" is increased by many orders of magnitude, or even vanishes entirely, when we search for macro-engineering artifacts instead of intentional messages. This reasoning was clear to thinkers of epochs long past. In a beautiful passage in Book V of the famous poem *De Rerum Natura* [On the Nature of Things], Roman poet and late-Epicurean philosopher Lucretius wrote the following intriguing verses (in translation of William E. Leonard, available via WWW Project Gutenberg; Lucretius 1997):

If there had been no origin-in-birth
Of lands and sky, and they had ever been
The everlasting, why, ere Theban war
And obsequies of Troy, have other bards
Not also chanted other high affairs?
Whither have sunk so oft so many deeds
Of heroes? Why do those deeds live no more,
Ingrafted in eternal monuments
Of glory? Verily, I guess, because
The Sun is new, and of a recent date
The nature of our universe, and had
Not long ago its own exordium.

Neglecting here its cosmological context of arguing for a finite past age of the universe, this passage indicates an oft-neglected aspect of Fermi's paradox—it is not enough to somehow remove all ATCs from our past light cone, but we need to erase their more durable and potentially detectable achievements as well, in order to reproduce empirical "Great Silence" (Brin, 1983). On Earth, the very existence of the fascinating discipline of archaeology tells us that cultures (and even individual memes) produce records significantly more durable than themselves. It is only to be expected that such trend will continue to hold even more forcefully for higher levels of complexity and more advanced cultures. There are even some factors related to the properties of our cosmic environment that enhance this trend; notably, it has already been repeatedly suggested that the traces of any hypothetical extraterrestrial visitations in the past of the Solar System would be easier to locate on the Moon than on Earth, due to the vastly supressed erosion there (e.g., McKay, 1996).

As an example, let us for the sake of discussion allow that a significant fraction of advanced technological civilizations evolves toward the Kardashev (1964) Type II, i.e. a community completely managing the energy output of its parent star; for the information-processing need of advanced communities, see Ćirković and Radujkov (2001). The straightforward way of achieving this is the construction of a Dyson shell. Once constructed, such an example of astroengineering, will be quite durable due to the properties of the interplanetary and interstellar space itself; like the Pyramids of Egypt, a Dyson shell is likely to outlive its creators for a vast period of time (on the physical eschatological scales), thus being an advanced analog of Lucretius' "eternal monuments". Some *very preliminary* searches (see §3 below) show the absence of such artifacts in the Galactic vicinity of the Sun.

Similar reasoning can be applied to the volume of space sampled by active searches. According to recent important studies by Lineweaver (2001) and

Lineweaver et al. (2004), Earth-like planets around other stars in the Galactic Habitable Zone (Gonzalez et al., 2001) are, on average, 1.8 ± 0.9 Gyr older than our planet. These calculations are based on chemical enrichment as the basic precondition for the existence of terrestrial planets, as well as on the rate of destructive processes (like supernovae). Applying the Copernican assumption naively, we would expect that correspondingly complex life forms on those others to be *on the average* 1.8 Gyr older. Intelligent societies, therefore, should also be older than ours by the same amount. In fact, the situation is even worse, since this is just the average value, and it is reasonable to assume that there will be, somewhere in the Galaxy, an inhabitable planet (say) 3 Gyr older than Earth. Since the set of intelligent societies is likely to be dominated by a small number of oldest and most advanced members (for an ingenious discussion in somewhat different context, see Olum, 2004), we are likely to encounter a civilization actually more ancient than 1.8 Gyr (and probably significantly more). It seems preposterous even to contemplate any possibility of communication between us and Gyr-older supercivilizations. Remember that 1 Gyr ago the appearance of even the simplest animals on Earth lay in the distant future. Some of the SETI pioneers have been very well aware of this on the qualitative level and warned about it (notably Sagan, 1975); these cautious voices have been consistently downplayed by the SETI community. All in all, we conclude that the conventional radio SETI assuming beamed broadcasts from a nearby Sun-like star (e.g., Turnbull and Tarter, 2003) is ill-founded and has almost no chance of success on the present hypothesis. Given the likely distances of ATCs that began their technological ascent tens of millions to billions of years ago, they are not likely to know of our development. While their astronomical capabilities probably allow them to observe the Solar System, they are looking at it before civilization developed. It is doubtful, to say at least, that they would waste resources sending messages to planetary systems possessing life, but quite uncertain (in light of the biological contingency) to develop a technological civilization. Dolphins and whales are quite intelligent and possibly even human-level conscious (e.g., Browne, 2004), but they do not have the ability to detect signals from ATCs, and it is unlikely, to say at least, that they will ever evolve such a capacity. By a mirror-image of such position, unless one has concrete evidence of an ATC at a given locale it would be wasteful to direct SETI resources towards them. Although this conclusion can offer a rationale to some of the SETI skeptics, it is based on the entirely different overall astrobiological picture and has different practical consequences.

Furthermore, locations of the original home and the bulk of technology of an ATC can be decoupled even at smaller spatial scales. It has already been repeatedly suggested that our descendants, in particular if they cease to be organic-based, may prefer low-temperature, volatile-rich outer reaches of the Solar System (if these ecological niches are not already filled; see Dyson, 2003). Thus, they could create what could be dubbed a "circumstellar technological zone" as different and complementary to the famous "circumstellar habitable zone" in which life is, according to most contemporary astrobiological views, bound to emerge. We propose to generalize this concept to the Galaxy (and other spiral galaxies) in complete analogy to the Galactic habitable zone.

In addition, the Dysonian approach allows us to re-assess *extragalactic* SETI, in the sense precluded by the orthodox paradigm. Extragalactic SETI has not been considered very seriously so far (for notable exceptions see Wesson, 1990; Annis, 1999b). The reason is, perhaps, the same old comforting prejudice that we should expect specific (and most conveniently radio) signals. Since these are not likely forthcoming over intergalactic distances and the two-way communication desired by orthodox SETI pioneers is here entirely senseless, there is no point in even thinking seriously about extragalactic SETI. Such view is fallacious: when we remove the cozy assumption of specific SETI signals (together with the second-order assumption of their radio nature), it collapses. On the contrary, extragalactic SETI would enable us to probe enormously larger part of physical space as well as the space of possible evolutionary histories of ATCs. (Of course, part of what we get ensemble-wise we lose time- and resolution-wise.) In fact, the definition of Kardashev's Type III civilization (i.e., those managing energy resources of its entire home galaxy; Kardashev, 1964) should prompt us to consider it more carefully, at least for a sample of nearby galaxies, visible at epochs significantly closer to us than the 1.8 Gyr difference between the average of Lineweaver (2001) and the age of Earth which is about 4.56 Gyr. In fact, it could be argued (although it is beyond the scope of the present study) that the null result of extragalactic SETI observations so far represents a strong argument against the viability of Kardashev's Type III civilizations. While it remains a possibility, in the formal sense of being in agreement with the known laws of physics, it seems that the type of pan-galactic civilization as envisaged by Kardashev and other early SETI pioneers is either much more difficult (suggesting that the sample of $\sim 10^4$ normal spiral galaxies close enough and observed in high enough detail is simply too small to detect even a single Type III civilization), or simply not worth striving to establish.

3. A PROLEGOMENA FOR "NEW" SETI

The list of both theoretical and observational SETI studies performed so far along the Dysonian guidelines is rather short; most of it is the following:

- Searches for artificial objects near Earth and anomalous spectral lines in stellar spectra performed by Robert A. Freitas, Jr. and Francisco Valdes in 1980s (Freitas and Valdes, 1980; Valdes and Freitas, 1983, 1986).
- Japanese program of searches for Dyson shells around nearby stars (Jugaku et al., 1995; Jugaku and Nishimura, 2003).
- Proposals for observational or archival searches for Dyson shells or related astro-engineering projects (Slysh, 1985; Tilgner and Heinrichsen, 1998; Timofeev et al., 2000).
- The detailed study of Sandberg (1999), fruitfully linking information-processing to macro-engineering.

- Investigation of gamma-ray signatures of antimatter burning by ATCs (Harris, 1986, 2002).
- A recent proposal for searching for transits of artificial objects across the observed stars (Arnold, 2005).
- The analysis of archival extragalactic data by Annis (1999b) suggesting the absence of star-powered Kardashev Type III civilizations among nearby galaxies.

To this disturbingly short list one may add several important theoretical studies showing either general feasibility of the astro-engineering feats detectable from afar (e.g., Suffern, 1977; Badescu, 1995; Badescu and Cathcart, 2000), or the necessity of taking a non-standard approach (e.g., Russell, 1983; Raup, 1992); the latter have been, significantly enough, often written by biologists interested in SETI, and have not been given due credit and attention in the orthodox SETI circles.

The proposed re-orientation of SETI projects is in a very deep sense independent of the favorite model for solution of puzzles related to the extraterrestrial intelligence, most notably Fermi's paradox. Although there may indeed be more than fifty solutions to Fermi's paradox (Webb, 2002), essentially all major solutions are compatible with, or indeed suggestive of the Dysonian approach.

Of course, even those projects or proposals put forward so far are limited in the sense of being often too conservative with respect to the full range of parameters. For instance, the controlling parameter for detection of a Dyson shell is, ultimately, its temperature (the differences in the intrinsic stellar output can be neglected in the first approximation). The searches thus far (including the mentioned studies of Jugaku et al.) relied on the original Dyson's proposal that the shell would be the size of Earth's orbit around the Sun, and that its working temperature would, thus, be close to the temperature of a solid body at 1 AU from a G2 dwarf. However, from a postbiological perspective, this looks to be quite wasteful, since computers operating at room temperature (or somewhat lower) are limited by a higher $kT \ln 2$ Brillouin limit, compared to those in contact with heat reservoir on lower temperature T (cf. Brillouin, 1962). Although it is not realistic to expect that efficiency can be increased by cooling to the cosmological limit of 3 K in the realistic model of the Galaxy (Ćirković and Bradbury, 2006), still it is considerable difference in practical observational terms whether one expects a Dyson shell to be close to a blackbody at 50 K, as contrasted to a blackbody at 300 K. This lowering of the external shell temperature is also in agreement with the study of Badescu and Cathcart (2000) on the efficiency of extracting work from the stellar radiation energy. In this sense, the "true" Dysonian approach needs to be even more radical than the intuitions of Dyson himself.

This is linked to another indicative practical issue: *parasitic* searches, which are now used by some of the ongoing SETI projects, are natural *modus operandi* for the observational search for the Galactic macro-engineering. This, of course, means a great increase in efficiency of operation, as well as a decrease in cost, especially when coupled with widely distributed processing, along the lines of ingenious *SETI@home*. However, this makes the role of creating solid theoretical groundwork for such projects much more delicate and important.

4. INSTEAD OF CONCLUSIONS

The Dysonian approach to search for other intelligent societies can be briefly summarized as follows. Even if they are not actively communicating with us, that does not imply that we cannot detect them and their astro-engineering activities. Their detection signatures may be much older than their communication signatures. Unless ATCs have taken great lengths to hide or disguise their IR detection signatures, the terrestrial observers should still be able to observe them at those wavelengths and those should be distinguishable from normal stellar spectra. (Ironically enough, surveys in the infrared have been proposed by one of the pioneers of microwave astronomy, Nobel-prize winner Charles H. Townes, although on somewhat different grounds; see Townes, 1983.) The same applies to other un-natural effects, like the antimatter-burning signatures (Harris, 1986, 2002), or recognizable transits of artificial objects (Arnold, 2005). Search for mega-projects such as Dyson shells, Jupiter Brains or stellar engines are most likely to be successful in the entire spectrum of SETI activities.

Some of the major differences between the orthodox and the Dysonian approach to SETI are laconically summarized in Table 1. It would not be too pretentious to claim that the comparison of the two approaches favors the Dysonian approach. However, this approach has yet to achieve its legitimacy in the circles of SETI researchers.

Bold and unconventional studies, such as Freitas', Harris', Arnold's, Slysh's, or the survey of Jugaku et al., represent still a small minority of the overall SETI research. We dare suggest that there is no real scientific reason for such situation: instead, it occurs due to excessive conservativeness, inertia of thought, overawe of the "founding fathers", or some combination of the three. Another, albeit extra-scientific, argument sometimes put forward in informal situations is that the massive pseudoscientific fringe surrounding SETI ("flying saucers" enthusiasts, archaeo-astronauts, and the like) would feel encouraged by relaxing the conservative tenets of the orthodox SETI. This argument is hard to evaluate due to its essentially social and extra-scientific nature. In any case, it gives far too much weight and influence

Table 1. A comparison between the orthodox and the Dysonian approach to SETI

	Orthodox SETI	Dysonian SETI
Main object of search	Intentional messages	Artifacts and traces
Working ATC model	Biological	Postbiological
Window of opportunity	Narrow	Wide
Prejudicates alien behavior	Yes	No
Communication	Yes	No
Interstellar travel	Irrelevant	Relevant
Main working frequencies	Radio (cm)	Infrared
Natural mode of work	Active	Parasitic
Extragalactic SETI	No	Yes

to lunatics and pseudo-scientists than is tolerable in any serious scientific discipline. The unconventional approach with emphasis on search for ATCs' manifestations would lose nothing of the advantages of conventional SETI before detection (Tough, 1998), but the gains could be enormous.

Since it is the success in search we are after, it goes without saying that this assessment has nothing to do with SETI skeptics such as Tipler, Mayr, Carter, Conway Morris and others. Insofar as some of their arguments cogently contribute to our astrobiological understanding, they are indeed welcome, but the overall interpretation along the lines of "we are alone in the Galaxy" is a dangerous anthropocentric pretension. That the early SETI optimism was unjustified has nothing to do with serious and realistic work which is being done and will, it is to be hoped, continue to be done in the field. The Dysonian approach should not be construed as some nebulous "search for miracles", albeit cosmic miracles. Instead, it can be regarded as operationalization of the old epistemological dictum of Heraclitus of Ephesos: *If you do not expect the unexpected, you will not find it; for it is hard to be sought out and difficult* (fragment B18). Doesn't history of science teach us that such was the attitude of great innovators, revolutionaries, and original thinkers in general?

What is the benefit for macroengineers arising from the proposed cooperation with astrobiologists? First and foremost, they would be offered another fresh outlet for the "frustration" Dyson wrote about. In this multidisciplinary enterprise, fruitful and liberal exchange of ideas between very different specialists can be only beneficial for all. Different types of macro-engineering projects will have different astrobiological impact, and will require different detection methods and procedures, thus opening a wide and still almost entirely empty field for theoretical studies and modeling. The same issue has another side: a civilization wishing for some reason to avoid detection will likely refrain from at least some macro-engineering projects; whether this can apply to future humanity/posthumanity is for future social sciences and decision-making processes to determine. But in order to do so, detailed studies of macro-projects *detectability* need to be performed, exactly of the same kind required by astrobiology. Finally, the very limits of the concept of macro-engineering itself, vis-à-vis materials, energy, time and sophistication constraints are possible to probe *only* through the Dysonian approach to SETI; this is similar to the ways an astrophysicist learns more about our Sun, its structure, evolution and final destiny by observing billions of other stars in the Milky Way.

Overall, the greatest beneficiary may be the long-term future of intelligence itself. As Tsiolkovsky famously wrote in a 1911 letter: "The Earth is the cradle of the mind, but we cannot live forever in a cradle." Neither can others.

ACKNOWLEDGEMENTS

Any eventual merits in this work are due to Irena Diklić, whose curiosity, tenderness and kind support have presented an inexhaustible source of inspiration and encouragement during the work on this project. I happily use this opportunity to express my gratitude to Srdjan Samurović, Branislav Nikolić, Marija Kotur, Olga Latinović, Nick Bostrom, Vesna Milošević-Zdjelar, Milan Bogosavljević, Nikola Milutinović, Saša Nedeljković, Vjera Miović, Larry Klaes, and Samir Salim for their help in finding some of the references, as well as to Richard B. Cathcart, Maja Bulatović and Alan Robertson for further technical help. The manuscript enormously benefited from discussions with Robert J. Bradbury, Nick Bostrom, Zoran Knežević, Petar Grujić, Robert A. Freitas, Slobodan Popović, Luc F. A. Arnold, and Fred C. Adams. This is an opportunity to thank *KoBSON* Consortium of Serbian libraries, which enabled overcoming of the gap in obtaining the scientific literature during the tragic 1990s in the Balkans.

REFERENCES

Adami C, Ofrla C, Collier TC (2000) Evolution of biological complexity. Proc Natl Acad Sci USA 97:4463–4468

Adams F (2003) The informational turn in philosophy. Minds Machines 13:471–501

Adams FC, Laughlin G (1997) A dying universe: the long-term fate and evolution of astrophysical objects. Rev Mod Phys 69:337–372

Annis J (1999a) An astrophysical explanation for the Great Silence. J Br Interplanet Soc 52:19–22

Annis J (1999b) Placing a limit on star-fed Kardashev type III civilisations. J Br Interplanet Soc 52:33–36

Arnold LFA (2005) Transit light-curve signatures of artificial objects. Astrophys J 627:534–539

Badescu V (1995) On the radius of Dyson's sphere. Acta Astronautica 36:135–138

Badescu V, Cathcart RB (2000) Stellar engines for Kardashev's type II civilization. J Br Interplanet Soc 53:297–306

Barrow JD, Tipler FJ (1986) The Anthropic cosmological principle. Oxford University Press, New York

Benton MJ (2003) When life nearly died: The greatest mass extinction of all time. Thames & Hudson, London

Birch P (1983) Orbital ring systems and Jacob's ladders – III. J Br Interplanet Soc 36:231–238

Birch P (1991) Supramundane planets. J Br Interplan Soc 44:169–182

Bostrom N (2002) Anthropic bias: Observation selection effects. Routledge, New York

Bostrom N (2005) A history of transhumanist thought. J Evol. Tech. 14(1):1–25 PDF version. http://www.jetpress.org/Vol, 14/bostrom.html

Bracewell RN (1975) The galactic club: Intelligent life in outer space. W.H. Freeman, San Francisco

Brillouin L (1962) Science and information theory. Academic Press, New York

Brin GD (1983) 'The Great Silence' – the controversy concerning extraterrestrial intelligent life. Q Jl R Astr Soc 24:283–309

Browne D (2004) Do dolphins know their own minds? Biol Philosophy 19:633–653

Carroll SB (2001) Chance and necessity: the evolution of morphological complexity and diversity. Nature 409:1102–1109

Carter B (1983) The anthropic principle and its implications for biological evolution. Philos Trans R Soc London A 310:347–363

Chaisson EJ (2001) Cosmic evolution: The rise of complexity in nature. Harvard University Press, Cambridge

Chaitin GJ (1987) Algorithmic information theory. Cambridge University Press, Cambridge

Ćirković MM (2003) Resource letter PEs-1: Physical eschatology. Am J Phys 71:122–133

Ćirković MM (2004) Forecast for the next eon: Applied cosmology and the long-term fate of intelligent beings. Found Phys 34:239–261

Ćirković MM (2005) 'Permanence' – an adaptationist solution to Fermi's paradox? J Br Interplanet Soc 58:62–70

Ćirković MM, Bradbury RJ (2006) Galactic gradients, postbiological evolution and the apparent failure of SETI. New Astronomy. 11:628–639 (ArXiv preprint astro-ph/0506110)

Ćirković MM, Radujkov M (2001) On the maximal quantity of processed information in the physical eschatological context. Serb Astron J 163:53–56

Clarke JN (1981) Extraterrestrial intelligence and galactic nuclear activity. Icarus 46:94–96

Conway Morris S (2003) Life's solution: Inevitable humans in a lonely universe. Cambridge University Press, Cambridge

Courtillot V (1999) Evolutionary catastrophes: the science of mass extinction. Cambridge University Press, Cambridge

Darling D (2001) Life everywhere. Basic Books, New York

Des Marais DJ, Walter MR (1999) Astrobiology: exploring the origins, evolution, and distribution of life in the universe. Annu Rev Ecol Syst 30:397–420

Dick SJ (1996) The biological universe: The twentieth century extraterrestrial life debate and the limits of science. Cambridge University Press, Cambridge

Dick SJ (2003) Cultural evolution, the postbiological universe and SETI. Int J Astrobiol 2:65–74

Drake F (1965) The radio search for intelligent extraterrestrial life. In: Mamikunian G, Briggs MH (eds) Current Aspects of Exobiology. Pergamon, New York, pp 323–345

Duric N, Field L (2003) On the detectability of intelligent civilizations in the Galaxy. Serb Astron J 167:1–11

Dyson FJ (1960) Search for artificial stellar sources of infrared radiation. Science 131:1667–1668

Dyson FJ (1966) The search for extraterrestrial technology. In: Marshak RE (ed) Perspectives in modern physics. Interscience Publishers, New York, pp 641–655

Dyson FJ (1979) Time without end: Physics and biology in an open universe. Rev Mod Phys 51:447–460

Dyson FJ (2003) Looking for life in unlikely places: reasons why planets may not be the best places to look for life. Int J Astrobiol 2:103–110

Fredkin E (2003) An introduction to digital philosophy. Int J Theor Phys 42:189–247

Freitas RA Jr (1985) Observable characteristics of extraterrestrial technological civilizations. J Br Interplanet Soc 38:106–112

Freitas RA Jr, Valdes F (1980) A search for natural or artificial objects located at the Earth-Moon libration points. Icarus 42:442–447

Garriga J, Mukhanov VF, Olum KD, Vilenkin A (2000) Eternal inflation, black holes, and the future of civilizations. Int J Theor Phys 39:1887–1900

Gold T (1998) The deep hot biosphere. Springer, New York

Gonzalez G, Brownlee D, Ward P (2001) The galactic habitable zone: galactic chemical evolution. Icarus 152:185–200

Gould SJ (1985) The paradox of the first tier: an agenda for paleobiology. Paleobiology 11:2–12

Gould SJ (1987) SETI and the wisdom of Casey Stengel. In: The Flamingo's Smile: Reflections in Natural History. W.W. Norton and Company, New York, pp 403–413

Grinspoon D (2003) Lonely planets: The natural philosophy of alien life. HarperCollins, New York

Harris MJ (1986) On the detectability of antimatter propulsion space-craft. Astrophys Space Sci 123:297–303

Harris MJ (2002) Limits from CGRO/EGRET data on the use of antimatter as a power source by extraterrestrial civilizations. J Br Interplan Soc 55:383–393

Hart MH (1975) An explanation for the absence of extraterrestrials on Earth. Q Jl R Astr Soc 16:128–135

Henry WP (2005) Artificial Intelligence, 4th edn. Addison-Wesley Co, Boston

Huggett R (1997) Catastrophism: Asteroids, comets and other dynamic events in Earth history. Verso, London

Hutchins E (1996) Cognition in the wild. MIT Press, Boston

Huxley J (1957) Transhumanism. In: New bottles for new vine. Chatto and Windus, London, pp 13–17

Jablonski D (1986) Background and mass extinctions: the alternation of macroevolutionary regimes. Science 231:129–133

Jones EM (1976) Colonization of the Galaxy. Icarus 28:421–422

Jugaku J, Nishimura S (2003) A search for Dyson spheres around late-type stars in the solar neighborhood. In: Norris RP, Stootman F (eds) Bioastronomy 2002: Life among the stars. Proceedings of IAU Symposium #213, ASP Conference Series, San Francisco, pp 437–438

Jugaku J, Noguchi K, Nishimura S (1995) A search for Dyson spheres around late-type stars in the solar neighborhood. In: Seth Shostak G (ed) Progress in the Search for Extraterrestrial Life. ASP Conference Series, San Francisco, pp 381–185

Kardashev NS (1964) Transmission of information by extraterrestrial civilizations. Sov Astron 8:217–220

Kauffman S (1995) At Home in the Universe. Oxford University Press, Oxford

Korykansky DG, Laughlin G, Adams FC (2001) Astronomical engineering: a strategy for modifying planetary orbits. Astrophys Space Sci 275: 349–366

Kurzweil R (2000) The age of spiritual machines: When computers exceed human intelligence. Penguin, New York

Lachmann M, Newman MEJ, Moore C (2004) The physical limits of communication, or why any sufficiently advanced technology is indistinguishable from noise. Am J Phys 72:1290–1293

Lem S (1977) Summa Technologiae. Nolit, Belgrade (in Serbian)

Lem S (1984) His master's voice. Harvest Books, Fort Washington

Lem, S. (1987) Fiasco. Harcourt, New York

Lineweaver CH (2001) An estimate of the age distribution of terrestrial planets in the universe: quantifying metallicity as a selection effect. Icarus 151:307–313

Lineweaver CH, Fenner Y, Gibson BK (2004) The galactic habitable zone and the age distribution of complex life in the Milky Way. Science 303:59–62

Livio M (1999) How rare are extraterrestrial civilizations, and when did they emerge?. Astrophys J 511:429–431

Locher JL et al (1992) Escher: The Complete Graphic Work. Thames & Hudson, London

Lucretius (1997) On the Nature of Things, translated by William E. Leonard, e-text version, Project Gutenberg, Urbana http://www.gutenberg.org/.

Maynard Smith J, Szathmary E (1997) The major transitions in evolution. Oxford University Press, Oxford

Mayr E (1993) The search for intelligence. Science 259:1522–1523

McDonough T (1995) Review of the planetary society's SETI program. In: Seth Shostak G (ed) Progress in the Search for Extraterrestrial Life. Astronomical Society of the Pacific, San Francisco, pp 419–422

McInnes CR (2002) Astronomical engineering revisited: planetary orbit modification using solar radiation pressure. Astrophys Space Sci 282: 765–772

McKay CP (1996) Time for intelligence on other planets. In: Doyle LR (ed) Circumstellar habitable zones, Proceedings of the First International Conference. Travis House Publications, Menlo Park, pp 405–419

Moravec HP (1988) Mind children: The future of robot and human intelligence. Harvard University Press, Cambridge

Olum KD (2004) Conflict between anthropic reasoning and observation. Analysis 64:1–8

Parkinson B (2005) The carbon or silicon colonization of the universe?. J Br Interplanet Soc 58:111–116

Pierson T (1995) SETI Institute: Summary of projects in support of SETI research. In: Seth Shostak G (ed) Progress in the Search for Extraterrestrial Life. Astronomical Society of the Pacific, San Francisco, pp 433–444

Raup DM (1991) Extinction: Bad genes or bad luck?. W.W. Norton, New York

Raup DM (1992) Nonconscious intelligence in the universe. Acta Astronautica 26:257–261

Raup DM (1994) The role of extinction in evolution. Proc Natl Acad Sci USA 91:6758–6763

Rose C, Wright G (2004) Inscribed matter as an energy-efficient means of communications with an extraterrestrial civilization. Nature 431:47–49

Roy KI, Kennedy RG III, Fields DE (2004) Shell worlds: an approach to making large moons and small planets habitable. AIP Conf Proc 699:1075–1084

Russell DA (1983) Exponential evolution: implications for intelligent extraterrestrial life. Adv Space Sci 3:95–103

Sagan C (1975) The recognition of extraterrestrial intelligence. Proc R Soc Lond B 189:143–153

Sagan C, Walker RG (1966) The infrared detectability of Dyson civilizations. Astrophys J 144:1216–1218
Sandberg A (1999) The physics of information processing superobjects: daily life among the Jupiter brains. J Transhumanism 5:1–34 PDF version, http://www.jetpress.org/Vol, 5/Brains2.pdf
Schroeder K (2002) Permanence. Tor Books, New York
Shostak S (1993) Ten-year SETI begins – NASA listens for cosmic intelligence. Spaceflight 35:116–118
Simpson GG (1964) The nonprevalence of humanoids. Science 143:769–775
Slysh VI (1985) A search in the infrared to microwave for astroengineering activity. In: Papagiannis MD (ed) The Search for extraterrestrial life: Recent developments. Reidel Publishing Co, Dordrecht, pp 315–319
Suffern KG (1977) Some thoughts on Dyson spheres. Proc Astron Soc Austral 3:177–179
Tarter J (2001) The search for extraterrestrial intelligence (SETI). Ann Rev Astron Astrophys 39:511–548
Tilgner CN, Heinrichsen I (1998) A program to search for Dyson spheres with the infrared space observatory. Acta Astronautica 42:607–612
Timofeev M Yu, Kardashev NS, Promyslov VG (2000) Search of the IRAS database for evidence of Dyson spheres. Acta Astronautica 46:655–659
Tipler FJ (1980) Extraterrestrial intelligent beings do not exist. Q Jl R Astr Soc 21:267–281
Tipler FJ (1981) Additional remarks on extraterrestrial intelligence. Q Jl R Astr Soc 22:279–292
Toffoli T (1998) How much of physics is just computation? Superlattices Microstructures 23:381–406
Tough A (1998) Positive consequences of SETI before detection. Acta Astronautica 42:745–748
Townes CH (1983) At what wavelengths should we search for signals from extraterrestrial intelligence?. Proc Natl Acad Sci USA 80:1147–1151
Turnbull MC, Tarter JC (2003) Target selection for SETI. I. A catalog of nearby habitable stellar systems. Astrophys J 145:181–198
Valdes F, Freitas RA Jr (1983) A search for objects near the Earth-Moon lagrangian points. Icarus 53:453–457
Valdes F, Freitas RA Jr (1986) A search for the tritium hyperfine line from nearby stars. Icarus 65:152–157
Ward PD, Brownlee D (2000) Rare earth: Why complex life is uncommon in the universe. Springer, New York
Webb S (2002) Where is everybody? Fifty solutions to the Fermi's paradox. Copernicus, New York
Wesson PS (1990) Cosmology, extraterrestrial intelligence, and a resolution of the Fermi-Hart paradox. Q Jl R Astr Soc 31:161–170
Wright R (2000) Nonzero: The logic of human destiny. Pantheon Books, New York

INDEX

COLOR PLATE SECTION

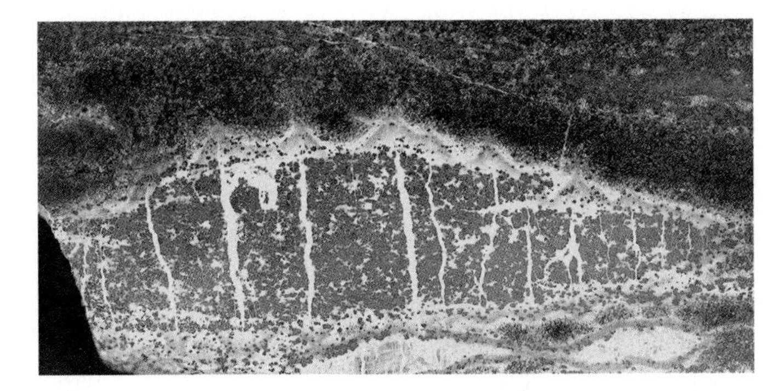

Figure 1 Chapter 2. Hand specimen of olivine rock (dunite) near Orhaneli/Turkey, showing magnesite veinlets inside a chromite lense, topped with serpentine where the veinlets pinch out into the surrounding dunite

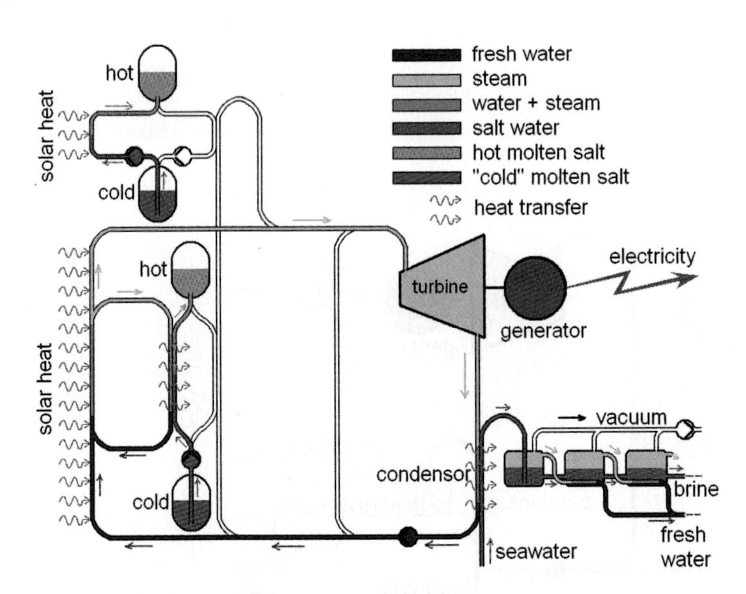

Figure 3 Chapter 3. Block diagram of a solar thermal power station equipped with heat storage, auxiliary gas heating and sea water desalination. Solar operation at daytime, two thirds of the solar heat is stored in tanks containing molten salts. The upper tanks are destined for superheating the steam

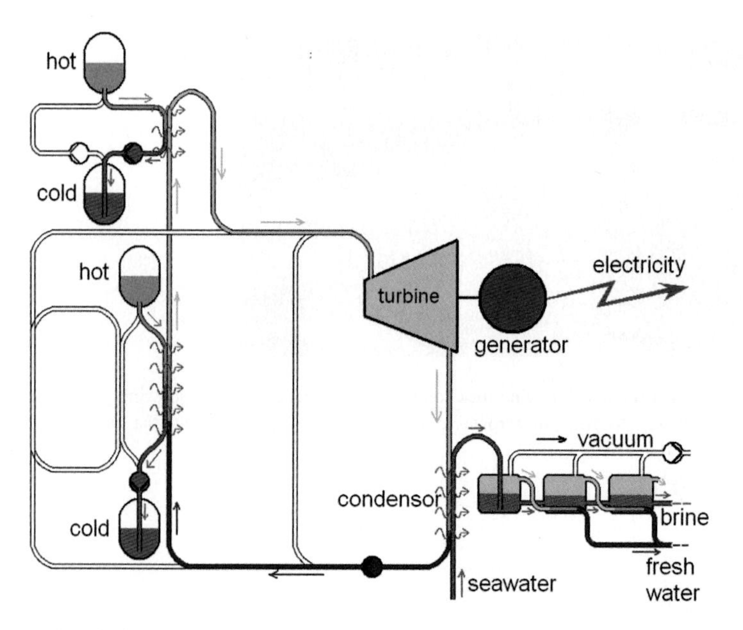

Figure 4 Chapter 3. Same as Fig. 3 *Chapter* 3. Operation at night, steam is produced out of the heat in the storage tanks

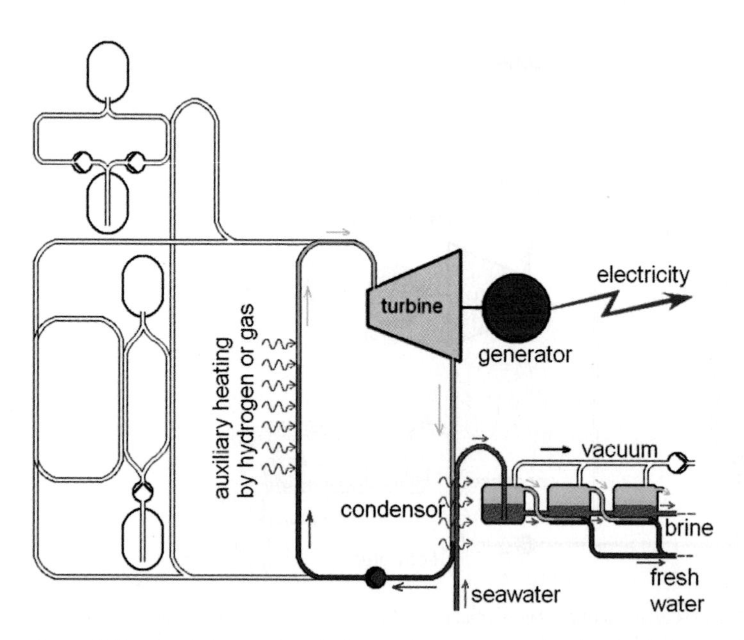

Figure 5 Chapter 3. Same as Fig. 3 *Chapter* 3. Operation on fossil fuel or hydrogen during cloudy days, when the heat storage tanks are empty

Figure 1 Chapter 4. Artist view of an offshore wind farm on a ring dike for better accessibility of wind turbines for maintenance and repair

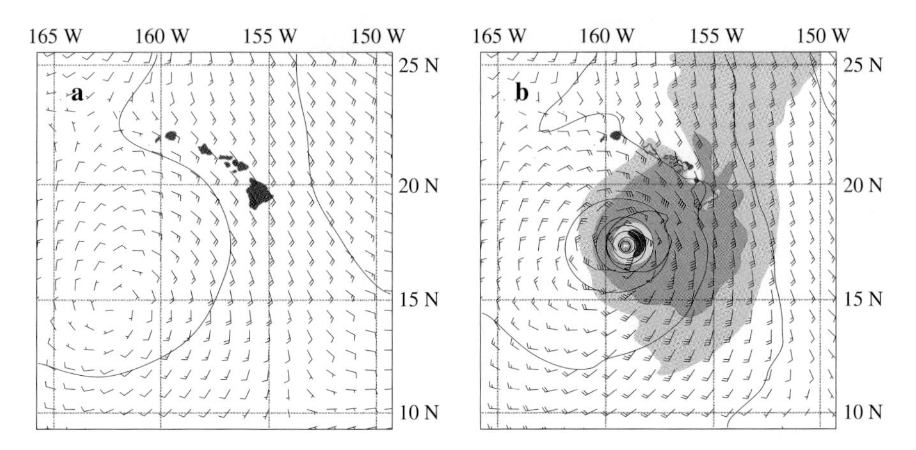

Figure 3 Chapter 7. Surface pressure and winds at the start of the 4d-VAR interval for Iniki as depicted by (a) the reanalysis and (b) the bogus procedure. Surface pressure is contoured in hPa; wind speed is color coded with green indicating tropical storm strength winds and with yellow, red, and blue indicating category 1 through 3 hurricane strength winds respectively; and a heavy contour indicates the extent of wind speeds above 25 m/s—the lower threshold of damaging winds. Wind symbols are in m/s with a full barb indicating 5 m/s

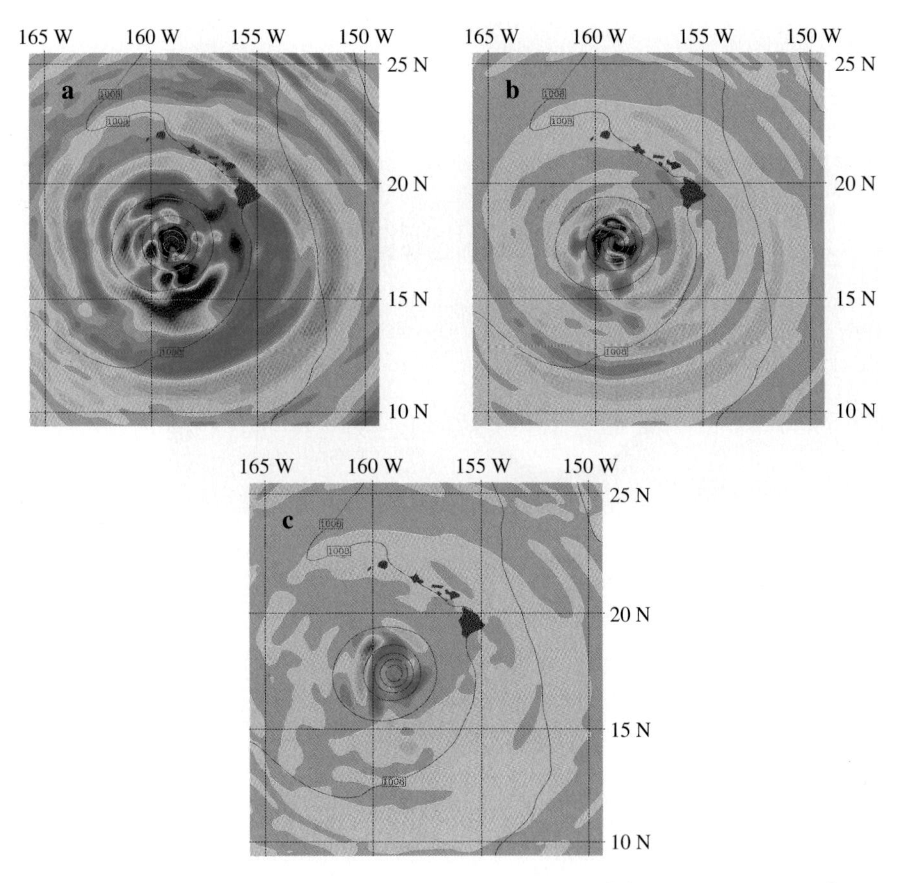

Figure 7 Chapter 7. Structure of the perturbation for experiment $C[T]$. Horizontal slices of δT are shown at (a) $\sigma = 0.350$, (b) $\sigma = 0.650$, and (c) $\sigma = 0.950$. Temperature from -4 to $+4\,°C$ is color coded from magenta to yellow, with green shades negative and red shades positive

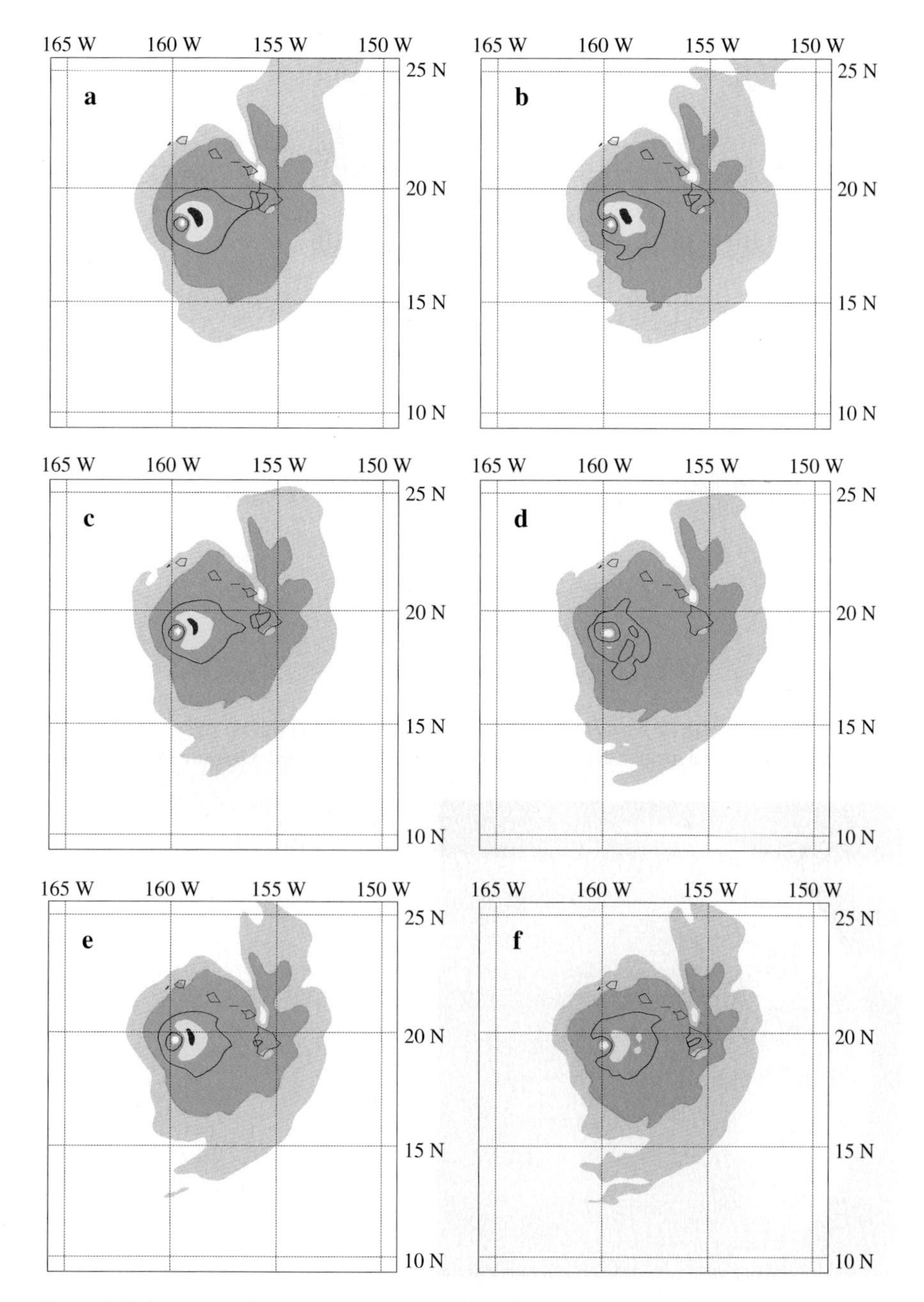

Figure 9 Chapter 7. Evolution of the surface wind field for the unperturbed simulation U (left) and for experiment $C[T]$ (right) at 4, 6, and 8 h (from top to bottom). Wind speed is color coded as in Fig. 7-3 and the damaging wind contour 25 m/s is plotted

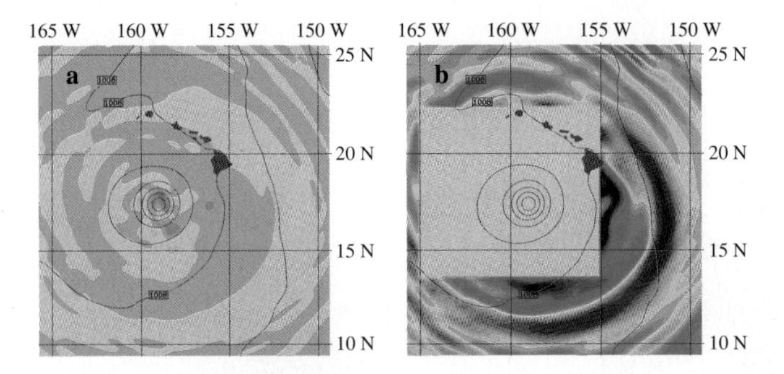

Figure 10 Chapter 7. Perturbation temperature δT at $\sigma = 0.350$ for experiments (a) $C[X]$ and (b) $C[T_d]$. Compare to Fig. 7-7a. Temperature is color coded as in Fig. 7-7

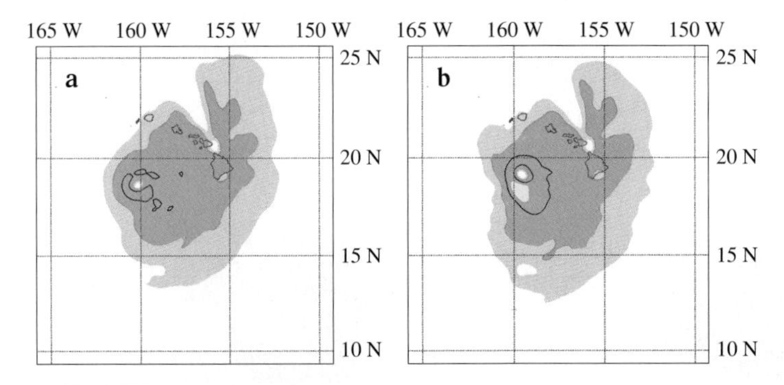

Figure 11 Chapter 7. Surface wind field at 6 h for experiments (a) $C[X]$ and (b) $C[T_d]$. Wind speed is color coded as in Fig. 7-3 and the damaging wind contour 25 m/s is plotted. Compare to Fig. 7-9c,d

Figure 14 Chapter 7. In the future solar power satellites may be also used to precisely heat parts of the atmosphere by transmitting at a frequency that is absorbed by water vapor. Image copyright by Pat Rawlings SAIC. Used with permission

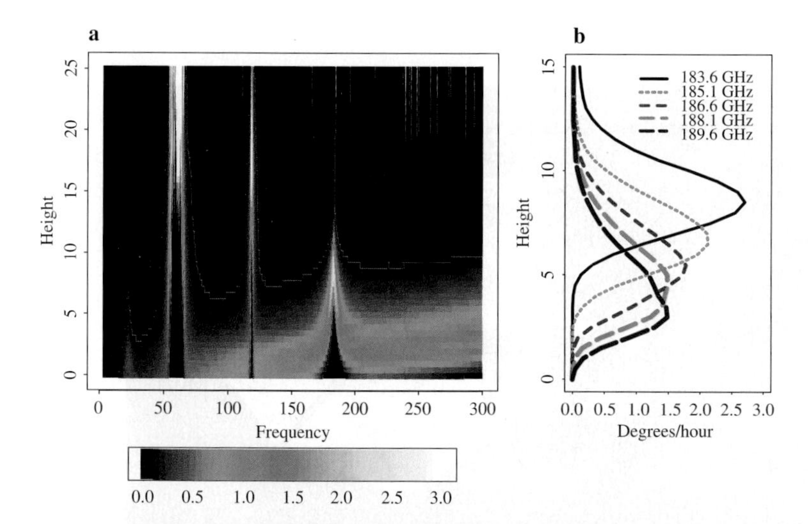

Figure 15 Chapter 7. The time rate of change of temperature (degrees Celsius/hour) as a function of frequency (GHz) and height (km). Calculations are for the standard tropical atmosphere assuming vertically incident radiation from an SSP satellite with power flux density of 1500 W m^{-2}. The whole microwave spectrum and heights to 25 km are shown in (a). The time rate of change of temperature is color coded (degrees/hour as shown by the color scale at the bottom). Values greater than $3\,°h^{-1}$ are reset to that value for plotting. This affects higher levels in the oxygen bands where absorption is strong and density is low. Selected profiles near the 183 GHz water vapor resonance are plotted in (b). Here as height increases, density decreases, so that peak values of *dT/dt* are larger for more opaque frequencies

Figure 1 Chapter 11. Large-scale deposition of aerosols by Mount Pinatubo (credit: Dave Harlow, United States Geological Survey)

Figure 8 Chapter 11. Assembly of a large occulting disk from a swarm of discrete elements (credit: Dario Izzo, ESA/ACT)

Figure 12 Chapter 11. Solar reflectors in Earth orbit (credit: Aerospace Systems Ltd)

Water Science and Technology Library

16/3: Water-quality hydrology ISBN 0-7923-3652-6
16/4: Water resources planning and management ISBN 0-7923-3653-4
Set 16/1–16/4 ISBN 0-7923-3654-2

17. V.P. Singh: *Dam Breach Modeling Technology*. 1996 ISBN 0-7923-3925-8
18. Z. Kaczmarek, K.M. Strzepek, L. Somlyódy and V. Priazhinskaya (eds.): *Water Resources Management in the Face of Climatic/Hydrologic Uncertainties*. 1996 ISBN 0-7923-3927-4
19. V.P. Singh and W.H. Hager (eds.): *Environmental Hydraulics*. 1996 ISBN 0-7923-3983-5
20. G.B. Engelen and F.H. Kloosterman: *Hydrological Systems Analysis*. Methods and Applications. 1996 ISBN 0-7923-3986-X
21. A.S. Issar and S.D. Resnick (eds.): *Runoff, Infiltration and Subsurface Flow of Water in Arid and Semi-Arid Regions*. 1996 ISBN 0-7923-4034-5
22. M.B. Abbott and J.C. Refsgaard (eds.): *Distributed Hydrological Modelling*. 1996 ISBN 0-7923-4042-6
23. J. Gottlieb and P. DuChateau (eds.): *Parameter Identification and Inverse Problems in Hydrology, Geology and Ecology*. 1996 ISBN 0-7923-4089-2
24. V.P. Singh (ed.): *Hydrology of Disasters*. 1996 ISBN 0-7923-4092-2
25. A. Gianguzza, E. Pelizzetti and S. Sammartano (eds.): *Marine Chemistry*. An Environmental Analytical Chemistry Approach. 1997 ISBN 0-7923-4622-X
26. V.P. Singh and M. Fiorentino (eds.): *Geographical Information Systems in Hydrology*. 1996 ISBN 0-7923-4226-7
27. N.B. Harmancioglu, V.P. Singh and M.N. Alpaslan (eds.): *Environmental Data Management*. 1998 ISBN 0-7923-4857-5
28. G. Gambolati (ed.): *CENAS. Coastline Evolution of the Upper Adriatic Sea Due to Sea Level Rise and Natural and Anthropogenic Land Subsidence*. 1998 ISBN 0-7923-5119-3
29. D. Stephenson: *Water Supply Management*. 1998 ISBN 0-7923-5136-3
30. V.P. Singh: *Entropy-Based Parameter Estimation in Hydrology*. 1998 ISBN 0-7923-5224-6
31. A.S. Issar and N. Brown (eds.): *Water, Environment and Society in Times of Climatic Change*. 1998 ISBN 0-7923-5282-3
32. E. Cabrera and J. García-Serra (eds.): *Drought Management Planning in Water Supply Systems*. 1999 ISBN 0-7923-5294-7
33. N.B. Harmancioglu, O. Fistikoglu, S.D. Ozkul, V.P. Singh and M.N. Alpaslan: *Water Quality Monitoring Network Design*. 1999 ISBN 0-7923-5506-7
34. I. Stober and K. Bucher (eds): *Hydrogeology of Crystalline Rocks*. 2000 ISBN 0-7923-6082-6
35. J.S. Whitmore: *Drought Management on Farmland*. 2000 ISBN 0-7923-5998-4
36. R.S. Govindaraju and A. Ramachandra Rao (eds.): *Artificial Neural Networks in Hydrology*. 2000 ISBN 0-7923-6226-8
37. P. Singh and V.P. Singh: *Snow and Glacier Hydrology*. 2001 ISBN 0-7923-6767-7
38. B.E. Vieux: *Distributed Hydrologic Modeling Using GIS*. 2001 ISBN 0-7923-7002-3
39. I.V. Nagy, K. Asante-Duah and I. Zsuffa: *Hydrological Dimensioning and Operation of Reservoirs*. Practical Design Concepts and Principles. 2002 ISBN 1-4020-0438-9

Water Science and Technology Library

40. I. Stober and K. Bucher (eds.): *Water-Rock Interaction*. 2002 ISBN 1-4020-0497-4
41. M. Shahin: *Hydrology and Water Resources of Africa*. 2002 ISBN 1-4020-0866-X
42. S.K. Mishra and V.P. Singh: *Soil Conservation Service Curve Number (SCS-CN) Methodology*. 2003 ISBN 1-4020-1132-6
43. C. Ray, G. Melin and R.B. Linsky (eds.): *Riverbank Filtration*. Improving Source-Water Quality. 2003 ISBN 1-4020-1133-4
44. G. Rossi, A. Cancelliere, L.S. Pereira, T. Oweis, M. Shatanawi and A. Zairi (eds.): *Tools for Drought Mitigation in Mediterranean Regions*. 2003 ISBN 1-4020-1140-7
45. A. Ramachandra Rao, K.H. Hamed and H.-L. Chen: *Nonstationarities in Hydrologic and Environmental Time Series*. 2003 ISBN 1-4020-1297-7
46. D.E. Agthe, R.B. Billings and N. Buras (eds.): *Managing Urban Water Supply*. 2003 ISBN 1-4020-1720-0
47. V.P. Singh, N. Sharma and C.S.P. Ojha (eds.): *The Brahmaputra Basin Water Resources*. 2004 ISBN 1-4020-1737-5
48. B.E. Vieux: *Distributed Hydrologic Modeling Using GIS*. Second Edition. 2004 ISBN 1-4020-2459-2
49. M. Monirul Qader Mirza (ed.): *The Ganges Water Diversion: Environmental Effects and Implications*. 2004 ISBN 1-4020-2479-7
50. Y. Rubin and S.S. Hubbard (eds.): *Hydrogeophysics*. 2005 ISBN 1-4020-3101-7
51. K.H. Johannesson (ed.): *Rare Earth Elements in Groundwater Flow Systems*. 2005 ISBN 1-4020-3233-1
52. R.S. Harmon (ed.): *The Río Chagres, Panama*. A Multidisciplinary Profile of a Tropical Watershed. 2005 ISBN 1-4020-3298-6
53. To be published
54. V. Badescu, R.B. Cathcart and R.D. Schuiling (eds): *Macro-Engineering: A Challenge for the Future,* 2006 ISBN 1-4020-3739-2